Armillaria Root Rot:
Biology and Control of Honey Fungus

Armillaria Root Rot: Biology and Control of Honey Fungus

Editor:
ROLAND T.V. FOX

The University of Reading, U.K.

Intercept

Andover

British Library Cataloguing in Publication Data
Armillaria Root Rot: Biology and Control of Honey Fungus

A CIP catalogue record for this book is available from the British Library
ISBN 1–898298–64–5

Published in January 2000 by Intercept Limited,
PO Box 716, Andover, Hants. SP10 1YG, England

Typeset in Times by
Ann Buchan (Typesetters), Shepperton, Middlesex.
Printed by Ashford Colour Press Limited, Gosport, Hampshire.

Contents

SECTION 5: FUTURE POSSIBILITIES

List of Colour Plates

Credits for photographs

Cover, courtesy of Dr. R.T.V. Fox

Figures 1.1a–d, 1.3a–b, 9.2 courtesy of Dr. J.S. West

Figure 1.2 reproduced from *Plant Pathology* (1949) edited by E.J. Butler and S.G. Jones courtesy of Macmillan

Figure 1.4 courtesy of Dr. J.A. Turner

Figures 3.1a–c courtesy of A. Lamour and Prof. M.J. Jeger

Figures 6.1, 9.1 courtesy of AstraZeneca Agrochemicals (now Syngenta)

Figures 10.1–10.4 courtesy of Dr. F. Raziq

Figure 10.5 courtesy of Dr. M. Dumas

Colour plates 1a, 3a, 6 courtesy of Dr. R.T.V. Fox

Colour plates 1b–d, 3c, 7 courtesy of Dr. J.S. West

Colour plate 2 courtesy of Dr. T.O. Popoola

Colour plate 4 courtesy of Dr. P.J. Herring

Colour plates 5a & b courtesy of Dr. H.A. van der Aa

Colour plate 3b courtesy of P. Johns and AstraZeneca Agrochemicals (now Syngenta)

Colour plate 8 courtesy of Dr. J.A. Turner

Colour plate 10a courtesy of the late Robin Buckley

Colour plate 10b courtesy of Dr. J.S. West

Contributors

ROLAND T.V. FOX, *The University of Reading, School of Plant Sciences, Department of Horticulture and Landscape (Crop Protection), 2 Earley Gate, Reading, Berkshire RG6 6AU, U.K.*

MICHAEL J. JEGER, *Wye College, University of London, Wye, Ashford, Kent TN25 5AH, U.K.*

ANGELIQUE LAMOUR, *Laboratory of Phytopathology, Wageningen Agricultural University, P.O. Box 8025, 6700 EE Wageningen, The Netherlands*

DAVID N. PEGLER, *The Herbarium, Royal Botanic Gardens, Kew, Richmond, Surrey TW9 3AB, U.K.*

ANA PÉREZ-SIERRA, *Royal Horticultural Society, Wisley, Woking, Surrey GU23 6QB, U.K.*

FAZLI RAZIQ, *Agricultural Research Station, Takhta Band Road, Mingora Swat, NWFP, Pakistan*

AAD J. TERMORSHUIZEN, *Laboratory of Phytopathology, P.O. Box 8025, 6700 EE Wageningen, The Netherlands*

JONATHAN S. WEST, *Institute of Arable Crops Research, Rothamsted, Harpenden, Herts. AL5 2JQ, U.K.*

DEBRA WHITEHEAD, *School of Applied Sciences, University of Wolverhampton, Wolverhampton WV1 1SB, U.K.*

MICHAEL WHITEHEAD, *School of Applied Sciences, University of Wolverhampton, Wolverhampton WV1 1SB, U.K.*

Editor's Foreword

In this book my colleagues and I seek to reveal the biology of *Armillaria* species and suggest ways to reduce the damage that the pathogenic species can cause.

It has often been said that the only sure way to be rid of *Armillaria mellea*, the Honey Fungus, is to move house. However, this is a rather unreliable answer as so many mature gardens are already affected, at least in my part of southern England.

Resist the temptation to flee and in your battle with this unseen virulent fungus you will experience a living organism as natural and fascinating as any fox, frog or finch that visits your garden. After all, the *Armillaria* living in your garden is a relic of the first wild wood that clothed the land after the ice age glaciers receded. It is difficult not to be impressed by an organism considered one of the largest and oldest on the planet.

Others just appreciate its mushroom as a tasty ingredient which is unusually sweet, meaty and, above all, abundant in its season. Some people who are also aware that the fungus is packed with biologically active molecules often shy away from eating it, unless well cooked. Not so the elderly sisters who rebuked me for trying to control it. This happened when I was manning a department stand which won a Lindley Medal at the 1987 Chelsea Flower Show. Apparently, these ladies had long ago abandoned their orchard to Honey Fungus in return for filling their freezers with harvested fruiting bodies in the autumn!

Armillaria, like most soil-borne pathogens, is in some ways just as mysterious as one of those fierce, deep-sea creatures hauled out of the ocean. Both exist largely unobserved in a murky, alien world. While *Armillaria* has no teeth, it nonetheless devours most plants, from giant forest trees to humble herbs and weeds. Regretfully, I can confirm this from bitter personal experience, as can many others.

Dr. Roland Fox
Reading, December 1999

About the Editor

Forests have always been important to Roland Fox, who was brought up close to Epping Forest. He attended Southampton University where he studied for his B.Sc. and Ph.D. (Plant Pathology) in the Department of Botany. This allowed easy access to the New Forest. While at Southampton, Roland Fox was secretary to the Science Faculty Society, an editor of its journal *Helios*, chairman of the British Universities North America Club, and Director of Fine Arts at the 1967 Southampton University Arts Festival.

After graduating, he moved to Wokingham, Berkshire (in the old Windsor Forest). He worked, first as a plant pathologist, then later as a biotechnologist, with ICI Plant Protection Ltd. (now merged into Syngenta) in the U.K. at Jealott's Hill Research Station near Bracknell, and in the U.S.A. at Goldsboro, North Carolina. He was a member of the ICI Interdivisional Biocides Committee, chairman of the ICI Plant Protection Gardening Club, and a founder member of the West Forest Round Table.

He now lives surrounded by pines in Crowthorne, travelling extensively around the world, and has been a Visiting Lecturer at the International Agricultural Centre, The Agricultural University, Wageningen in the Netherlands. In 1984, he joined the Department of Agriculture and Horticulture at The University of Reading as a lecturer in plant pathology. At present, he is a lecturer in crop protection in the School of Plant Sciences and the Course Director and Admissions Tutor for a B.Sc. in Crop Protection. His current research activities encompass the diagnosis of plant pathogens and improving the control of soil-borne diseases of plants (particularly honey fungus root rot), especially by developing microbiological methods.

He is the author of *Principles of Diagnostic Techniques in Plant Pathology* (CAB International) and *The Gardener's Book of Pests and Diseases* (B.T. Batsford). He has been awarded two Lindley Medals for Science and Education by the Royal Horticultural Society (Gold, 1987; Bronze, 1993), is an adviser to *Gardening from Which?*, and lectures to groups of amateur and professional gardeners.

Plate 1a. Clump of fruiting bodies of *Armillaria mellea* above decayed root of Turkey oak (*Quercus cerris).*

Plate 1b–d. Clump of fruiting bodies of *Armillaria mellea* showing deposition of white basidiospores. Note: fruiting bodies possess a ring and gills (white with basidiospores) beneath the honey-coloured caps.

Plate 2. Fruiting body of *Armillaria ostoyae* produced on slope of 3% Malt Extract Agar after several weeks exposure to light.

Plate 3a. Network of rhizomorphs of *Armillaria mellea* under bark of horse chestnut (*Aesculus hippocastanum*).

Plate 3b. Cut lengths of large rhizomorphs of *Armillaria lutea* showing resprouting.

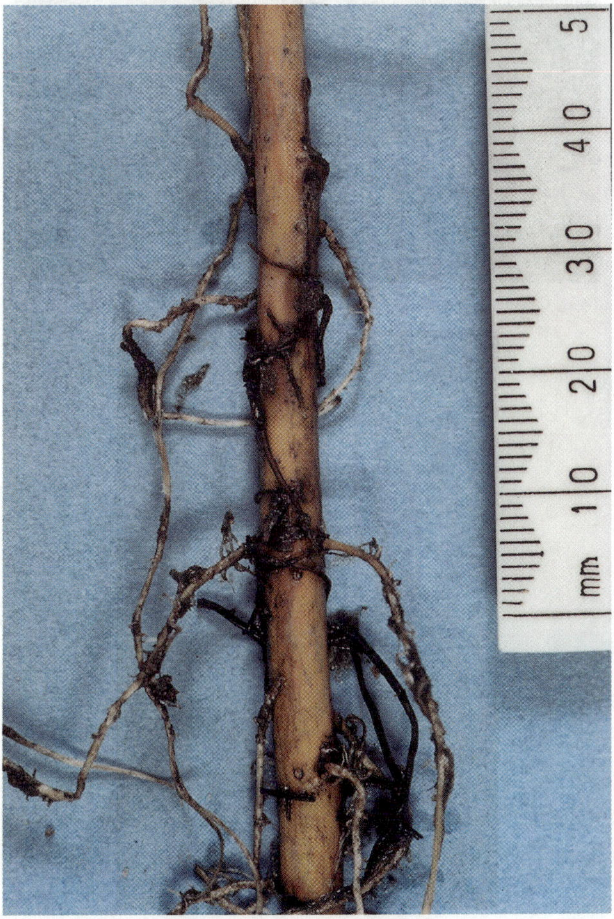

Plate 3c. Rooted cutting of willow showing attached rhizomorphs of *Armillaria mellea*.

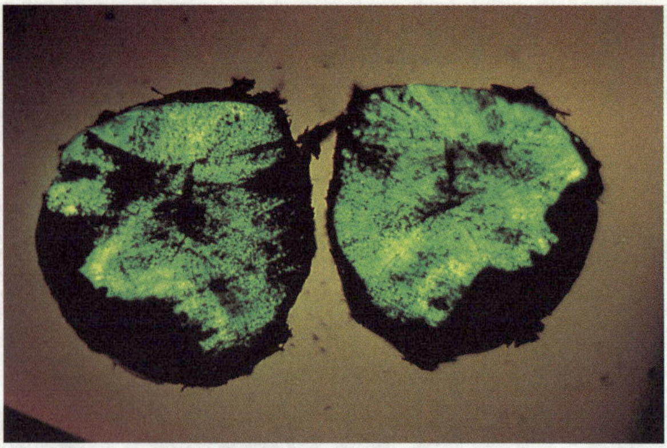

Plate 4. Bioluminescent mycelium of *Armillaria mellea* on decaying root of rowan (*Sorbus aucuparia)*.

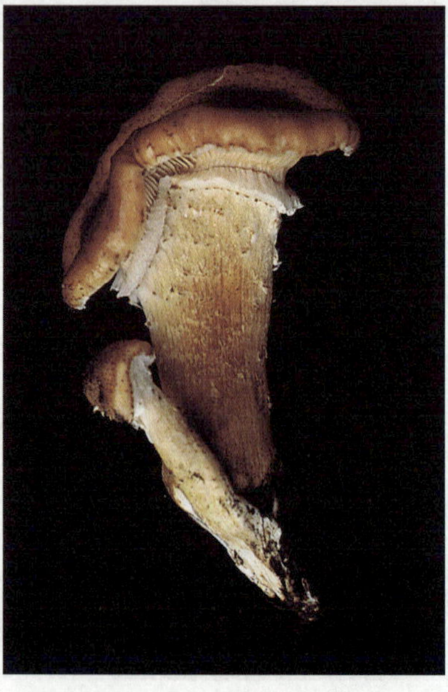

Plate 5a & b. Fasciated basidiocarps of *Armillaria ostoyae* (possibly as a result of virus infection).

Plate 6. Many climbers (like ivy) and herbaceous plants (like chrysanthemum) are highly tolerant to *Armillaria mellea*.

Plate 7. Infected strawberry plant, three months after inoculation with *Armillaria mellea,* dissected to show its colonized root system and advanced fruit production.

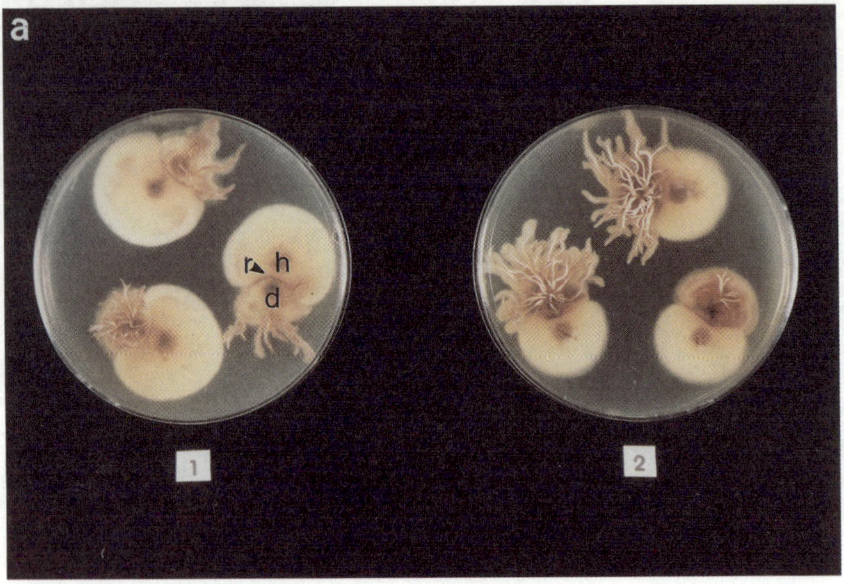

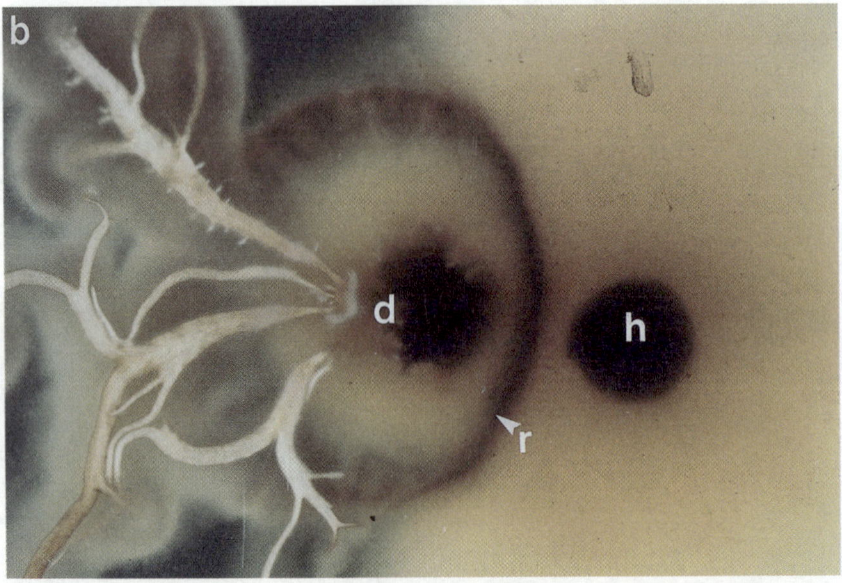

Plate 8. Incompatible reaction between the mycelium of a haploid tester (*Armillaria borealis*) on a diploid isolate on malt extract agar after 2–3 weeks incubation at 25°C in darkness (*A. mellea*).

Plate 9a. Privet hedge being killed by *Armillaria mellea* in Reading garden (Note fruiting bodies).

Plate 9b. Fruiting bodies beneath privet hedge being killed by *Armillaria mellea* in Reading garden (Note deposits of white basidiospores).

Plate 10a. Individual blackcurrant (*Ribes nigrum*) killed by *Armillaria* root rot while neighbouring bushes are apparently unaffected.

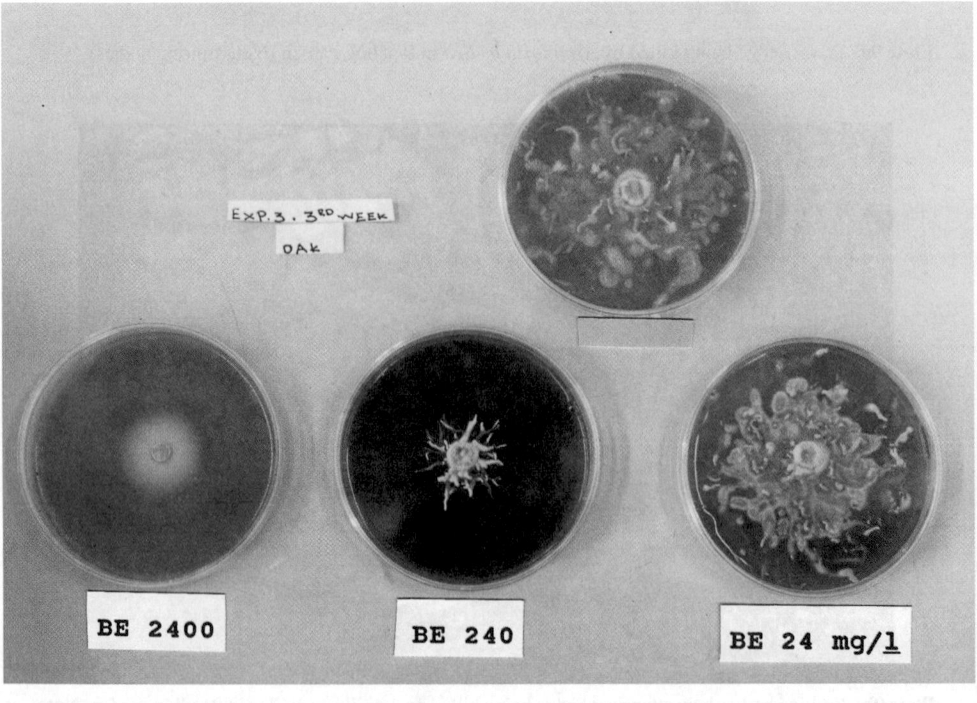

Plate 10b. Response of *Armillaria mellea* to Bray's Emulsion at different concentrations of product.

SECTION 1

Biology

1
Biology and life cycle

ROLAND T.V. FOX

The University of Reading, School of Plant Sciences, Department of Horticulture and Landscape (Crop Protection), 2 Earley Gate, Reading, Berkshire RG6 6AU, U.K.

Synopsis

Many species of *Armillaria* are facultative pathogens but others are obligate saprobes or symbionts. Although the basidiomes are commonly gathered for food, they may be allergenic and are also the source of many medicinal compounds. Apart from a few days in autumn when these fruit bodies occur, the hyphae and rhizomorphs have to be identified by other means, including molecular methods. *Armillaria* assembles hyphae into rhizomorphs which slowly aggregate into huge networks, lasting over hundreds of years if there continue to be sufficient sources of nutrition for them to absorb and translocate. Rhizomorphs sprout the white fan-shaped mats of mycelium which extend the decay up infected tree's phloem and cambium, separating the wood from the bark, killing the host, and thus encouraging windthrow. Although rhizomorphs permit tree-to-tree spread by mechanical force, healthy roots can be infected directly if they contact the mycelium on diseased roots. Infection depends on the enzymes and toxins produced by the mycelium. Injured roots are biochemically predisposed to infection. After a rot has been established, providing there is a sufficiently large food base, the mycelium may persist for many years. The bulk of the hyphae are protected within large pseudosclerotia, which are protected by a melanized rind relatively impenetrable to moulds or predators, similar to that covering the rhizomorphs. The majority of rhizomorphs are produced during the terminal stages of decay but the amount they grow depends on the species and is influenced by its habitat and environmental conditions. High temperatures and certain soils may suppress the development of rhizomorphs, although some species form rhizomorphs readily only *in vitro*, not in field soil. Rhizomorphs consist of a melanized rind over a densely packed cortex outside the loosely intertwined wide-diameter hyphae of the medulla that transports water and nutrients, with a central canal able to translocate oxygen. The rhizomorphs are surrounded by mucilage and a loose network of hyphae. While the structure and functions of a rhizomorph are comparable in complexity to a plant root, less is known of the biochemistry of solute transport and gas diffusion within the

Armillaria *Root Rot: Biology and Control of Honey Fungus*

hyphae or its genetic control. Rhizomorphs meet the criteria for pressure-driven flow as a substantial gradient of water and turgor potential extends from their tips to their base through conducting channels which are relatively impermeable to water laterally but very permeable to solutes along their lengths. Rhizomorphs of different species of *Armillaria* show varying degrees of branching and elasticity, so it is possible to differentiate some species by the branching patterns of their rhizomorphs. However, the rhizomorphs used for this method of diagnosis are not dug from soil but are 'cultivated' quite slowly *in vitro* in mist chambers from inoculated food base substrates. Rhizomorph elongation, originally thought to be due to a meristematic apical centre actively dividing off cells to form the other layers, might be analogous to the mechanism proposed for hyphal extension, where a plasticized apical dome is driven forward by the pressure generated within a tube with rigid side walls, compensated for by branching and growth of the intercalated apical hyphae. Most infections are on the root collar, possibly because rhizomorphs grow towards the soil surface due to the oxygen gradient, when not limited by its seasonal desiccation. Rhizomorph production may be stimulated by the presence of some other fungi. The basidiomes of *Armillaria* species are gilled toadstools which release millions of basidiospores. They can be produced in culture, light is necessary. Although basidiomes are responsible for long range dispersal and important for taxonomic studies, they are of less use in routine identification as their appearance is usually highly seasonal and erratic. There is genetic diversity in cultural characteristics, virulence, reaction to environmental stimuli, nutritional responses to low-molecular-weight alcohols, gallic acid and light. The precise physiological and biochemical causes for this are still obscure. Although virus-like particles have been found within the genus *Armillaria* which might result in some of the differences seen among isolates, true interspecific variation occurs in rhizomorph branching patterns, basidiome and vegetative morphology, pathogenicity, and isoenzyme profiles. The variation observed in the bioluminescence of *Armillaria* in different areas may depend on the prevailing environmental factors. There may be some cytological differences as there has been some debate as to whether the haploid monokaryotic hyphae in *A. mellea* fuse into diploids or dikaryons. At least 10 different antibiotics and many other metabolites are bioactive, and exudates may be released. *Armillaria* can detoxify mycotoxins produced by some other fungi. Chitinolytic enzymes are also produced. Phenoloxidizing enzyme production affects rhizomorph morphogenesis and deserves further investigation. In plant tissues, laccase is involved in lignin oxidation, degradation and detoxificiation of antifungal phenols. Isolates of *A. ostoyae* with low laccase activity have low rhizomorph production, whereas *A. lutea* has a broader range of laccase activities and rhizomorph production. Glucose is the favourite carbohydrate but *Armillaria* species utilize other carbohydrates, lipids, phenols, alcohols and fix carbon dioxide. *Armillaria* requires an adequate source of nitrogen in a suitable form, in addition to carbon, for adequate growth and development. Casein hydrolyzate is superior to individual amino acids, some of which outperform inorganic nitrogen sources such as ammonium and nitrate. The optimum C:N ratio for rhizomorph development varies between different isolates. The availability of inorganic ions also appears to influence the behaviour of *Armillaria* in soil. Some organic compounds, like the alcohols, auxins, fatty acids, and phenols, stimulate the growth and development of *Armillaria* at concentrations far below those of inorganic nutrients, such as carbon and nitrogen, but substantially above those normal for vitamins, Hence, undefined substrate supplements, such as fungal or plant extracts, may be vital for optimal growth and development of *Armillaria* in defined media.

General biology

Several species of basidiomycete fungi in the genus *Armillaria* are collectively known in England, and some places elsewhere, as the 'Honey Fungus' (Fox, 1990; Shaw and Kile, 1991). For a few days in the autumn, these fungi may be identified by their characteristic fruit bodies, which can often be abundant on patches of soil above buried wood (Grieg and Strouts, 1983; Grieg *et al.*, 1991). Apart from this brief period, the fruit bodies may also be grown *in vitro* (Fox and Popoola, 1990; Intini, 1993; Togashi, 1996; Togashi and Takizawa, 1996). During other periods in their life history (Lamoure and Guillaumin, 1985; Rishbeth, 1985; Guillaumin and Lung, 1985; Guillaumin *et al.*, 1991), mycelium can be detected and diagnosed in other ways, including molecular methods (Wargo and Shaw, 1985; Guillaumin, 1988; Fox and Hahne, 1988a, 1989; Fox, 1993; Fox *et al.*, 1993; Priestley *et al.*, 1994; Smith *et al.*, 1992; Pérez-Sierra *et al.*, 1999) any time of the year.

Although the attractive epithet, 'Honey Fungus', like the specific name, *mellea*, was probably adopted because of the tawny colour of the caps of the mushrooms, it is also appropriate because of their sweetish taste (Yoshida *et al.*, 1984). Despite this, first time consumers frequently consider it unpalatable, and occasionally even toxic (Herink, 1983; Bocchi *et al.*, 1995), probably as the whitish flesh of the fructifications tastes acrid when fresh or undercooked and is full of a variety of biologically active chemicals, some of pharmaceutical interest (Junhua *et al.*, 1990; Rehman and Thurston, 1992; Lasota and Florczak, 1984; Wang *et al.*, 1996; Sonnenbichler *et al.*, 1994, 1997; O'Neil *et al.*, 1990; Liengswangwong *et al.*, 1987; Piepp and Sonnenbichler, 1992; Obuchi *et al.*, 1990; Donnelly *et al.*, 1985, 1990, 1997; Donnelly and Hutchinson, 1990; Kiho *et al.*, 1992a,b; Yang *et al.*, 1984, 1989a,b, 1990a,b, 1991; Watanabe, 1997; Watanabe *et al.*, 1990; Yi *et al.*, 1998; Xue and Xu, 1991). Like many other fungi, *Armillaria* is allergenic (O'Neil *et al.*, 1990), there is evidence of cross reactivity with some Deuteromycetes (Liengswangwong *et al.*, 1987). Enzymes from *Armillaria tabescens* have been reported to detoxify the mycotoxin, aflatoxin B_1 (Liu *et al.*, 1998). In Chinese medicine, tablets containing cultured mycelium of *Armillaria* are prescribed for a variety of ailments including dizziness, headache, neurasthenia, insomnia, numbness in limbs, and infantile convulsions (Jungshan *et al.*, 1984). Indirectly, *Armillaria* also helps to provide another Chinese medicine as it is symbiotic with *Galeola septentrionalis* and some other orchids (Terashita, 1996) which are themselves used to treat an assortment of complaints. However, several other aspects of the behaviour of the living honey fungus are also far from pleasant (Hagle and Shaw, 1991).

Rhizomorphs

Armillaria is one of the few genera of fungi able to construct rhizomorphs from vegetative hyphae which, over the centuries, can form huge networks of relatively stable genets (Smith *et al.*, 1992). Ions and other substances in solution are absorbed

Figure 1.1a. Fruiting bodies produced for a few days in the autumn from infected roots of an old tree stump.

Figure 1.1b. Mycelial fans of *Armillaria mellea* beneath loose bark of a side root of a windthrown tree (horse chestnut).

Figure 1.1c. Cross section of infected horse chestnut tree showing rotted area and zone lines of pseudosclerotia of *Armillaria mellea*.

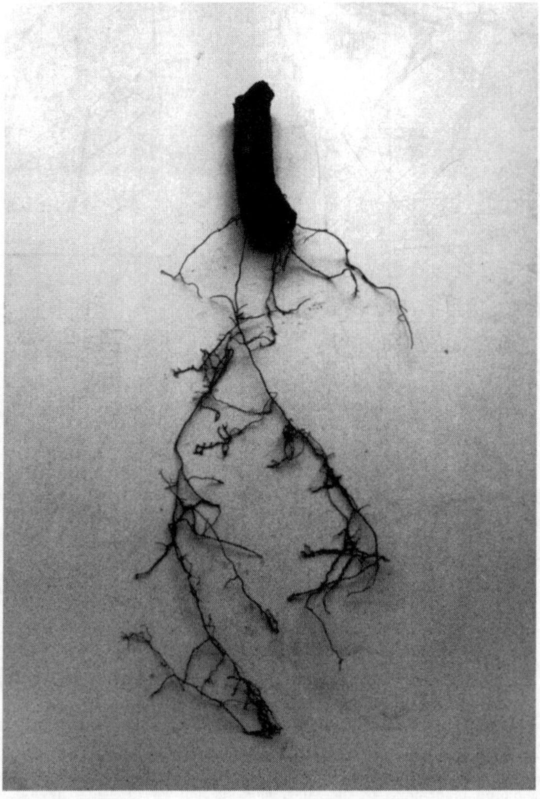

Figure 1.1d. Network of rhizomorphs of *Armillaria mellea* produced from artificial inoculum (hazel billet) in 12 weeks.

(Rizzo *et al.*, 1992) and translocated along these rhizomorphs (Granlund *et al.*, 1985; Cairney *et al.*, 1988a; Gray *et al.*, 1996) under turgor pressure (Eamus and Jennings, 1984), and they also enable the white fan-shaped mats of mycelium to extend the decay upwards in infected trees through the region of the phloem and cambium, thus separating the wood from the bark, usually with lethal consequences. Hyphae also grow down inside the trunk, often spreading between the ray cells, rotting, softening and weakening the heartwood of its host's roots and butt (Whitney, 1997). Subsequent windthrow may also result in the premature death of the tree.

The linear rhizomorphs grow apically, enabling them to translocate resources and intensifying their inoculum potential (Woeste, 1956), as well as directing the growth of their individual, highly differentiated autonomous hyphae to external stimuli by maintaining and channelling the expansion of the fungus from the decaying roots, even traversing patches of unsuitable substrate (Thompson, 1984). The cord-like rhizomorphs are 1–3 mm in diameter with a compact reddish brown to black outer cortex layer and a core of whitish mycelium in the medulla (Zhang and Dong, 1986; Cairney *et al.*, 1988a). They usually form a branched, flattened network within the roots, under the bark, or in thoroughly decayed wood, with similar, but more rounded,

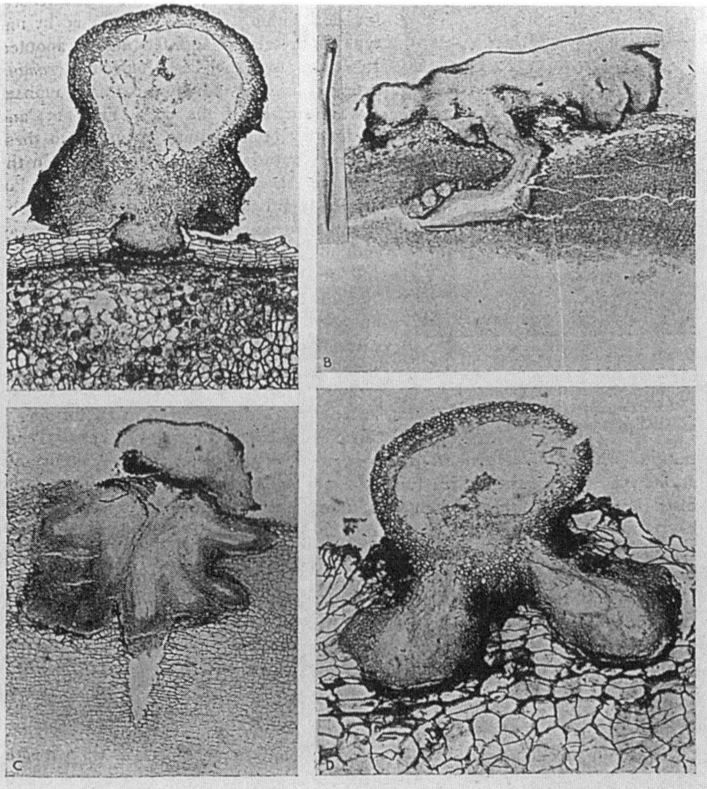

FIG. 112.—Penetration of cork by rhizomorphs of *Armillaria mellea*. *A*, cross-section of root of Persian walnut, the rhizomorph showing almost complete break-through of the cork layers. *B*, top corner, rhizomorph tip; deeper penetration and disruption of the host tissues in root of pear, the rhizomorph entering the wood below cambium. *C*, the same, invading carrot, the rhizomorph branching. *D*, the same, entering a tuber of dahlia (photos by Thomas, *J. Agric. Res.*)

Figure 1.2. Penetration into various hosts by the rhizomorphs of *Armillaria mellea*.

strands spreading out into the soil surrounding the rotten substrate, such as the infected roots or bases of dead or dying trees.

Contact with such rhizomorphs can result in tree-to-tree spread of the fungus, even when direct contact between diseased and healthy roots is not possible. The rhizomorph becomes firmly attached after the mucilagenous substance covering its growing tip hardens. Hyphae then emerge from the rhizomorph at the points of attachment. These penetrate beneath the bark scales by mechanical force and chemical secretions (Morrison *et al.*, 1991). Therefore, the rhizomorphs of *A. mellea* need neither wounds nor points of weakness to attack healthy, vigorously growing host plants (Day, 1927; Thomas, 1934; Woeste, 1956). Nonetheless, root injuries can serve as infection courts. Injured roots are also biochemically more predisposed to infection than those that remain intact (Popoola and Fox, 1996).

Infection of a healthy root begins when its bark is acted upon by toxic substances produced by *A. mellea* mycelium (Zeller, 1926). Shallow brown spots appear in the outer parenchyma of the bark, and these eventually coalesce. Flakes of dead cork are sloughed as new cork layers are formed. Eventually, the fungus reaches the cambium and a canker develops. Mycelium may persist for years after a rot has been established in a large food base, often producing rhizomorphs, chiefly during the terminal stages of decay.

The growth of rhizomorphs that develop from wooden inocula in soil can be assessed by measuring their total length or dry weight. Rishbeth (1968) investigated the effect of temperature by repeatedly measuring the growth of individual rhizomorphs. The amount they grow is influenced by their habitat as well as their environment (Pearce and Malajczuk, 1990; Turcsanyi *et al.*, 1996). Although rhizomorphs develop from inocula in many soils in which they are placed (Gramss, 1983; Morrison, 1976; Redfern, 1970, 1973, 1975; Rishbeth, 1985b), certain soils in the tropics appear to suppress their development (Dade, 1927; Fox, 1964; Rishbeth, 1980; Swift, 1968; Pearce and Malajczuk, 1990; Podger *et al.*, 1978). Pure sand restricts their production (Garrett, 1956; Redfern, 1973; Rykowski, 1984), whereas peat encourages their growth and branching (Redfern, 1973).

Armillaria species can be found in a wide variety of forest and agricultural soils (Ono, 1965, 1970; Rhoads, 1956; Ritchie, 1932; Shields and Hobbs, 1979). However, there may be inconsistent results where there is a patchy distribution of inoculum in the soil or if morphologically similar species differing widely in virulence are misidentified. The abundance of rhizomorph production varies between the species. Some species of *Armillaria* form rhizomorphs readily, but not every species develops rhizomorphs in field soil, even if they do when cultured *in vitro*.

Among the North Temperate species, *A. lutea* and *A. cepistipes* produce more rhizomorphs than *A. ostoyae* and *A. mellea* (Gregory, 1985; Guillaumin *et al.*, 1989; Redfern, 1975; Rishbeth, 1985a). *Armillaria sinapina* produces abundant rhizomorphs in the field (Bérubé and Dessureault, 1988). However, rhizomorphs are rare or absent in *A. tabescens* (Rhoads, 1956; Rishbeth, 1982; Ross, 1970).

In Australasia, *A. limonea* and *A. novae-zelandiae* form rhizomorphs readily (Hood and Sandberg, 1987) but in *A. hinnulea*, they are restricted to the surface of host roots (Kile, 1980; Kile and Watling, 1983) and are absent in *A. luteobubalina* (Kile, 1981; Shearer and Tippett, 1988; Pearce *et al.*, 1986; Podger *et al.*, 1978).

Different species of *Armillaria* produce rhizomorphs with either a monopodial or

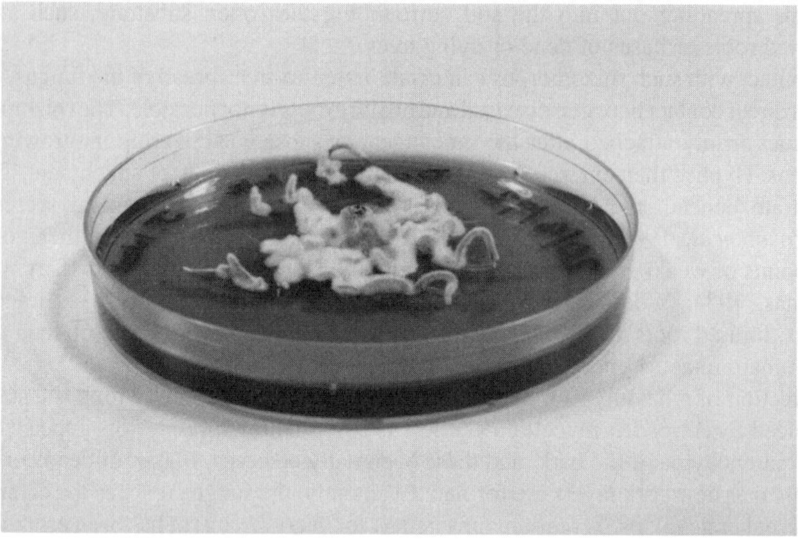

Figure 1.3a. *Armillaria mellea* growing on 3% malt extract agar culture plate showing rhizomorphs.

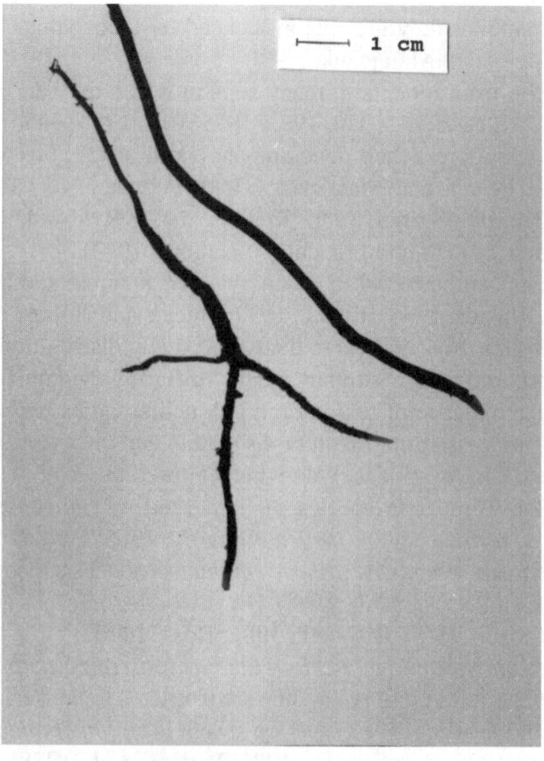

Figure 1.3b. Pale tips of freshly exposed rhizomorphs from soil.

dichotomous branching pattern (Morrison, 1982b, 1989; Mihail *et al.*, 1995). Frequently, the former tend to be less virulent than the latter (Morrison, 1989). However, under experimental conditions rhizomorph production is often abnormal (Gregory, 1985; Redfern, 1975; Rishbeth, 1985a,b; Morrison, 1989; Podger *et al.*, 1978).

When they were first described, rhizomorphs were considered to be an independent fungus species, *Rhizomorpha fragilis* Roth, which was subdivided into *R. subterranea* found in the soil and *R. subcorticalis* found underneath tree bark. After finding rhizomorphs in decayed timber in mine shafts, Schmitz (1848) realised that the fungus must have entered the standing trees prior to felling (Ainsworth and Sussman, 1965; Hüttermann, 1987). Hartig (1873) established that rhizomorphs are produced by the majority of species of *Armillaria* and that they extend out through the soil to colonize and utilize woody substrates, including living trees in the case of the pathogens. Hartig (1874) also considered that the differences between the conditions in the soil or beneath the bark are responsible for the morphological differences between the subterranean and subcortical forms of rhizomorphs. He also observed that rhizomorphs exposed to the air are darker, with a denser rind which restricts additional lateral growth, than those concealed under tree bark. Both types of rhizomorphs are also produced *in vitro* (Lopez-Real, 1975); ribbon-shaped rhizomorphs form underneath the cultures, whereas round, pigmented rhizomorphs develop at the surface.

The structure of rhizomorphs in *Armillaria* described by De Bary (1869, 1887) was illustrated in detail by Hartig (1870, 1874), who divided the thallus into the three layers; cortex, subcortex, and medulla. He also reported that the tip of the rhizomorph, which Brefeld (1877) described as a meristem, is mucilagenous and there are differences between the walls of the hyphae in each layer of the thallus. This differentiation of the rhizomorphs can be observed more clearly using an electron microscope to reveal that they consist of a melanized, densely packed cortex outer layer surrounded by mucilage and a loose network of hyphae (Turner, 1991). The melanin in the outer walls of the cortex of the rhizomorph shields the hyphae inside from microbial antagonists (Bloomfield and Alexander, 1967; Khuo and Alexander, 1967). The subcortical layer which forms the transition to the medulla, lies below the cortex. The loosely intertwined, wide-diameter hyphae of the medulla transport water and nutrients (Jennings, 1984). The biochemistry of the hyphal walls of *Armillaria mellea* has been characterized by Sanchez-Hernandez *et al.* (1990). Near the middle of the rhizomorph, the medullary hyphae become less abundant so that the central canal is able to translocate oxygen (Smith and Griffin, 1971). The rhizomorphs may also release an exudate (Mallett and Colotelo, 1984).

While its impermeable rind permits *Armillaria* to flourish in hostile environments, 'breathing pores' may form when a rhizomorph grows at the substrate-air interface. These tufts of intertwining hyphae burst through the rind enabling oxygen to diffuse into its central canal (Smith and Griffin, 1971). Breathing pores may derive from aborted side branches as their apices are loosely intertwined hyphae with no organized meristem.

The structure and functions of a rhizomorph are comparable in complexity to a plant root, yet little is known of the biochemistry of solute transport and gas diffusion within the hyphae or its genetic control. According to Garraway *et al.* (1991), a rhizomorph consists of:

1. A gelatinous sheet and mucilage layer at the apex which protect its growth in the soil.
2. A central region of the apex associated with mucilage production – surrounding the meristem responsible for the growth of the rhizomorph.
3. Circum-medullary cells of the apex – nearby supply of growth material for apical dome.
4. Lateral meristem – initiates lateral growth behind the apex.
5. Melanized cortex, outer rind of the rhizomorph – protects against fungi and bacteria.
6. Subcortical layer, secondary meristem – permits lateral growth.
7. Medulla – large cells allow solute-mediated transport of nutrients.
8. Breathing pores – regions in the rhizomorph which ensure oxygen uptake.
9. Central canal – cavity within the rhizomorph which enables it to translocate gases.

In *Armillaria* rhizomorphs a substantial gradient of water and turgor potential extends from the tip to the base of the rhizomorphs (Eamus and Jennings, 1984). Cytological investigations confirm that *Armillaria* rhizomorphs meet the criteria that are considered necessary for pressure-driven flow to be accepted as a translocation mechanism in plants (Zimmermann, 1971). These are conducting channels which are relatively impermeable to water laterally but very permeable to solutes and water longitudinally, with turgor gradients between source and sink.

Measurements of the internal structure and hydraulic conductivity of rhizomorphs imply that long-distance transport occurs predominantly by solutes moving along the vessel hyphae of the medulla (Eamus *et al.*, 1985). Estimates of the velocity of translocation are between 0.55–10.8 $cm.h^{-1}$ and the flux of carbon and phosphate are between 0.07–3.8[$nMcm^{-2}s^{-1}$] (Granlund *et al.*, 1985). When the source-sink relations are altered, transport can be shown to be both basipetal and bidirectional. However, the chemical form in which the labelled carbon is translocated within the rhizomorph is uncertain as it is transferred, metabolized and metabolically compartmented away from the stream.

Phosphate uptake by the rhizomorphs of *A. mellea* is biphasic with the two different carrier systems having different Km and Vmax values, resulting in most of the orthophosphate becoming located in the vacuole, but with a significant portion of the cytoplasmic phosphorus present in the rhizomorph as polyphosphate (Cairney *et al.*, 1988b).

Thin sections of rhizomorph tips examined by transmission electron microscopy confirm that there is a primary meristem where new hyphal elements are formed from apical initials, and secondary meristems in the lateral regions, in which secondary cross walls form (Motta, 1969, 1982; Schmid and Liese, 1970). The apical initials resemble typical basidiomycete hyphae with denser cytoplasm, more nuclei, ribosomes and non-membrane-bound fibrous bundles, but fewer and smaller vacuoles. They differentiate by synchronous nuclear divisions, and segment in many planes. The walls are probably synthesized continuously because they are able to maintain the same thickness while they expand. The rhizomorph apical region is rich in protein and nucleic acids, especially RNA, with glycogen nearby, indicating that it is a meristem (Motta, 1971).

This basic morphology was also confirmed when rhizomorphs were studied by scanning electron microscopy (Wolkinger *et al.*, 1975; Granlund *et al.*, 1984; Turner, 1991). A loose network of hyphae, the peripheral cover, can be seen over the mature parts of the rhizomorphs when it is preserved by critical point drying (Granlund *et al.*, 1984). The same technique also permits the diameter of hyphae in different regions of the rhizomorph to be measured in order to estimate their resistance to solution flow. As a result, it is most likely that solute flow-mediated transport is through the hyphae of the medulla.

If rhizomorphs growing from infected logs are incubated inside polyethylene bags, their apices enlarge sufficiently for the morphology and formation of the mucilagenous outer layers and the apical region to be investigated. Ultrastructural studies showed they were densely packed with cells (Powell and Rayner, 1983). They also confirmed the presence in the mucilage of both the long hyphae and swollen cells with a dense interior described by Hartig (1874), and revealed that mucilage originated in vesicles which coalesced with the plasma membrane in cells clustered between the mucilagenous and cellular regions of the rhizomorphs. This established a mucilage-filled space between the membrane and all parts of the cell wall. The septal plate is bridged by membrane-bound protoplasmic protuberances. After the cell wall is partially or completely digested, this material is released outside the cells.

In the apical dome, Powell and Rayner (1984) found there is a biochemically active specialized layer of several cells which may provide a short-term supply of growth materials. Each of these cells is full of mitochondria and several axial bundles of the microfilaments described by Motta (1969).

The mechanisms for rhizomorph elongation which permit some species to grow 19 mm per day may be due to a meristematic apical centre containing actively dividing cells that form the other layers (Brefeld, 1877; Motta, 1969). Rayner *et al.* (1985) hold the opposing view that rhizomorph extension is analogous to the balanced lysis mechanism proposed for hyphal extension, mediated by a plasticized apical dome driven forward by the pressure generated within a tube with rigid side walls (rind) and compensated for by branching and growth of the intercalated apical hyphae (Bartnicki-Garcia, 1973). The latter view supposes that plasticization would be bolstered by mucilage production as this disrupts the continuity of the hyphal mesh that covers the dome.

Melanization and compaction of the outer (rind) crust may provide sufficient rigidity for forward pressure to be maintained through osmotically driven flow through the medullary region, but the mechanisms underlying growth and extension of rhizomorphs are far from being completely understood (Garraway *et al.*, 1991).

Although the pattern of soil moisture influences the vertical distribution of rhizomorphs in soil, they grow towards the soil surface, possibly due to the oxygen gradient in soil (Morrison, 1976). While oxygen and carbon dioxide concentrations may be critical deeper in the soil, the seasonal desiccation of the upper soil layers probably governs vertical distribution (Morrison, 1976; Rishbeth, 1978). Since the greatest concentration of rhizomorphs are stratified only 10–20 cm deep (Hartig, 1873; Lawrence, 1910; Day, 1927; Pearce *et al.*, 1986; Ono, 1970; Redfern, 1973), trees are often more susceptible to infections initiated on the root collar and proximal part of the root system than the deeper, more peripheral roots (Bliss, 1946; Hintikka, 1974; Patton and Riker, 1959; Shaw, 1980).

The relatively long rhizomorphs of a number of species, like *Armillaria lutea*, extend from the food bases from which they originated at almost 1 metre per year (Redfern, 1973) to branch and anastomose as a 'complete network' over the surface of living roots (Lawrence, 1910; Childs and Zeller, 1929; Rishbeth, 1985a). In such situations, the profusion of rhizomorphs in soil is remarkable (Hood and Sandberg, 1989; Ono, 1965, 1970; Rykowski, 1984). Hintikka (1974) recovered 121 cm of rhizomorphs from soil with a surface area of only 100 cm^2.

Rhizomorph production in *Armillaria mellea in vitro* is stimulated by *Macrophoma* sp. and several other fungi (Watanabe, 1986), possibly by their amino acids (Behboudi *et al.*, 1987).

Basidiomes

The fruit bodies of *Armillaria* species are the gilled toadstools described in detail in Chapter 2. Basidiomes may eventually develop from old rotten trunks, stumps, or on the ground near affected roots. Some species emerge singly on the ground, others in clumps. The speckled, honey-coloured, fruiting bodies of *Armillaria mellea* typically grow in dense tufts on stumps, at the bases, or sometimes even several metres up, the trunks of dead or dying trees.

The basidiome of *Armillaria mellea* is typically about 7 cm or more tall, with a yellowish or brownish honey coloured cap about 5–15 cm in diameter. Initially, the cap is rounded and may be covered with dark brown fibrillose scales, but these frequently disappear later as it becomes flattened or depressed, with a striate margin. Their radial gills are adnate or slightly decurrent, opening pale pinkish or yellowish brown, but becoming whitish after they have released millions of colourless, elliptical, smooth, 8–9 × 5–6 μm basidiospores. These appear white *en masse*, dusting the caps of the lower fruiting bodies and anything else beneath them. The yellowish or brownish, flexible stipe is about 10–17 cm long and about 7–12 mm across, striate above the thick, white, membranous ring. The fructifications soon turn to a dark brown, slimy mass before the onset of hard frosts.

Basidiome ontogeny has been described in *Armillaria* (Hoffman, 1861; Hartig, 1874; Beer, 1911; Atkinson, 1914). When Reijnders (1963) and Watling (1985) studied the basidiome developmental pattern in some exannulate species, they found it to be monovelangiocarpic (only a universal veil encloses the hymenial primordium), but bivelangiocarpic (when the hymenium is enclosed by a partial and a universal veil) in some annulate species, a few of which might also be metavelangiocarpic (hyphae from various tissues proliferate to grow over and cover the developing hymenium). Hymenophore development is probably ruptohymenial, differentiated from the background material, and the overall development pattern is stipitocarpic in which the young primordium is a stipe-like group or bundle (fascicle) of hyphae lacking an apical area of differentiated cells.

The ultrastructure of *Armillaria* basidiomes (Singh and Bal, 1973) has been studied, as has the cytology (Guillaumin, 1986) of basidiospore production. Despite some abnormalities, meiosis and basidiospore formation appears to be similar to other hymenomycetes, the four nuclei resulting from meiosis migrate to four spores formed on the basidium. However, additional mitotic divisions may occur in the basidium. Hence, there may be more than four nuclei but, nonetheless, only four nuclei move to the top of the basidia to enter the developing basidiospores and the other nuclei degenerate. There may be two, three, or five sterigmata, not four. A few basidiospores (1%–5%) are binucleate. The haploid chromosome number (n) in the basidia of *A. lutea, A. mellea,* and *A. ostoyae* is four (Guillaumin, 1986a; Nguyen, 1980).

Basidiomes were first produced in culture by Molisch (1904), who used autoclaved bread as a substrate. Falck (1907) used basidiospore inoculum to grow basidiomes and found that light is necessary before the basidiomes develop (Falck, 1909). Since then, the basidiomes of several *Armillaria* species have been produced *in vitro* using a variety of other substrates, including autoclaved wood or woodchips, filter paper soaked in nutrients, orange fruits, maize kernels, as well as nutrient solutions or agars with various amendments, including fruit or plant extracts like malt (an extensive list is given in Guillaumin *et al.*, 1991). However, none of the existing techniques is sufficiently dependable to enable studies on the basidiome out of season (Ullrich and Anderson, 1978). Without a satisfactory synthetic culture medium, the role of inorganic nutrients, vitamins or other compounds in stimulating basidiome production is, as yet, poorly understood. Nonetheless, a complex carbohydrate source appears essential to sustain mycelial growth and basidiome development. Basidiome development is also stimulated by low concentrations of the fungicide sodium pentachlorophenolate (Rykowski, 1974; Shaw *et al.*, 1981).

Although less effective than vegetative spread from an established source like an old stump, *Armillaria ostoyae* and *A. mellea* can presumably be introduced by basidiospores, since some infections do eventually also occur in forests or plantations planted on former grassland where there was previously no soil-borne inoculum. Generally, such colonies appear in the proximity of decayed woody or leaf material, which seems an essential medium for the initial growth and union of the haploid mycelium from the basidiospores to establish before rhizomorphs radiate from the heterozygous mycelium to attack living roots.

Although basidiomes are responsible for long range dispersal and important for taxonomic studies, they are of less use in routine identification as their appearance is usually highly seasonal and can be even somewhat erratic. For example, whereas fruiting bodies of *A. mellea* tend to emerge abundantly only for a few days in October in southern England, in some years they may fail to appear at all.

Pseudosclerotia

Pseudosclerotia form within the zone lines of melanized cells after the mycelium interacts with other fungi and with the tissues of infected hosts (Radzievskaya and Bobko, 1985a,b). Although, in some respects, they resemble massive sclerotia or greatly swollen rhizomorphs, unlike them, they contain wood in the process of being digested. The relationship between these analogous structures can be demonstrated by comparing the black crust of the pseudosclerotial plate with the rind of the round rhizomorphs (Campbell, 1934; Lopez-Real, 1975). Both shield the hyphae within against microbes in the soil or rotten wood which might attack the mycelium and permit *Armillaria* species to survive and complete their life cycles in a variety of habitats. Whereas the production of a rhizomorph allows the fungus to reach out to infect and colonize beyond the roots of its hosts, there are parallels between the function of a pseudosclerotium and a 'stomach' enclosing substrates that continue to be 'digested', despite environmental fluctuations. This phenomenon differs from the compartmentalization of decayed wood in living trees (Falck, 1924; Shigo and Tippett, 1981). Wood that has been incubated with a single isolate of *Armillaria* under sterile conditions to produce characteristic dark zone-lines (Hansson and Seifert, 1987a) is used commercially for special decorative veneers (Hartig, 1894; Bavendamm, 1939; Rayner and Todd, 1979). Zone-lines can also be produced in sawdust cultures (Hopp, 1938), as well as wood (Campbell, 1934). They also form during the intra- and interspecific pairings of different isolates in agar culture (Mallett and Hiratsuka, 1986) and in wood (Hood and Morrison, 1984). Hopkin *et al.* (1989) used L-DOPA to intensify the 'black line' between species of *Armillaria*.

Basidiomycetes that decay wood often need an external stimulus to induce their genetic ability to form pseudosclerotial plates. Garraway *et al.* (1991) list three distinct mechanisms that can stimulate pseudosclerotial plate or zone-line formation:

1. Mechanical and physical factors. These include:–
   a. fluctuating moisture content (Campbell, 1934; Lopez-Real and Swift, 1975; Radzievskaya and Bobko, 1985a);
   b. gas phase composition (Lopez-Real and Swift, 1977);
   c. wounding respiration-induced damage to hyphae (Lopez-Real and Swift, 1977).
2. Antagonistic interaction of different mycelia (incompatibility reactions).
3. Genetic factors within a species.

The differentiation and development of the structure of the rhizomorphs and pseudosclerotia involves a succession of intracellular reorganizations. During these processes of differentiation and development, metabolic pathways become redirected, organelles redistributed and structural materials rearranged in ways that have not yet been fully investigated. The brittle pseudosclerotial plate of *Armillaria* consists of a layer of bladder-like melanized cells, formed in three distinct phases (Campbell, 1934; Lopez-Real, 1975):–

1. The proliferation of hyphae.
2. Hyphal swelling and aggregation.
3. Pigmentation and melanization of hyphae.

Similar morphological changes in hyphal structure result, whatever the differences in the mode of induction, the species or substrate, whether inside decayed wood or in culture (Hopp, 1938; Lopez-Real, 1975; Mallett and Hiratsuka, 1986; Rayner, 1976). Improved knowledge of this regulation process might eventually allow *Armillaria* to be managed without adversely affecting its host or other organisms.

Uptake and transport of nutrients and water

The uptake and transport of nutrients and water by mycelial cords was established (Falck, 1912) about the same time as the role the rhizomorphs play in conducting oxygen to the growing fungus was also recognized (Munch, 1909; Reitsma, 1932).

The translocation of water through a rhizomorph can be demonstrated by applying fluorescein to its base and watching it move to the tip (Schütte, 1956). When applied in a similar way to the base of a rhizomorph, radioactively labelled chloride and phosphate were translocated acropetally to its tip. However, the ions were not translocated in the reverse direction from the tip to the base, even though the ions were readily taken up by rhizomorph tips (Morrison, 1975). Morrison (1975) also found that the production of amino acids was stimulated by dipping the tips of rhizomorphs into a medium containing ammonium. Transport in rhizomorphs depends on aerobic respiration, not diffusion. Rhizomorphs living under aerobic conditions both absorb and transport C-14 labelled glucose and P-32 labelled phosphate as isotopic markers, whereas those surviving under anaerobic conditions are unable to transport the nutrients that they absorb (Anderson and Ullrich, 1982).

Although fungal nutrition and physiology may explain how *Armillaria* behaves in soil and on infected hosts, our ignorance of the biochemical factors regulating its growth and development in response to nutritional and environmental influences has limited its control *in vitro*, in soil and on its hosts (Garraway *et al*., 1991).

The growth of the mycelium and rhizomorphs of *Armillaria* on nutrient media is characteristically slower than for most other fungi. Many different agar media have been used for culture of *Armillaria* species (Hansson and Seifert, 1987b). Agar media based on roots or tubers such as carrots or cassava are suitable (Togashi and Takizawa, 1994; Popoola, 1991), but malt extract agar (MEA) supports the fastest rate of growth. Rishbeth (1976) used 3% malt agar in combination with 2.5 N lactic acid and sodium o-phenylphenate to avoid the growth of bacteria and fungal moulds which are prevalent because of the slow growth of *Armillaria* species in culture.

Rhizomorph development is associated with mineral nutrient status. Some of the nutrients that are absorbed from its food base by the growing tips of rhizomorphs may be supplemented from the soil (Morrison, 1975). Application of potassium can stimulate growth, whereas it is inhibited by nitrogen and phosphorus (Rykowski, 1984). Organic matter like peat (Redfern, 1973; Rykowski, 1984; Guillaumin and Leprince, 1979) and pine bark compost (Rykowski, 1984) are stimulatory, and rhizomorphs are more common where there is a humus layer than in the mineral soil below (Hintikka, 1974; Singh, 1981).

The induction of rhizomorph initials in *Armillaria* is controlled by nutritional factors *in vitro* (Garrett, 1953). Garraway *et al.* (1991) considered the underlying processes and mechanisms that govern the growth and development of morphological structures in *Armillaria* would be revealed once the key physiological and biochemical processes in fungi are understood. Cell-wall polysaccharides and other macromolecules, and phenoloxidizing enzymes, could be implicated, but the biochemistry of *Armillaria* is either obscure or still unknown, even though the mechanisms regulating biochemical changes entailed in fungal development have been reviewed (Burnett and Trinci, 1979; Moore *et al.*, 1985; Smith and Berry, 1978).

In fungi, cell-wall polysaccharide biosynthesis is often associated with growth in response to various stimuli (Stewart and Rogers, 1978; Sietsma and Wessels, 1977; Wang *et al.*, 1968; Wessels, 1966). Ethanol concentrations that boost the growth and development of rhizomorphs, increase the incorporation of glucose into cell-wall polysaccharides by over 50% (Garraway and Weinhold, 1968a, 1970).

During growth and development there are changes in other macromolecules which are not associated with the cell walls. A few days after ethanol is added to *Armillaria* its DNA and RNA content rises three times as much as the dry weight (Sortkjaer and Allermann, 1973). Similar rises in protein content are observed in response to ethanol concentrations which encourage the growth and development of *Armillaria* and initially boost the constituents necessary for both nuclear division and protein synthesis (Garraway *et al.*, 1991).

Rhizomorph development is strongly stimulated when ortho- and para-aminobenzoic acids (Garraway, 1970) or lipids and fatty acids (Moody and Weinhold, 1972a,b) are added to a defined basal medium. Possibly all of these organic growth factors, and some other natural substrates including lecithin, oleic acid, and linoleic acid, may employ a common mechanism (Entry *et al.*, 1992). Ortho- and para-aminobenzoic acids are linked metabolically to auxin. Ethanol is linked metabolically to lipids and fatty acids, C-14 labelled ethanol is preferentially incorporated into lipids (Garraway and Weinhold, 1968a).

Fluctuations in various enzymes associated with fungal morphogenesis are probably secondary to other more fundamental changes (Bromberg and Schwalb, 1978; Ullrich, 1977). Analysis of alcohol dehydrogenase during ethanol-induced rhizomorph formation revealed a significant increase in both the activity of the enzyme and the number of its isoenzymes in the rhizomorphs, but not in the mycelium (Mallett and Colotelo, 1984). Although alcohol dehydrogenase is essential for the metabolism of ethanol, its relevance to rhizomorph morphogenesis is uncertain.

Exposure to oxygen at partial pressures above 0.04 atm stimulates the enzymatic polymerization of phenols by p-diphenol oxidase to an impermeable layer of brown pigment, visible under the TEM, that accumulates between the hyphae in the rhizomorphs. This may inhibit their development by preventing the uptake of nutrients or the disposal of waste products (Smith and Griffin, 1971).

Laccase (phenol oxidase) stimulates rhizomorph initiation and development in *Armillaria* spp. (Robene-Soustrade and Lung-Escarmant, 1997). Its activity accumulates during rhizomorph development (Rehman and Thurston, 1992) and is also induced by ethanol and other substances that induce rhizomorphs. Curir *et al.* (1997) characterized a laccase secreted by a pathogenic *Armillaria mellea*. Laccase can be detected just before the first rhizomorphs appear, peaking when their growth is greatest and disappearing when their growth stops. Laccase is not detected in cultures that do not form rhizomorphs. The production of rhizomorphs, but not mycelium, is decreased by enzyme inhibitors active against laccase (Worrall *et al.*, 1986). Garraway *et al.* (1991) have suggested that using a different species of *Armillaria* might account for the apparent contradictory explanation of the role of phenoloxidizing enzymes by Smith and Griffin (1971). Both nitrogen and carbon sources can affect lignin and cellulose degradation (Entry *et al.*, 1993). Also, rhizomorph formation and phenoloxidizing enzyme (presumably laccase) activity increase on a synthetic medium with casein hydrolyzate as the nitrogen source when guaiacol (200 mg/1) is added, but there was no rhizomorph development when L-asparagine was substituted as the nitrogen source, though enzyme activity was maintained. Adding ethanol and guaiacol, however, increased both, suggesting they are associated with the response to ethanol and other substances but not its cause (Garraway and Edwards, 1983). Higher laccase activity was associated with greater rhizomorph production among those isolates of five species of *Armillaria* that produced them, even though some isolates that produced none had some laccase activity (Marsh and Wargo, 1989). Since phenoloxidizing enzymes appear to control the morphogenesis and differentiation of sporulating and resting structures in many different fungi, including other basidiomycetes (e.g., Leonard, 1971; Chet *et al.*, 1972; Wong and Willetts, 1974), it is probable they do so in *Armillaria* as well.

Garraway *et al.* (1991) have reviewed the nature of the phenoloxidizing enzymes produced by *Armillaria.* Mayer (1987) proposed the general terms of 'catechol oxidase' and 'laccase'. Catechol oxidase can oxidize monophenols (tyrosinase or cresolase activity) or o-diphenols (catecholase activity) but cannot oxidize p-diphenols, so this is diagnostic (Mayer and Harel, 1979). Laccase can oxidize a wide range of substrates including mono-, di-, and tri-phenols and both o- and p-diphenols, and its ability to oxidize p-diphenols is diagnostic (Mayer and Harel, 1979). Since catechol oxidase (tyrosinase) is primarily an intracellular enzyme in fungi, it may be involved in melanin formation. Laccase is excreted by fungi and involved in lignin oxidation and the degradation and detoxificiation of antifungal phenols in plant tissues (Mayer and Harel, 1979).

Peroxidase also catalyzes the oxidation of phenols by hydrogen peroxide but is non-specific. Its activity can be detected in exudates from *Armillaria* rhizomorphs using 4-amino antipyrine, a specific substrate or catechol, also oxidized by tyrosinase and laccase, so peroxidase is only responsible for a fraction of the phenol oxidase activity detected; beta-glucosidase, acid protease, and alkaline protease were also present (Mallett and Colotelo, 1984). Although peroxidase activity in *Armillaria* rhizomorph extracts had been reported by Lanphere (1934) and Lyr (1955), no specific substrate was used, nor was catalase added to destroy peroxide and eliminate peroxidase activity. Generally, the production of laccase is greatest in those species of *Armillaria* that produce rhizomorphs (Jacques-Felix, 1968; Smith and Griffin, 1971; Worrall *et al.*, 1986; Marsh and Wargo, 1989). However, in the latter, laccase activity was also present in some isolates that produced no rhizomorphs.

An unusual protease enzyme from *Armillaria* (Broadbent *et al.*, 1972) cleaves peptide bonds N-terminal to lysine residues in proteins (Hunneyball and Stanworth, 1975; Lewis *et al.*, 1978) and is neutral in charge (Barry *et al.*, 1981). Its role during basidiome formation or the rest of the developmental cycle, and in the environment, are unknown. Proteins which are insoluble in water can be fragmented because the enzyme is very stable in the presence of denaturing detergents such as sodium dodecylsulfate and can be solubilized (Barry and Doonan, 1987). Chitinolytic enzymes may originate from *Armillaria*, associated pathogenic and ectomycorrhizal fungi, or from trees (Hodge *et al.*, 1995, 1996).

Antibiotics and other metabolites

Armillaria produces antibiotics against other fungi, as well as bacteria, when cultivated on wood, solid, and liquid media (Oppermann, 1952; Richard, 1971), particularly with other fungi. Sonnenbichler *et al.* (1994) found that *A. ostoyae* produces four hundred times more toxic metabolites when grown in culture with an antagonist. Stimulation occurs before any cell contact and the inducing signals are the same toxins at low concentrations. Toxin synthesis is stimulated by the enhanced *de novo* sythesis of enzymes. *Armillaria* can also detoxify mycotoxins produced by some other fungi (Liu *et al.*, 1998). Nonetheless, the waste media remaining after the production of cultures of *Armillaria* species can act as a substrate for the cultivation of the oyster mushroom (*Pleurotus ostreatus*) (Togashi, 1996). Also, perithecium and ascospore production in *Sordaria fimicola* were stimulated by *Armillaria* (Watanabe, 1997).

At least 10 different antibiotic compounds have been reported (Ayer and MacCauley, 1987; Donnelly *et al.*, 1982; Jungshan *et al.*, 1984; Midland *et al.*, 1982; Obuchi *et al.*, 1990). Most are complicated sesquiterpenoid esters, some belonging to the protolludane group, containing the antifungal and antibacterial active Sparassol or orsellinic acid (Cwielong, 1986) bound to the same organic acid as in *Sparassis crispa* (Falck, 1907, 1909, 1924). In *Armillaria*, the sesquiterpenoids are attached to the aromatic group, allowing them to penetrate more easily through membranes and, hence, they may be more toxic. The chloroform-soluble antibiotics produced by *Armillaria* isolates are active against *Cladosporium cucumerinum* or a *Bacillus* sp. isolated from fumigated citrus roots naturally infected by *Armillaria* (Oduro *et al.*, 1976).

The various highly active antibiotics *Armillaria* produces might partly account for its success as a soil-borne pathogen. However, antibiotic production is inhibited by fumigation with methyl bromide even at sub-lethal concentrations (Ohr and Munnecke, 1974), thus exposing it to *Trichoderma* species and other potential antagonists that are normally suppressed (see also Chapter 3).

Antibiotics might also explain some of its traditional medicinal attributes. In the U.S.A., *Armillaria* fans were once wrapped round wounds caused during logging accidents, both as protection and as a styptic to enhance healing (Garraway *et al.*, 1991).

Bioluminescence

Although the bioluminescent basidiomycetes include *Armillaria* (Guyot, 1927), different isolates vary considerably in their ability, and may be inactive. One emitted light throughout its mycelium (Berliner and Hovnanian, 1963). One 15-day culture of *Armillaria* was sufficiently luminescent for photomicrography (Airth and Foerster, 1960). The light emitted by *Armillaria* has an emission maximum of 528 nm, similar to that of other fungi rather than bacteria, with an energy of activation for emission of 17,500 calories with a temperature optimum of 26°C and is present more in younger, marginal hyphae rather than the older, central area (Airth and Foerster, 1960). Possibly, bioluminescent fungi may be releasing waste energy of oxidation as light not heat. Although *Armillaria* can sustain luminescence for 10 weeks equal or longer than many other fungi, it takes longer to attain maximum light emission values (Berliner, 1961).

The luminescence of *Armillaria* and other fungi is affected by environmental factors such as temperature, exposure to X-rays and ultraviolet light, nutrition and other growth factors. Light emission is low at -10°C and negligible above 40°C, with the optimum around 18–26°C (Airth and Foerster, 1960; Berliner, 1961) and is inhibited by ultraviolet light (Berliner, 1963; Berliner and Brand, 1962), like several other bioluminescent basidiomycetes, the effect depending on the fungus, the wave-length of incident radiation, and extent of time. The bioluminescence of *Armillaria* may be enhanced by X-rays if it functions like another basidiomycete (Berliner, 1961). Nutrition also affects light emission (Harvey, 1952), a specific pH and nitrogen source ensure optimum light emission by another basidiomycete and presumably *Armillaria* (Airth and Foerster, 1965). At certain concentrations, some growth factors, such as biotin (0.75 mg/1), can intensify the luminescence of *Armillaria* by more than 150%, whereas others cannot (Berliner and LaRochelle, 1964). The effects of antibiotics on light emission have also been studied (Berliner, 1965).

Airth and Foerster (1962) proposed that the biochemical reaction leading to bioluminescence of *Armillaria* and other fungi involves either reduced nicotinamide adenine dinucleotide (NADH) or reduced nicotinamide adenine dinucleotide phosphate (NADPH), an electron acceptor found in hot water extracts, soluble dehydrogenases, molecular oxygen and the particulate enzyme luciferase. Fungal and bacterial bioluminescence, notwithstanding some apparent differences, may even share some fundamental chemistry in common with chemoluminescence (Airth *et al.*, 1966).

Nutrition

CARBON SOURCES

Confirmation that numerous sources of carbon can be used by *Armillaria* species can be demonstrated *in vitro* by their ability to utilize carbohydrates (Wargo, 1981; Weinhold and Garraway, 1966), lipids (Moody and Weinhold, 1972a,b), phenols (Cheo, 1982; Shaw, 1985; Wargo, 1983, 1984), alcohols (Weinhold, 1963; Weinhold and Garraway, 1966) and even fix carbon dioxide (Schinner and Concin, 1981). This miscellany is not unexpected considering their diverse range of hosts (Raabe, 1962a, 1979; Rishbeth, 1983; Singh and Carew, 1983), and their exploitation of organic substrates in soil (Garrett, 1960; Morrison, 1982a) and plants (Rishbeth, 1972; Wargo, 1980b).

When glucose, fructose, and sucrose were used as different sources of carbohydrates, the subsequent growth of hyphae was so poor it is likely that these carbohydrates were predominately utilized to provide the energy to stay alive, rather than for growth. However, if ethanol is added to media as the sole carbon source, or in addition to glucose, fructose, or sucrose, the hyphae and rhizomorphs grow abundantly. They are most profuse if ethanol is added to glucose, less with fructose, with sucrose the least affected correlating to the different rates of uptake and utilization of C-14 labelled sugars (Garraway, 1975). This could explain certain aspects of the ecological behaviour of pathogenic species of *Armillaria* and their hosts. While there may be limited access to host nutrients and growth promoters when the interaction is quiescent, aggressive colonization of the host is likely to require increased access to them (Wargo, 1972).

NITROGEN SOURCES

Armillaria requires an adequate source of nitrogen in a suitable form, in addition to carbon for adequate growth and development. While nitrate cannot be used as the sole source of nitrogen (Garrett, 1953), *Armillaria* can grow on ammonium tartrate, but amino acids support greater growth. When growth and development of *Armillaria* was assessed using a variety of nitrogen sources in culture, together with glucose (0.5%) as a carbon source and ethanol (0.05%) as a growth stimulant, casein hydrolyzate, which is composed of a mixture of amino acids, including glutamic acid and leucine, is superior to individual amino acids. The latter outperformed inorganic nitrogen sources such as ammonium and nitrate (Weinhold and Garraway, 1966). The efficacy of casein hydrolyzate may be correlated to amino acid uptake, which is governed by specific transport systems in fungi (Pateman and Kinghorn, 1976). Once system-specific amino acids accumulate inside the hyphae, transinhibition or transport system shut-down occurs (Horak *et al.*, 1977). Since substrates like casein hydrolyzate supply an assortment of amino acids, they enable more transport systems to operate, hence nitrogen uptake is greater. Nonetheless, the extent to which fungi can exploit an accessible nitrogen source depends on the type and size of the source of carbon. The concentration of nitrogen optimally necessary to induce rhizomorphs increases with their supply of carbohydrates (Garrett, 1953). The optimum C:N ratio for rhizomorph development varies between different isolates. Although more carbon boosts the dry weight of mycelium when added at some levels of nitrogen, growth is inhibited at some carbon levels if nitrogen is raised above a certain level (Rykowski, 1976).

OTHER INORGANIC NUTRIENTS

Although it has not attracted experimentation, the availability of inorganic ions appears to influence the behaviour of *Armillaria* in soil (Morrison, 1975). No doubt inorganic nutrients perform the same physiological functions in *Armillaria* as in other fungi (Garraway and Evans, 1984). Hence, more substantial amounts of magnesium, phosphorus, potassium, sulphur, and to a lesser extent, calcium, are likely to be required than the minute quantities of copper, iron, magnesium, zinc, and possibly molybdenum.

VITAMINS

Certain vitamins are important for growth. Fungi are more likely to be auxoheterotrophic for thiamine than for any other vitamin (Garraway and Evans, 1984) as it is essential for growth, unlike biotin (Garrett, 1953; Garraway, 1966). Like many other decay fungi, *Armillaria* is able to synthesize the vitamins it requires from simple precursors with the exception of thiamine which, as thiamine pyrophosphate, serves as the coenzyme necessary for several enzymes of intermediate metabolism. These catalyze the removal or transfer of aldehyde groups, including pyruvate carboxylaylase, transketolase, pyruvate dehydrogenase, and alpha-ketoglutarate dehydrogenase.

ORGANIC GROWTH FACTORS

Major effects on the growth and development of *Armillaria* are also stimulated by some organic compounds like the alcohols, auxin and related compounds, fatty acids, and phenols and related compounds. These operate at concentrations far below those of inorganic nutrients, such as carbon and nitrogen, but substantially above those normal for vitamins. Initially, undefined substrate supplements, such as yeast or figwood extract, were considered vital to support optimal growth and development of *Armillaria* in defined media (Raabe, 1962; Weinhold *et al.*, 1962). Subsequently, this was guaranteed by the addition of low-molecular-weight alcohols and related compounds, thus permitting critical studies on its nutrition and physiology (Weinhold, 1963). Low-molecular-weight alcohols, as well as being sources of carbon, act as organic growth factors. Growth and rhizomorph formation were greatly stimulated by the addition of 0.05% of either ethanol, 1-propanol, or 1-butanol to a synthetic medium containing 0.5% glucose as the sole carbon source (Weinhold, 1963; Weinhold and Garraway, 1966). Different isolates of *Armillaria* can differ significantly in their reaction to different alcohols (Allermann and Sortkjaer, 1973). Low-molecular-weight alcohols produced either by soil micro-organisms (Pentland, 1965, 1967) or in tree roots (Coutts and Armstrong, 1976; Crawford and Baines, 1977) may influence the ecology of *Armillaria* in the soil.

The majority of the growth occurred after the ethanol supplement had been depleted from the medium, implying that *Armillaria* utilized glucose as a carbon source after it had adapted to the ethanol supplement (Garraway and Weinhold, 1968b). The growth rate also rose significantly when ethanol was added to synthetic media that had already been incubated for 7 days (Garraway and Weinhold, 1970) or 15 days (Sortkjaer and Allermann, 1972). Increases in growth in response to ethanol coincided with decreases in the short-term uptake and utilization of glucose (Garraway and Weinhold, 1968a, 1970) and increased uptake of nitrogen and phosphate while the rate of DNA and RNA accumulation increased (Sortkjaer and Allermann 1973).

The production of *Armillaria* rhizomorphs is also significantly stimulated by some compounds with auxin activity, like indole-3-acetic acid above 10 mg/l in synthetic media (Garraway, 1970, 1975) and 2,4-dichlorophenoxyacetic acid (2,4-D) which also boosted their growth rate (Pronos and Patton, 1979). The stimulation of *Armillaria* rhizomorph growth by auxins in higher plants (Key, 1969; Key *et al.*, 1967; Rayle, 1973) may result from the formation of new proteins in the cytoplasm from MRNA synthesized after the stimulation of RNA polymerase in the nucleus by a factor formed when auxin interacts with the plasma membrane which moves through the cytoplasm into the nucleus (Key, 1969).

When added to a defined basal medium, lipids and fatty acids (Moody and Weinhold, 1972a,b) and ortho- and para-aminobenzoic acid (Garraway, 1970) encourage development of rhizomorphs, conceivably a common mechanism is involved since there are metabolic links between them and ethanol (Garraway and Weinhold, 1968a) and auxin (Garraway *et al.*, 1991).

Phenoloxidizing enzymes

The nature of the phenoloxidizing enzymes and their production by *Armillaria* needs to be evaluated as they appear to affect rhizomorph morphogenesis. Mayer (1987) and Mayer and Harel (1979) devised the terminology to describe the phenoloxidizing enzymes and their substrates. Laccase can oxidize a wide range of substrates, including mono-, di-, and tri-phenols. As it can oxidize both o- and p-diphenols, its ability to oxidize p-diphenols is diagnostic (Mayer and Harel, 1979). In the fungi, catechol oxidase (tyrosinase) is essentially an intracellular enzyme with a possible function in melanin formation. Although fungi excrete laccase, in plant tissues it is involved in lignin oxidation, degradation and detoxificiation of antifungal phenols (Mayer and Harel, 1979).

The oxidation of phenols by peroxidase is also catalyzed by hydrogen peroxide, but is non-specific for phenols. If the hydrogen peroxide in cell-free preparations is not removed, it could be mistaken for polyphenol oxidase activity (Mallett and Colotelo, 1984). Robene-Soustrade *et al.* (1992) have identified and partially characterized an extracellular manganese-dependent peroxidase in *Armillaria ostoyae* and *A. mellea.* Although peroxidase was also reported in extracts of *Armillaria* rhizomorphs (Lanphere, 1934; Lyr, 1955), no specific substrate was used, nor was catalase added to extract and destroy hydrogen peroxide and eliminate peroxidase activity. Also, while tyrosinase and laccase activities have been reported in extracts of *Armillaria* mycelium (Kaarik, 1965), laccase can oxidize both tyrosine and guaiacol (1 and o-diphenols), the two substrates used. Genuine laccase activity in culture liquid by species of *Armillaria* using 2,6 dimethoxyphenol and p-phenylenediamine as substrates revealed a positive correlation between high production and the number of rhizomorphs produced. For example, while isolates of *A. ostoyae* recorded low laccase activity and low rhizomorph production, and *A. lutea* had a broad range of laccase activities and rhizomorph production (Worrall *et al.*, 1986).

Phenol oxidase in five North American *Armillaria* species were measured (Marsh and Wargo, 1989). Although peroxidase was not detected in all isolates, some isolates had detectable levels. There was some intracellular tyrosinase activity. Generally, high laccase activity again correlated positively with rhizomorph production, but some isolates that produced no rhizomorphs had some laccase activity.

Cytogenetics

The precise basis of much of the major genetic diversity within the genus *Armillaria* is still obscure. There are substantial differences in the physiological and biochemical processes, cultural characteristics (Raabe, 1966), virulence (Raabe, 1967), reaction to environmental stimuli, nutritional responses to low-molecular-weight alcohols (Allermann and Sortkjaer, 1973; Hong *et al.*, 1990), gallic acid (Cheo, 1982; Shaw, 1985; Wargo, 1980a), and light (Benjamin, 1983; Doty and Cheo, 1974). Virus-like particles might also be the cause of some of the differences seen among *Armillaria* isolates (Reaves *et al.*, 1988). Nonetheless, like other genera, true intraspecific variation occurs in rhizomorph branching patterns, basidiome and vegetative morphology, pathogenicity, and physiological and biochemical characteristics polymorphism, isoenzyme profiles (Lin *et al.*, 1989; Morrison *et al.*, 1985b; Robene-Soustrade and Lung-Escarmant, 1997) and restriction fragment patterns in nuclear (Anderson *et al.*, 1987, 1989; Anderson and Smith, 1989) and mitochondrial DNA (Jahnke *et al.*, 1987; Anderson and Smith, 1989). There has been some debate whether the hyphae formed by the fusion of the haploid monokaryotic hyphae of *Armillaria mellea* from basidiospores becomes diploid or remains dikaryotic. The evidence points to a diploid mycelium in *Armillaria mellea* (Lamoure and Guillaumin, 1985) but this is not necessarily the case, even in closely related species. The effects of this, if any, on their behaviour as pathogens possibly deserves more attention.

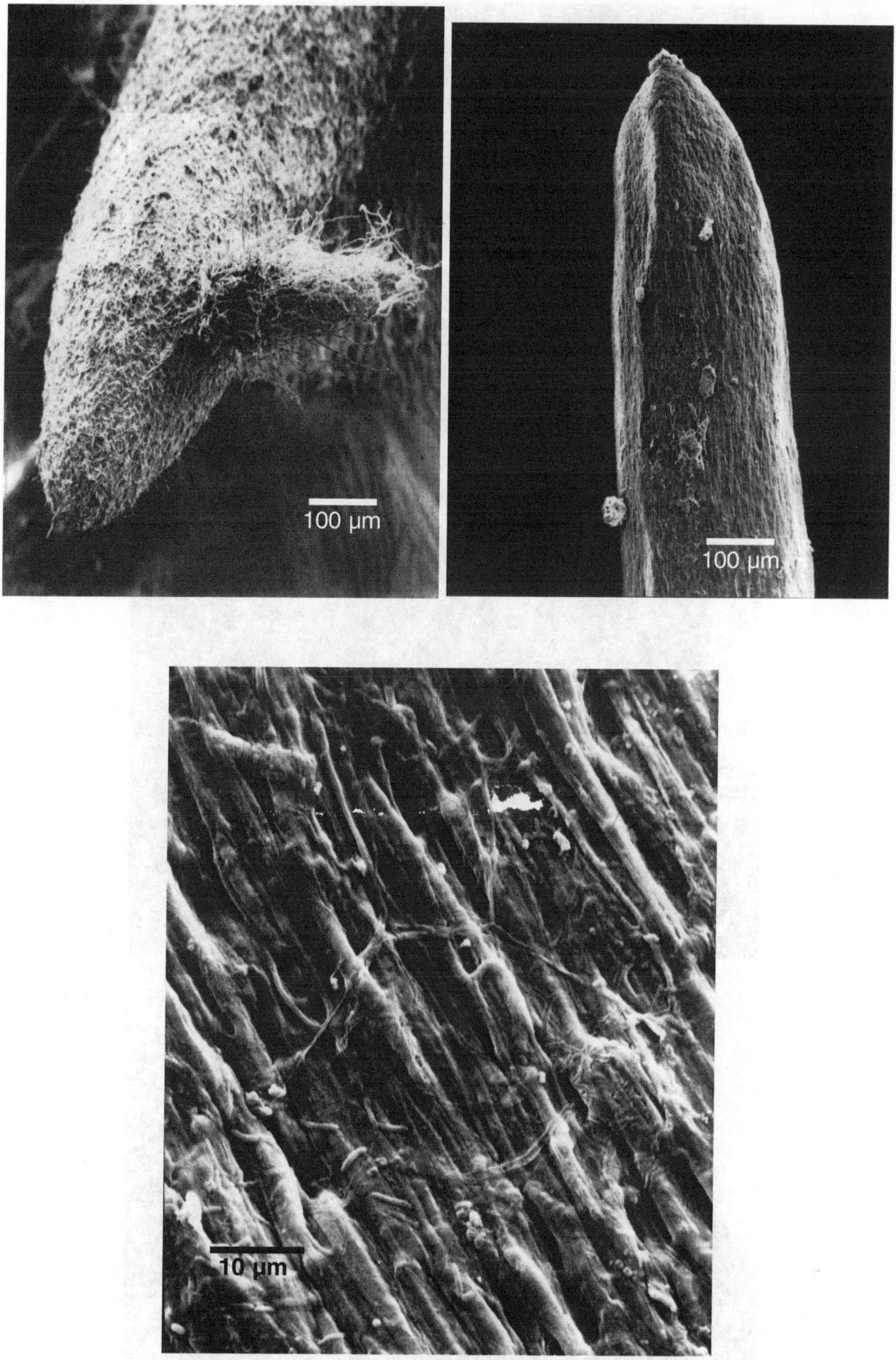

Figure 1.4a–c. Stereo electron micrographs of rhizomorphs of *Armillaria lutea.* These photographs illustrate the appearance of the external surface of the rhizomorph and its general shape: (a) mycelium which is extruded through the outer cortex of the rhizomorph from the core appears to constitute a primary contact between *Armillaria* and its host; (b) tip of rhizomorph; (c) surface of rhizomorph.

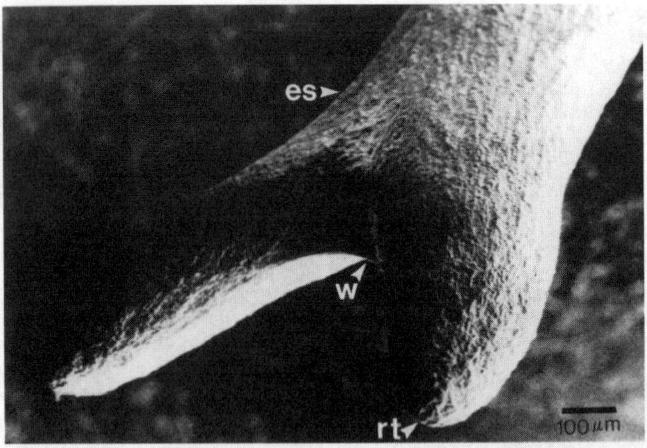

Figure 1.4d. Ultramicroscopic examination of rhizomorphs of *Armillaria lutea.* External surface: es, external surface; w, web of tissue a point of branching of rhizomorph; rt, rhizomorph tip.

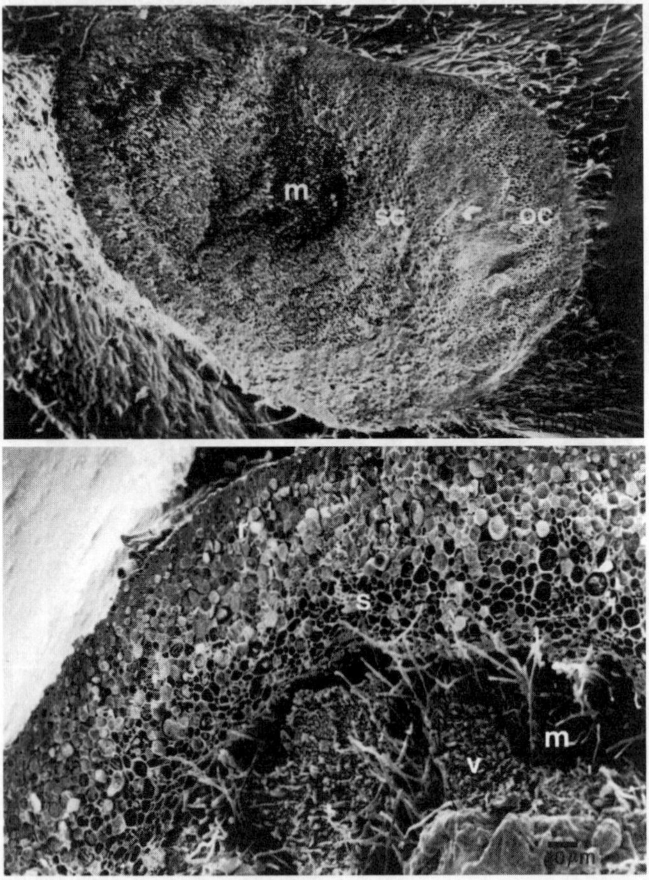

Figure 1.4e and 1.4f. Ultramicroscopic examination of freeze fractured rhizomorph of *Armillaria lutea.* **Figure 1.4e.** Transverse section of rhizomorph tip. **Figure 1.4f.** Transverse section 7 cm behind tip. r, rind; oc, outer cortex; sc, sub-cortex; v, vessel hyphae; m, medulla.

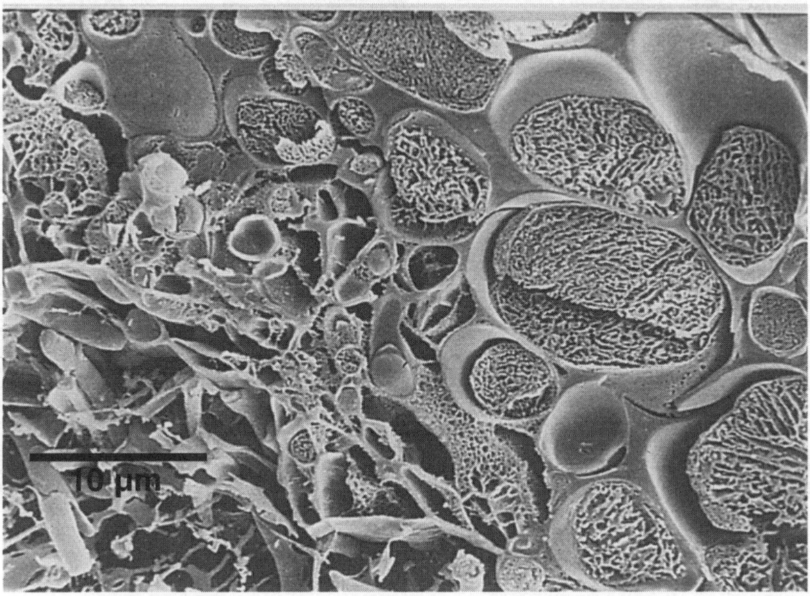

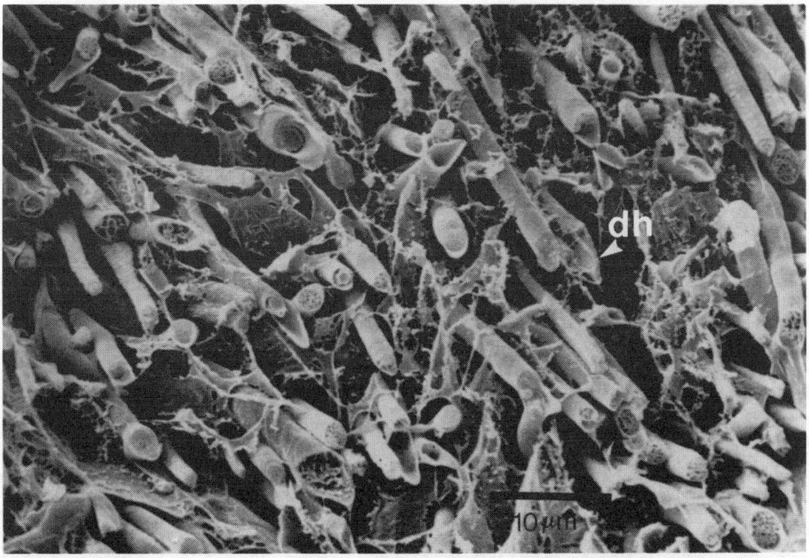

Figure 1.4g and 1.4h. Ultramicroscopic examination of a transverse section 7 cm behind tip of freeze fractured rhizomorph of *Armillaria lutea*. **Figure 1.4g.** Outer cortex. **Figure 1.4h.** Central core or medulla, dh, densely packed hyphae.

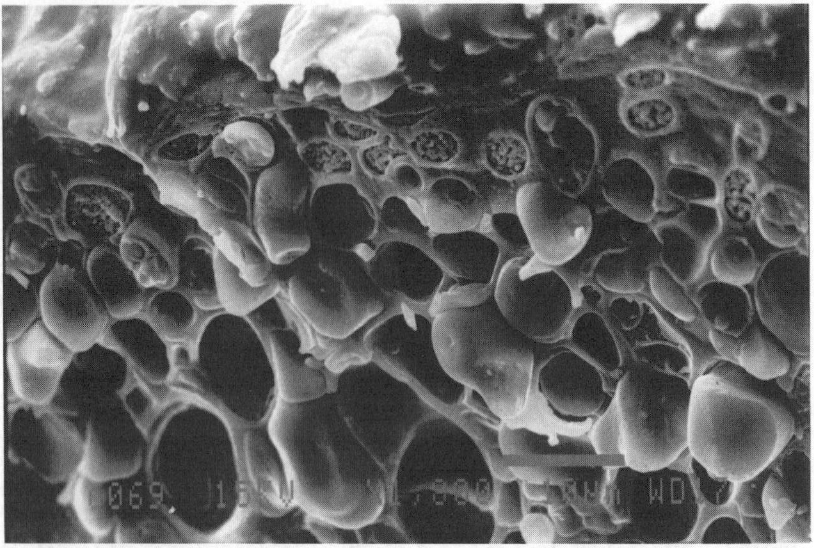

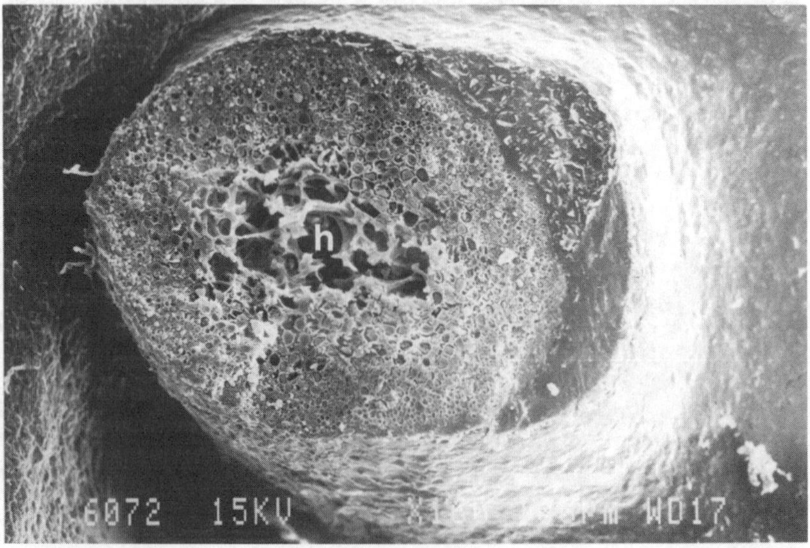

Figure 1.4i and 1.4j. Ultramicroscopic examination of a transverse section of freeze fractured rhizomorph of *Armillaria mellea.* **Figure 1.4i.** Outer cortex. **Figure 1.4j.** Central core or medulla, h, central hole.

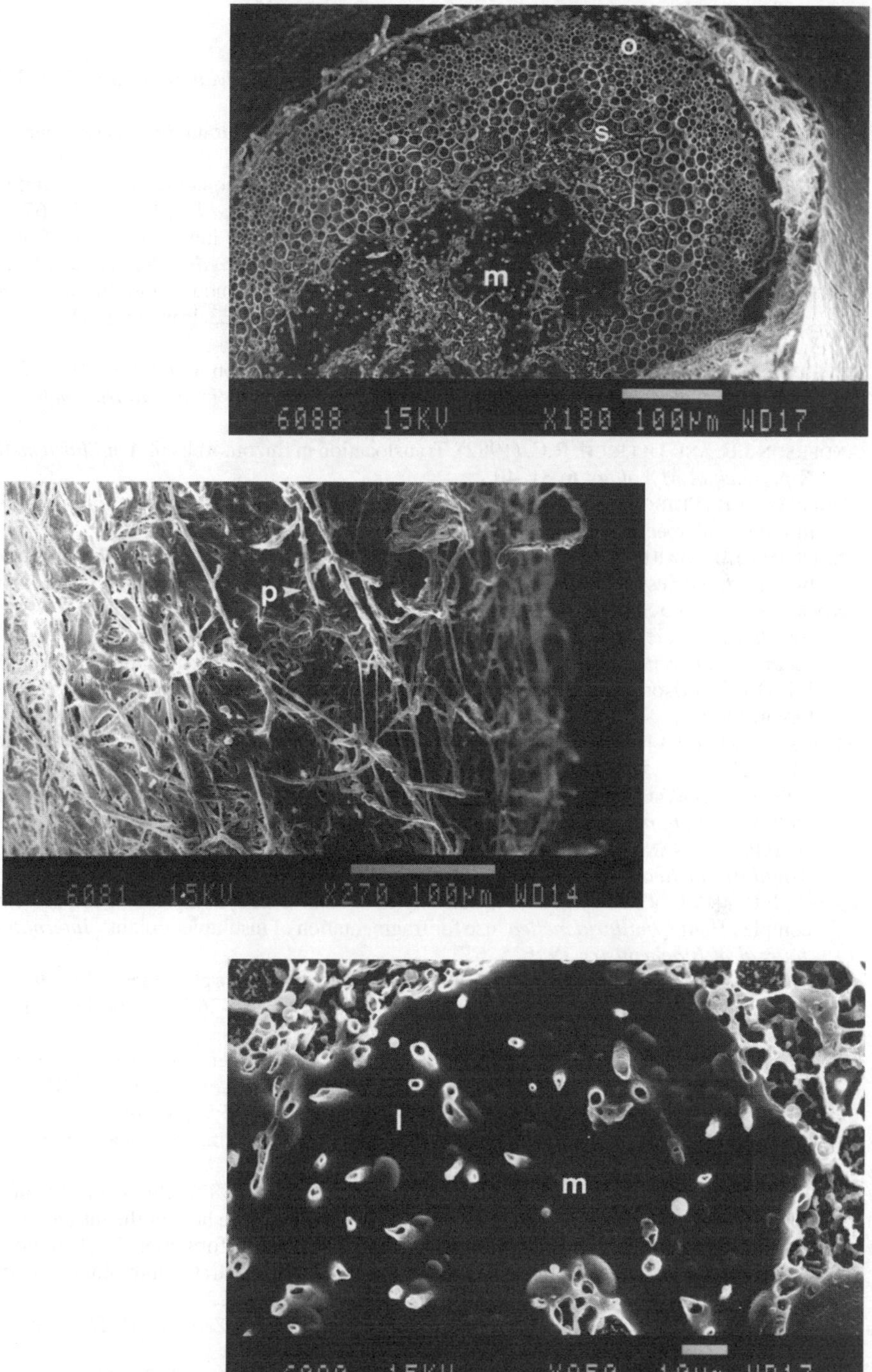

Figure 1.4k–m. Stereo electron micrographs of a transverse section 1 cm behind tip of freeze fractured rhizomorph of *Armillaria ostoyae*. p, peripheral hyphae; oc, outer cortex; sc, sub-cortex; m, medulla; l, liquid or mucilage filling medulla.

References

AINSWORTH, G.C. AND SUSMAN, A.S., eds. (1965). *The fungi: an advanced treatise. Vol. 3. The fungal population*, 738pp. New York: Academic Press.

AIRTH, R.L. AND FOERSTER, G.E. (1960). Some aspects of fungal bioluminescence. *Journal of Cellular and Comparative Physiology* **56**, 173–182.

AIRTH, R.L. AND FOERSTER, G.E. (1962). The isolation of catalytic components required for cell-free fungal bioluminescence. *Archives of Biochemistry and Biophysics* **97**, 567–573.

AIRTH, R.L. AND FOERSTER, G.E. (1965). Light emission from the luminous fungus *Collybia velutipes* under different nutritional conditions. *American Journal of Botany* **52**, 495–505.

AIRTH, R.L., FOERSTER, G.E. AND BEHRENS, P.Q. (1966). The luminous fungi. In: *Bioluminescence in progress*. Eds. F.H. Johnson and Y. Haneda, pp 203–223. Princeton, NJ: Princeton University Press.

ALLERMANN, K. AND SORTKJAER, O. (1973). Rhizomorph formation in fungi, H. The effect of 12 different alcohols on growth and rhizomorph formation in *Armillaria mellea* and *Clitocybe geotropa. Physiologia Plantarum* **28**, 51–55.

ANDERSON J.B. AND ULLRICH, R.C. (1982). Translocation in rhizomorphs of *Armillaria mellea. Experimental Mycology* **6**, 31–40.

ANDERSON, J.B., PETSCHE, D.M. AND SMITH, M.L. (1987). Restriction fragment polymorphisms in biological species of *Armillaria mellea. Mycologia* **79**, 69–76.

ANDERSON, J.B., BAILEY, S.S. AND PUKKILA, P. (1989). Variation in ribosomal DNA among biological species of *Armillaria*, a genus of root-infecting fungi. *Evolution* **43**, 1652–1662.

ANDERSON, J.B. AND SMITH, M.L. (1989). Variation in ribosomal and mitochondrial DNAs in *Armillaria* species. Sex and evolution in *Armillaria*. In: *Proceedings of the 7th international conference on root and butt rots*, August 9–16, 1988, Vernon and Victoria, B.C. Ed. D.J. Morrison, pp 60–71. Victoria, B.C.: International Union for Forestry Research Organizations.

ATKINSON, G.F. (1914). The development of *Armillaria mellea. Mycologisches Zentrablatt* **4**, 113–121.

AYER, W.A. AND MACCAULAY, J.B. (1987). Metabolites of the honey mushroom, *Armillaria mellea. Canadian Journal of Chemistry* **65**, 7–14.

BARRY, F.P., DOONAN, S. AND ROSS, C.A. (1981). Cleavage of trypsin by the proteinase from *Armillaria mellea* N-formyl-lysine residues. *Biochemical Journal* **193**, 737–742.

BARRY, F.P. AND DOONAN, S. (1987). Stability to sodium dodecyl sulphate of the proteinase complex from *Armillaria mellea*: use for fragmentation of insoluble proteins. *International Journal of Biochemistry* **19**, 625–632.

BARTNICKI-GARCIA, S. (1973). Fundamental aspects of hyphal morphogenesis. In: *Microbial differentiation*. Eds. J.W. Ashworth and J.E. Smith, pp 245–267. Cambridge, England: Cambridge University Press.

BAVENDAMM, W. (1939). Erkennen, nachweis und kultur der holzverfar-benden und holzzersetzenden pilze. In: *Handbuch der biologischen arbeitsmethoden*, Abt. XII, Part 2, Vol. 3. Ed. E. Abderhalden, pp 927–1134. Berlin: Urban and Schwarzenberg.

BEER, R. (1911). Notes on the development of the carpophore of some Agaricaceae. *Annals of Botany* **25**, 683–689.

BEHBOUDI, B.C., EBRAHIMZADEH, H. AND HADADTCHI, G. (1987). Study of rhizomorph formation by *Armillariella mellea* (Vahl ex Fr.) Karst. with the help of the nitrate medium alternative ammonium and certain amino acids. [Controle de la formation des rhizomorphes d'*Armillariella mellea* (Vahl ex Fr.) Karst. par l'alternative nitrate-ammonium et certains acides amines]. (In French.) *Cryptogamie, Mycologie* **8** (3), 227–234.

BENJAMIN, M. (1983). *Studies on the biology of* Armillaria *in New Zealand*. Ph.D. dissertation. 135pp. Auckland, New Zealand: University of Auckland.

BERLINER, M.D. (1961). Studies in fungal luminescence. *Mycologia* **53**, 84–90.

BERLINER, M.D. (1963). The action of monochromatic ultraviolet radiation on luminescence in *Armillaria mellea. Radiation Research* **19**, 392–401.

BERLINER, M.D. (1965). Effects of physiologically active chemical and antibiotics on light emission by three basidiomycetes. *Canadian Journal of Microbiology* **11**, 291–295.

BERLINER, M.D. AND BRAND, P.B. (1962). Effects of monochromatic ultraviolet light on luminescence in *Panus stipticus*. *Mycologia* **54**, 415–421.

BERLINER, M.D. AND HOVNANIAN, H.P. (1963). Autophotography of luminescent fungi. *Journal of Bacteriology* **86**, 339–341.

BERLINER, M.D. AND LAROCHELLE, M.F. (1964). Effects of plant growth substances on light emission and growth of luminescent fungi. *Developments in Industrial Microbiology* **5**, 390–398.

BÉRUBÉ, J.A. AND DESSUREAULT, M. (1988). Morphological characterisation of *Armillaria ostoyae* and *Armillaria sinapina* sp. nov. *Canadian Journal of Botany* **66**, 2027–2034.

BLISS, D.E. (1946). The relation of soil temperature to the development of Armillaria root rot. *Phytopathology* **36**, 302–318.

BLOOMFIELD, B.J. AND ALEXANDER, M. (1967). Melanins and resistance of fungi to lysis. *Journal of Bacteriology* **93**, 1276–1280.

BOCCHI, A., BRACCHI, P.G., DELBONO, G. AND CADONICI, O. (1995). Report on mushroom poisoning in the Parma area. (In Italian.) *Annali della Facolta di Medicina Veterinaria, Universita di Parma* **15**, 251–256.

BREFELD, O. (1877). *Botanische untersuchungen ber Schimmelpilze, Vol. III*. Lepzig: Felix.

BROADBENT, D., TURNER, R.W. AND WALTON, P.L. (1972). BR Patent 1263956.

BROMBERG, S.K. AND SCHWALB, M.N. (1978). Sporulation in *Schizophyllum commune*: Changes in enzyme activity. *Mycologia* **70**, 481–486.

BURNETT, J.H. AND TRINCI, A.P.J., eds. (1979). Fungal cell walls and hyphal growth. In: *Proceedings of the symposium of the British Mycological Society*, April, 1978, p 418. Cambridge, England: Cambridge University Press.

CAIRNEY, J.W.G., JENNINGS, D.H., RATCLIFFE, R.G. AND SOUTHON, T.E. (1988a). The physiology of basidiomycete linear organs. II. Phosphate uptake by rhizomorphs of *Armillaria mellea*. *New Phytologist* **109** (3), 327–333.

CAIRNEY, J.W.G., JENNINGS, D.H. AND VELTKAMP, C.J. (1988b). Structural differentiation in maturing rhizomorphs of *Armillaria mellea* (Tricholomatales). *Nova Hedwiga* **46** (1–2), 1–25.

CAMPBELL, A.H. (1934). Zone lines in plant tissues, II. The black lines formed by *Armillaria mellea* (Vahl) Quél. *Annals of Applied Biology* **21**, 1–22.

CHEO, P.C. (1982). Effects of tannic acid on rhizomorph production by *Armillaria mellea*. *Phytopathology* **72**, 676–679.

CHET, I., RETIG, N. AND HENIS, Y. (1972). Changes in total soluble proteins and in some enzymes during morphogensis of *Sclerotium rolfsii*. *Journal of General Microbiology* **72**, 451–456.

CHILDS, L. AND ZELLER, S.M. (1929). Observations on Armillaria root rot of orchard trees. *Phytopathology* **19**, 869–873.

COUTTS, M.P. AND ARMSTRONG, W. (1976). Role of oxygen transport in the tolerance of trees to waterlogging. In: *Tree physiological yield improvement*. Eds. M.G.R. Cannell and F.T. Last, pp 361–385. New York: Academic Press.

CRAWFORD, R.M.M. AND BAINES, M.A. (1977). Tolerance of anoxia and the metabolism of ethanol in tree roots. *New Phytologist* **79**, 519–526.

CURIR, P., THURSTON, C.F., D'AQUILA, F., PASINI, C. AND MARCHESINI, A. (1997). Characterization of a laccase secreted by *Armillaria mellea* pathogenic for *Genista*. *Plant Physiology & Biochemistry (Paris)* **35** (2), 147–153.

CWIELONG, P. (1986). *Mechanismen der Resistenz und Pathogenität von fungizid wirksamen Naturstoffen gegenüber dem Erreger der Rotfäule* Heterobasidion annosum *(Fr.) Bref.* Ph.D. dissertation. 210pp. Gottingen, West Germany: Universität Gottingen.

DADE, H.A. (1927). 'Collar crack' of Cacao [*Armillaria mellea* (Vahl) Fr.]. *Gold Coast: Department of Agriculture; Bulletin* **5**, 7–21.

DAY, W.R. (1927). The parasitism of *Armillaria mellea* in relation to conifers. *Quarterly Journal of Forestry* **21**, 9–21.

DE BARY, A. (1869). Morphologie und physiologie de pilze, flechten und mycetozoa. In: *Handbuch der physiologischen botanik: heft II*. Ed. W.F.B. Hoffmeister. Leipzig: Engelmann.

DE BARY, A. (1887). *Comparative morphology and biology of the Fungi, Mycetozoa and bacteria*, 525pp. Oxford: Clarendon Press.

DONNELLY, D.M.X., SANADA, S. AND O'REILLY, J. (1982). Isolation and structure (x-ray analysis) of the orsellinate of Armillol, a new antibacterial metabolite from *Armillaria mellea*. *Journal of the Chemical IV, Society: Chemical Communications* **2**, 135–137.

DONNELLY, D.M.X., ABE, F., COVENEY, D., FUKADA, N., O'REILLY, J., POLONSKY, J. AND PRANGE, T. (1985). Antibacterial sesquiterpene aryl esters from *Armillaria*. *Journal of Natural Products* **48**, 10–16.

DONNELLY, D.M.X. AND HUTCHINSON, R.H. (1990). Armillane, a saturated sesquiterpene ester from Armillaria mellea. *Phytochemistry* **29** (1), 179–182.

DONNELLY, D.M.X., HUTCHINSON, R.M., COVENEY, D. AND YONEMITSU, M. (1990). Sesquiterpene aryl esters from *Armillaria mellea*. *Phytochemistry* **29** (8), 2569–2572.

DONNELLY, D.M.X., KONISHI, T., DUNNE, O. AND CREMIN, P. (1997). Sesquiterpene aryl esters from *Armillaria tabescens*. *Phytochemistry* **44** (8), 1473–1478.

DOTY, J.E. AND CHEO, P.C. (1974). Light inhibition of thallus growth of *Armillaria mellea*. *Phytopathology* **64**, 763–764.

EAMUS, D. AND JENNINGS, D.H. (1984). Determination of water, solute and turgor potentials of mycelium of various basidiomycete fungi causing wood decay. *Journal of Experimental Botany* **35** (161), 1782–1786.

EAMUS, D., THOMPSON, W., CAIRNEY, J.W.G. AND JENNINGS, D.H. (1985). Internal structure and hydraulic conductivity of basidiomycete translocating organs. *Journal of Experimental Botany* **36** (168), 1110–1116.

ENTRY, J.A., DONNELLY, P.K. AND CROMACK, K., JR. (1992). The influence of carbon nutrition on *Armillaria ostoyae* growth and phenolic degradation. *European Journal of Forest Pathology* **22** (2–3), 149–156.

ENTRY, J.A., DONNELLY, P.K. AND CROMACK, K., JR. (1993). Effect of nitrogen and carbon sources on lignin and cellulose degradation by Armillaria ostoyae. *European Journal of Forest Pathology* **23** (3), 129–137.

FALCK, R. (1907). Wachstumsgesetze, wachstumsfaktoren und temperaturwerte der holzzerstörenden mycelien. *Hausschwamm Forschungen* **1**, 53–154.

FALCK, R. (1909). Die lenzitesfäule des coniferenholzes. *Hausschwamm Forschungen* **3**, 1–234.

FALCK, R. (1912). Die merulius-fäule des bauholzes. *Hausschwamm Forschungen* **6**, 1–405.

FALCK, R. (1924). Über das eichensterben im regierungsbezirk stralsund nebst beiträgen zur biologie des hallimaschs und eichenmehltaus. In: *Festschrift zur feier der einführung der neuen hochschulverfassung an der seitherigen forstakademie hann.* Münden am 3. Mai 1923. Ed. L. Rhumbler, pp 57–75. Frankfurt, West Germany: Sauerlanders.

FALCK, R. (1930). Neue Mitteilungen über die Rotfaule. *Mitt. Forstwirtsch. Forstwiss.* **1**, 525–566.

FEILER, S., TESCHE, M., ZENTSCH, W., SCHMIDT, P.A. AND BELLMANN, C. (1992). Investigations on water relations of spruce infected by Armillaria sp. [Untersuchungen zum Wasserhaushalt der Fichte bei Befall durch Armillaria sp.] (In German.) *European Journal of Forest Pathology* **22** (6–7), 329–336.

FOX, R.A. (1964). A report on a visit to Nigeria (9–30 May, 1963) undertaken to make a preliminary study of root diseases of rubber. Document Research Archives, Rubber Research Institute of Malaya 27. 34pp. [*Review of Applied Mycology* **43**, 3003.]

FOX, R.T.V. (1990). Diagnosis and control of Armillaria honey fungus root rot of trees. *Professional-Horticulture* **4** (3), 121–127.

FOX, R.T.V. (1993). *Principles of Diagnostic Techniques in Plant Pathology*. Wallingford: CAB International.

FOX, R.T.V. AND HAHNE, K. (1988a). Prospects for the rapid diagnosis of *Armillaria* by monoclonal antibody ELISA. In: *Proceedings of the 7th International Conference IUFRO Working Party S2.06.01 on Root and Butt Rots of Forest Trees*, August 9–16, 1988, Vernon and Victoria, British Columbia, Canada. Ed. D.J. Morrison, pp 458–468. British Columbia, Canada: International Union of Forestry Research Organizations.

FOX, R.T.V. AND HAHNE, K. (1988b). A procedure for the diagnosis of *Armillaria* by monoclonal antibody ELISA. *Abstracts of Papers, The 5th International Congress of Plant Pathology*. Kyoto, Japan. August 20–27, 1988, p 379.

FOX, R.T.V. AND HAHNE, K. (1989). Separation of the *Armillaria* complex using monoclonal antibodies. *Abstracts of Papers, The British Society for Plant Pathology/British Crop Protection Council Conference on Techniques for the Rapid Diagnosis of Plant Disease.* University of East Anglia, Norwich. July 11–13, 1989, p 15.

FOX, R.T.V., AND POPOOLA, T.O.S. (1990). Induction of fertile basidiocarps in *Armillaria bulbosa. Mycologist* **4** (2), 70–72.

FOX, R.T.V., MANLEY, H.M., CULHAM, A., HAHNE, K. AND TIFFIN, A.I. (1993). Methods for detecting *Armillaria mellea.* In: *Ecology of Plant Pathogens.* Ed. J.P. Blakeman. Oxford: Blackwells/BSPP.

GARRAWAY, M.O. (1966). *Nutrition and metabolism of* Armillaria mellea *(Vahl ex Fr.) Quél in relation to growth and rhizomorph formation.* Ph.D. dissertation. 113pp. Berkeley: University of California.

GARRAWAY, M.O. (1967). Nutrition and metabolism of *Armillaria mellea* (Vahl ex Fr.) in relation to growth and rhizomorph formation. *Diss. Abstr. Serv.* B.E.28., 16B.

GARRAWAY, M.O. (1970). Rhizomorph initiation and growth in *Armillaria mellea* promoted by o-aminobenzoic and p-aminobenzoic acids. *Phytopathology* **60**, 861–865.

GARRAWAY, M.O. (1973). Promotion of rhizomorph growth in *Armillaria mellea* by plant growth hormones in relation to the concentration and type of carbohydrate. *Abstract No 872, 2nd International Congress on Plant Pathology*. Minneapolis, Minn.

GARRAWAY, M.O. (1975). Stimulation of *Armillaria mellea* growth by plant hormones in relation to the concentration and type of carbohydrate. *European Journal of Forest Pathology* **5** (1), 35–43.

GARRAWAY, M.O. AND WEINHOLD, A.R. (1968a). Influence of ethanol on the distribution of glucose ^{14}C assimilated by *Armillaria mellea. Phytopathology* **58**, 1652–1657.

GARRAWAY, M.O. AND WEINHOLD, A.R. (1968b). Period of access to ethanol in relation to carbon utilization and rhizomorph initation and growth in *Armillaria mellea. Phytopathology* **58**, 1190–1191.

GARRAWAY, M.O. AND WEINHOLD, A.R. (1970). *Armillaria mellea* infection structures: rhizomorphs. In: *Root diseases and soil-borne pathogens: Proceedings of the symposium: July 1968; London: Imperial College.* Eds. T.A. Toussoun, R.V. Bega and P.E. Nelson, pp 122–124. Berkeley: University of California Press.

GARRAWAY, M.O. AND EDWARDS, D.F. (1983). Casein hydrolyzate enhances rhizomorph production and polyphenoloxidase activity in *Armillaria mellea. Phytopathology* **73**, 815. Abstract.

GARRAWAY, M.O. AND EVANS, R.C. (1984). *Fungal nutrition and physiology.* 401pp. New York: Wiley.

GARRAWAY, M.O., HÜTTERMANN, A. AND WARGO, P.M. (1991). Ontogeny and Physiology. In: *Armillaria Root Disease.* Eds. C.G. Shaw III and G.A. Kile, pp 21–47. *Forest Service Handbook No. 691.* Washington, D.C.: USDA.

GARRETT, S.D. (1953). Rhizomorph behaviour in *Armillaria mellea* (Vahl) Quél. I. Factors controlling rhizomorph initiation by *Armillaria mellea* in pure culture. *Annals of Botany, London* **17**, 63–79.

GARRETT, S.D. (1956). Rhizomorph behaviour in *Armillaria mellea* (Vahl) Quél. I. Logistics of infection. *Annals of Botany (n.s)* **20**, 193–209.

GARRETT, S.D. (1960). Rhizomorph behaviour in *Armillaria mellea* (Vahl) Quél. III. Saprophytic colonization of woody substrates in the soil. *Annals of Botany (n.s)* **24**, 275–285.

GRAMSS, G. (1983). Examination of low-pathogenicity isolates of *Armillaria mellea* from natural stands of *Picea abies* in Middle-Europe. *European Journal of Forest Pathology* **13**, 142–151.

GRANLUND, H.I., JENNINGS, D.H. AND VELTKAMP, K. (1984). Scanning electron microscope studies of rhizomorphs of *Armillaria mellea. Nova Hedwiga* **39** (1–2), 85–100.

GRANLUND, H.I., JENNINGS, D.H. AND THOMPSON, W. (1985). Translocation of solute along rhizomorphs of *Armillaria mellea. Transactions of the British Mycological Society* **84** (1), 111–119.

Gray, S.N., Dighton, J. and Jennings, D.H. (1996). The physiology of basidiomycete linear organs. III. Uptake and translocation of radiocaesium within differentiated mycelia of *Armillaria* spp. growing in microcosms and in the field. *New Phytologist* **132** (3), 471–482.

Gregory, S.C. (1985). The use of potato tubers in pathogenicity studies of *Armillaria* isolates. *Plant Pathology* **34**, 41–48.

Gregory, S.C. (1989). *Armillaria* species in northern Britain. *Plant Pathology* **38**, 93–97.

Greig, B.J.W. and Strouts, R.G. (1983). Honey fungus. *Arboricultural Leaflet 2*. London: HMSO.

Greig, B.J.W., Gregory, S.C. and Strouts, R.G. (1991). Honey fungus. *Forestry Commission Bulletin* No. 100 (Ed. 8), vii + 11pp.

Guillaumin, J.J. (1986). *Contribution à l'étude des Armillaires phytopathogènes, en particulier du groupe* Mellea: *cycle caryologique, notion d'espèce, rôle biologique des espèces.* Thesis. 270pp. Univ. Claude Bernard-Lyon I.

Guillaumin, J.J. and Leprince, S. (1979). Influence de divers types de matière organique sur l'initiation et la croissance des rhizomorphes d'*Armillaria mellea* (Vahl) Karst. dans le sol. [Influence of different organic materials on the initiation and growth of rhizomorphs of *Armillaria mellea* in soil.] *European Journal of Forest Pathology* **9**, 355–366.

Guillaumin, J.J. and Lung, B. (1985). Study of the specialization of *Armillaria mellea* (Vahl) Kumm. and *Armillaria obscura* (Secr.) Herink in the saprophytic phase and in the parasitic phase. [Etude de la spécialisation d'*Armillaria mellea* (Vahl) Kumm. et *Armillaria obscura* (Secr.) Herink en phase saprophytique et en phase parasitaire.] (In French.) *European Journal of Forest Pathology* **15** (5–6), 342–349.

Guillaumin, J.J., Mohammed, C. and Berthelay, S. (1989). *Armillaria* species in the northern temperate hemisphere. In: *Proceedings of the 7th international conference on root and butt rots*, August 9–16, 1988, Vernon and Victoria, British Columbia. Ed. D.J. Morrison, pp 27–43. Victoria, British Columbia, Canada: International Union of Forestry Research Organizations.

Guillaumin, J.J., Anderson, J.B. and Korhonen, K. (1991). Life Cycle, Interfertility, and Biological Species. In: *Armillaria Root Disease.* Eds. C.G. Shaw III and G.A. Kile, pp 10–20. *Forest Service Handbook No. 691*. Washington, D.C.: USDA.

Hagle, S.K. and Shaw, C.G., III (1991). Avoiding and reducing losses from *Armillaria* root disease. In: *Armillaria Root Disease.* Eds. C.G. Shaw III and G.A. Kile, pp 157–173. *Forest Service Handbook No. 691*. Washington, D.C.: USDA.

Hansson, G. and Seifert, G. (1987a). Production of zone lines (Pseudosclerotia) in veneers. *Material und Organismen* **22**, 87–102.

Hansson, G. and Seifert, G. (1987b). Effects of cultivation techniques and media on yields and morphology of the basidiomycete *Armillaria mellea. Applied Microbiology and Biotechnology* **26** (5), 468–473.

Hansson, G. and Seifert, G. (1987). Production of zone lines (Pseudosclerotia) in veneers. *Material und Organismen* **22**, 87–102.

Hartig, R. (1870). Pas Auftreten der Rhizomorpha in Nadelholzkulturen. *Zeitschrift fur Forst- und Jagdwesen* **2**, 359–361.

Hartig, R. (1873). Vorlaufige Mittheilung uber den Parasitismus von *Agaricus melleus* und dessen Rhizomorphen. [Preliminary report on the parasitism of *Agaricus melleus* and its rhizomorphs.] *Botanische Zeitung* **31**, 295–297.

Hartig, R. (1874). Wichtige Krankheiten der Waldbaume. Beitrage zur Mycologie und Phytopathologie fur Botaniker und Forstmanner. [Important Diseases of Forest Trees. Contributions to mycology and phytopathology for botanists and foresters. Phytopathological Classics No. 12, 1975. St. Paul, MN: American Phytopathological Society.] 127pp. Berlin: Springer.

Harvey, E.N. (1952). *Bioluminescence*. 649pp. New York: Academic Press.

Herink, J. (1983). Toxicology of honey fungus, *Armillaria mellea* (Vahl ex Fr.) Kumm. Translation, Environment Canada. No. OOENV TR–2386, 22pp; Transl. from Sympozium o vaclavce obecne – *Armillaria mellea* (Vahl ex Fr.) Kumm. *Sbornik referatu* (1972, publ. 1973). 153–170.

HINTIKKA, V. (1974). Notes on the ecology of *Armillariella mellea* in Finland. *Harstenia* **14**, 12–31.

HODGE, A., ALEXANDER, I.J. AND GOODAY, G.W. (1995). Chitinolytic enzymes of pathogenic and ectomycorrhizal fungi. *Mycological Research* **99** (8), 935–941.

HODGE, A., GOODAY, G.W. AND ALEXANDER, I.J. (1996). Inhibition of chitinolytic activities from tree species and associated fungi. *Phytochemistry* **41** (1), 77–84.

HOFFMANN, H. (1861). Icones analyticae fungorum. *Abbildungen und beschreibungen von pilzen mit besonderer rucksicht auf anatomie und entwicklungsgeschichte*. 105pp.

HONG, J.S., KIM, M.K., LEE, J.H. AND KIM, H.M. (1990). Alcohols and volatile organic acids as stimulants of rhizomorph production by *Armillaria mellea* (in Korean). *Korean Journal of Mycology* **18** (3), 158–163.

HOOD, I.A. AND MORRISON, D.J. (1984). Incompatibility testing of *Armillaria* isolates in a wood substrate. *Canadian Forestry Service Research Notes* **4**, 8–9.

HOOD, I.A. AND SANDBERG, C.J. (1987). Occurrence of *Armillaria* rhizomorph populations in the soil beneath indigenous forests in the Bay of Plenty, New Zealand. *New Zealand Journal of Forestry Science* **17**, 83–99.

HOOD, I.A. AND SANDBERG, C.J. (1989). Changes in soil populations of *Armillaria* species following felling and burning of indigenous forest in the Bay of Plenty, New Zealand. In: *Proceedings of the 7th international conference on root and butt rots*; August 9–16, 1988; Vernon and Victoria, B.C. Ed. D.J. Morrison, pp 288–296. Victoria, B.C.: International Union of Forestry Research Organizations.

HOOD, I.A., SANDBERG, C.J. AND KIMBERLEY, M.O. (1989). A decay study of windthrown indigenous trees. *New Zealand Journal of Botany* **27**, 281–297.

HOPKIN, A.A., MALLETT, K.I. AND BLENIS, P.V. (1989). The use of L-DOPA to enhance visualization of the 'black line' between species of the *Armillaria mellea* complex. *Canadian Journal of Botany* **67** (1), 15–17.

HOPP, H. (1938). The formation of colored zones by wood-destroying fungi in culture. *Phytopathology* **28**, 601–620.

HORAK, J., HOTYK, A. AND RIHOVA, L. (1977). Specificity of transinhibition of amino acid transport in baker's yeast. *Folia Microbiologica* **22**, 360–362.

HUGHES, C.N.G., FROUD-WILLIAMS, R.J. AND FOX, R.T.V. (1993). The effects of fragmentation and defoliation on *Rumex obtusifolius* and its implications for grassland management. *Proceedings 1993 Brighton Crop Protection Conference – Weeds*, 767–772.

HUGHES, C.N.G., WEST, J.S., FROUD-WILLIAMS, R.J. AND FOX, R.T.V. (1996). Control of broad-leaved docks by *Armillaria mellea*, pp 531–534. In: *Proceedings of the IX International Symposium on Biological Control of Weeds*. Eds. V.C. Moran and J.H. Hoffmann, 563pp. Cape Town, South Africa: University of Cape Town.

HUNNEYBALL, I.M. AND STANWORTH, D.R. (1975). Fragmentation of human IgG by a new protease isolated from the basidiomycete *Armillaria mellea*. *Immunology* **29**, 921–931.

HÜTTERMANN, A. (1987). History of forest botany (forstbotanik) in Germany from the beginning in 1800 until 1940 – Science in the tension field between university and professional responsibility. *Per Deutsche Botanische Gesellschaft* **100**, 107–141.

INTINI, M.G. (1993). Observations on the *in vitro* development of *Armillaria* carpophores. *Mycologist* **7** (1), 18–24.

JACQUES-FELIX, M. (1968). Recherches morphologiques, anatomiques, morphogénétiques et physiologiques sur des rhizomorphes de champignons supérieurs et sur le déterminisme de leur formation. Deuxième partie. *Bulletin de la Société Mycologique de France* **84**, 166–307.

JAHNKE, H.-D., BAHNWEG, G. AND WORRALL, J.J. (1987). Species delimitation in the *Armillaria mellea* complex by analysis of nuclear and mitochondrial DNAs. *Transactions of the British Mycological Society* **88**, 572–575.

JENNINGS, D.H. (1984). Water flow through mycelia. In: *The ecology and physiology of the fungal mycelium.* Eds. D.H. Jennings and A.D.M. Rayner, pp 143–163. Cambridge: Cambridge University Press.

JUNGSHAN, Y., YUWU, C. AND XIAOZHANG, F. (1984). Chemical constituents of *Armillaria mellea* mycelium, I. Isolation and characterization of Armillarin and Armillaridin. *Planta Medica* **50**, 288–290.

JUNHUA, H., DECHAO, Y., XIANYU, C., ZEMIN, H., XIAOZHANG, F., ARAKI, H., TSUCHIDA, K., ASAMI, Y., AIHARA, H., TAMAI, M., WATANABE, N., OBUCHI, T. AND OMURA, S. (1990). Effects of Mi Huan Jun (Armillaria mellea) on central nervous and vascular system. *Fitotherapia* **61** (3), 207–214.

KAARIK, A. (1965). The identification of the mycelia of wood decay fungi by their oxidation reactions with phenolic compounds. *Studia Forestalia Suecica* **31**, 1–80.

KEY, J.L. (1969). Hormones and nucleic acid metabolism. *Annual Review of Plant Physiology* **20**, 449–474.

KEY, J.L., BARNETT, M.N. AND LEN, C.Y. (1967). RNA and protein biosynthesis and the regulation of cell elongation by auxin. *Annals of the New York Academy of Sciences* **144**, 49–62.

KHUO, M.J. AND ALEXANDER, M. (1967). Inhibition of the lysis of fungi by melanins. *Journal of Bacteriology* **94**, 624–629.

KIHO, T., SHIOSE, Y., NAGAI, K., SAKUSHIMA, M. AND UKAI, S. (1992a). Polysaccharides in fungi. XXIX. Structural features of two antitumor polysaccharides from the fruiting bodies of *Armillariella tabescens*. *Chemical and Pharmaceutical Bulletin* **40** (8), 2110–2112.

KIHO, T., SHIOSE, Y., NAGAI, K. AND UKAI, S. (1992b). Polysaccharides in fungi. XXX. Antitumor and immunomodulating activities of two polysaccharides from the fruiting bodies of *Armillariella tabescens*. *Chemical and Pharmaceutical Bulletin* **40** (8), 2112–2114.

KILE, G.A. (1980). Behaviour of an *Armillaria* in some *Eucalyptus obliqua – Eucalyptus regnans* forests in Tasmania and its role in their decline. *European Journal of Forest Pathology* **10**, 278–296.

KILE, G.A. (1981). *Armillaria luteobubalina*: a primary cause of decline and death of trees in mixed species eucalypt forests in central Victoria. *Australian Forest Research* **11**, 63–77.

KILE, G.A. AND WATLING, R. (1983). *Armillaria* species from south-eastern Australia. *Transactions of the British Mycological Society* **81**, 129–140.

LAMOURE, D. AND GUILLAUMIN, J.J. (1985). The life cycle of the *Armillaria mellea* complex. Le cycle caryologique des Armillaires du groupe mellea (in French). *European Journal of Forest Pathology* **15** (5–6), 288–293.

LANPHERE, W.M. (1934). Enzymes of the rhizomorphs of *Armillaria mellea*. *Phytopathology* **24,** 1244–1249.

LASOTA, W. AND FLORCZAK, J. (1984). Chemical composition of Armillaria mellea. OT: Badania skladu chemicznego opienki miodowej [Armilariella mellea (Vahl In. Fl. Dan. ex Fr.) P. Karst] (in Polish). *Bromatologia-i-Chemia-Toksykologiczna* **17** (4), 287–291.

LAWRENCE, W.H. (1910). Root diseases caused by *Armillaria mellea* in the Puget Sound Country. *Bulletin 3* [special series], 1–16. Puyallup, WA: State College of Washington, [Western Washington] Agricultural Experiment Station.

LEONARD, T.J. (1971). Phenoloxidase and fruiting body formation in *Schizophyllum* commune. *Journal of Bacteriology* **106**, 162–167.

LEWIS, W.G., BASFORD, J.M. AND WALTON, P.L. (1978). Specificity and inhibition studies on *Armillaria mellea* protease. *Biochimica et Biophysica Acta* **522**, 551–560.

LIENGSWANGWONG, V., SALVAGGIO, J.E., LYON, F.L. AND LEHRER, S.B. (1987). Basidiospore allergens: determination of optimal extraction methods. *Clinical Allergy* **17** (3), 191–198.

LIN, P., PUMAS, M.T. AND HUBBES, M. (1989). Isozyme and general protein patterns of *Armillaria* spp. collected from the boreal mixed wood forest of Ontario. *Canadian Journal of Botany* **67**, 1143–1147.

LIU, D., YAO, D., LIANG, R., MA, L., CHENG, W. AND GU, L. (1998). Detoxification of aflatoxin B1 by enzymes isolated from *Armillariella tabescens*. *Food & Chemical Toxicology* **36**, 563–574.

LOPEZ-REAL, J.M. (1975). The formation of pseudosclerotia ('zone lines') in wood decayed by *Armillaria mellea* and *Stereum hirsutum*, I. Morphological aspects. *Transactions of the British Mycological Society* **64**, 465–471.

LOPEZ-REAL, J.M. AND SWIFT, M.J. (1975). The formation of pseudosclerotia ('zone lines') in wood decayed by *Armillaria mellea* and *Stereum hirsutum*, II. Formation in relation to the moisture content of the wood. *Transactions of the British Mycological Society* **64**, 473–481.

LOPEZ-REAL, J.M. AND SWIFT, M.J. (1977). The formation of pseudosclerotia ('zone lines') in wood decayed by *Armillaria mellea* and *Stereum hirsutum*, III. Formation in relation to composition of atmosphere in wood. *Transactions of the British Mycological Society* **68**, 321–325.

LYR, H. (1955). Occurrence of peroxidase in wood-destroying basidiomycetes. *Planta* **46**, 408–413.

MALLETT, K.I. AND COLOTELO, N. (1984). Rhizomorph exudate of *Armillaria mellea. Canadian Journal of Microbiology* **30**, 1247–1252.

MALLETT, K.I. AND HIRATSUKA, Y. (1985). The 'trap-log' method to survey the distribution of *Armillaria mellea* in forest soils. *Canadian Journal of Forest Research* **15**, 1191–1193.

MALLETT, K.I. AND HIRATSUKA, Y. (1986). Nature of the 'black line' produced between different biological species of the *Armillaria mellea* complex. *Canadian Journal of Botany* **64** (11), 1588–1590.

MARSH, S.F. AND WARGO, P.M. (1989). Phenol oxidases of five *Armillaria* biospecies. *Phytopathology* **79**, 1150. Abstract.

MAYER, A.M. (1987). Polyphenol oxidases in plants – recent progress. *Phytochemistry* **26** (1), 11–20.

MAYER, A.M. AND HAREL, E. (1979). Polyphenol oxidases in plants. *Phytochemistry* **18**, 193–215.

MIDLAND, S.L., IZAC, R.R. AND WING, R.M. (1982). Melleolide, a new antibiotic from *Armillaria mellea. Tetrahedron Letters* **23**, 2515–2518.

MIHAIL, J.D., OBERT, M., BRUHN, J.N. AND TAYLOR, S.J. (1995). Fractal geometry of diffuse mycelia and rhizomorphs of Armillaria species. *Mycological Research* **99** (1), 81–88.

MOLISCH, H. (1904). *Leuchtende Pflanzen*, pp 34–38. Jena: Fischer.

MOODY, A.R. AND WEINHOLD, A.R. (1972a). Fatty acids and naturally occurring plant lipids as stimulants of rhizomorph production in *Armillaria mellea. Phytopathology* **62**, 264–267. Abstract.

MOODY, A.R. AND WEINHOLD, A.R. (1972b). Stimulation of rhizomorph production by *Armillaria mellea* with lipid from tree roots. *Phyopathology* **62**, 1347–1350.

MOORE, D., CASSELTON, L.A. AND WOOD, D.A., eds. (1985). *Developmental biology of higher fungi*. 615pp. Cambridge: Cambridge University Press.

MORRISON, D.J. (1975). Ion uptake by rhizomorphs of *Armillaria mellea. Canadian Journal of Botany* **53**, 48–51.

MORRISON, D.J. (1976). Vertical distribution of *Armillaria mellea* rhizomorphs in soil. *Transactions of the British Mycological Society* **66**, 393–399.

MORRISON, D.J. (1982a). Effect of soil organic matter on rhizomorph growth by *Armillaria mellea. Transactions of the British Mycological Society* **78**, 201–208.

MORRISON, D.J. (1982b). Variation among British isolates of *Armillaria mellea. Transactions of the British Mycological Society* **78**, 459–464.

MORRISON, D.J. (1989). Pathogenicity of *Armillaria* species is related to rhizomorph growth habit. In: *Proceedings of the 7th international conference on root and butt rots*, August 9–16, 1988, Vernon and Victoria, B.C. Ed. D.J. Morrison, pp 584–589. Victoria, B.C.: International Union for Forestry Research Organizations.

MORRISON, D.J., CHU, D. AND JOHNSON, A.L.S. (1985a). Species of *Armillaria* in British Columbia. *Canadian Journal of Plant Pathology* **7**, 242–246.

MORRISON, D.J., THOMSON, A.J. AND CHU, D. (1985b). Isozyme patterns of *Armillaria* intersterility groups occurring in British Columbia. *Canadian Journal of Microbiology* **31**, 651–653.

MORRISON, D.J., THOMSON, A.J., CHU, D., PEET, F.G. AND SAHOTA, T.S. (1989). Variation in isozyme patterns of esterase and polyphenol oxidase among isolates of *Armillaria ostoyae* from British Columbia. *Canadian Journal of Plant Pathology* **11**, 229–234.

MORRISON, D.J., WILLIAMS, R.E. AND WHITNEY, R.D. (1991). Infection, disease development, diagnosis, and detection. In: *Armillaria Root Disease.* Eds. C.G. Shaw III and G.A. Kile, pp 62–75. *Forest Service Handbook No. 691*. Washington, D.C.: USDA.

MOTTA, J.J. (1969). Cytology and morphogenesis in the rhizomorph of *Armillaria mellea. American Journal of Botany* **56**, 610–619.

MOTTA, J.J. (1971). Histochemistry of the rhizomorph meristem of *Armillaria mellea. American Journal of Botany* **58**, 80–87.

MOTTA, J.J. (1982). Rhizomorph cytology and morphogenesis in *Armillaria tabescens. Mycologia* **74**, 671–674.

MUNCH, E. (1909). *Untersuchungen uber Immunitat und Krankheitsempfanglichkeit der Holzpflanzen.* 81pp. Ludwigsburg: Ungeheuer und Ulmer.

NGUYEN, T.H.H. (1980). *Etude morphologique, morphogénétique et cytologique de quatre espèces de Basidiomycetes* Armillariella mellea, A. bulbosa, A. ostoyae, Clitocybe tabescens, *et de leur pouvoir pathogène.* Thèse de Dr. Ingénieur. 136pp. INA, Paris Grignon, No. 157.

O'NEIL, C.E., HORNER, W.E., REED, M.A., LOPEZ, M. AND LEHRER, S.B. (1990). Evaluation of basidiomycete and deuteromycete (fungi imperfecti) extracts for shared allergenic determinants. *Clinical and Experimental Allergy* **20** (5), 533–538.

OBUCHI, T., KONDOH, H., WATANABE, N., TAMAI, M., OMURA, S., YANG, J.S. AND LIANG, X.T. (1990). Armillaric acid, a new antibiotic produced by Armillaria mellea. *Planta Medica* **56**, 198–201.

ODURO, K.A., MUNNECKE, D.E. AND SIMS, J.J. (1976). Isolation of antibiotics produced in culture by *Armillaria mellea. Transactions of the British Mycological Society* **66**, 195–199.

OHR, H.D. AND MUNNECKE, D.E. (1974). Effects of methyl bromide on antibiotic production by *Armillaria mellea. Transactions of the British Mycological Society* **62**, 65–72.

ONO, K. (1965). [Armillaria root rot in plantations of Hokkaido. Effects of topography and soil conditions on its occurrence.] *Meguro Bulletin of the Government Forest Experiment Station* **179**,1–62. (In Japanese.) [*Review of Applied Mycology* **45**, 635.]

ONO, K. (1970). Effect of soil conditions on the currence of Armillaria root rot of Japanese larch. *Meguro Bulletin of the Government Forest Experiment Station* **229**, 123–219. (In Japanese.) [*Review of Plant Pathology* **50**, 2001.]

OPPERMANN, A. (1952). Das antibiotische Verhalten einiger holzzersetzender Basidiomyceten zueinander and zu Bakterien. *Archiv Mikrobiologie* **16**, 364–409.

PATEMAN, J.A. AND KINGHORN, J.R. (1976). Nitrogen metabolism. In: *The filamentous fungi* Vol. 2. Eds. J.E. Smith and D.R. Berry, pp 159–237. New York: Wiley.

PATTON, R.F. AND RIKER, A.J. (1959). Artificial inoculations of pine and spruce trees with *Armillaria mellea. Phytopathology* **49**, 615–622.

PEARCE, M.H., MALAJCZUK, N. AND KILE, G.A. (1986). The occurrence and effects of *Armillaria luteobubalina* in the karri (*Eucalyptus diversicolor* F. Muell.) forests of Western Australia. *Australian Forest Research* **16**, 243–259.

PEARCE, M.H. AND MALAJCZUK, N. (1990). Factors affecting growth of *Armillaria luteobubalina* rhizomorphs in soil. *Mycological Research* **94** (1), 38–48.

PENTLAND, G.D. (1965). Stimulation of rhizomorph development of *Armillaria mellea* by *Aureobasidium pullullans* in artificial culture. *Canadian Journal of Microbiology* **11**, 345–350.

PENTLAND, G.D. (1967). Ethanol produced by *Aureobasidium pullullans* and its effect on the growth of *Armillaria mellea. Canadian Journal of Microbiology* **13**, 1631–1639.

PÉREZ-SIERRA, A., WHITHEAD, D. AND WHITEHEAD, M. (1999). Molecular methods for the detection and identification of *Armillaria.* In: *Armillaria root rot: biology and control of honey fungus.* R.T.V. Fox. Andover, U.K.: Intercept Limited.

PIEPP, H. AND SONNENBICHLER, J. (1992). Secondary metabolites and their biological activities II. Occurrence of antibiotic compounds in cultures of Armillaria ostoyae growing in the presence of an antagonistic fungus or host plant cells. *Biological Chemistry Hoppe-Seyler* **373**, 675–683.

PODGER, F.D., KILE, G.A. AND WATLING, R. (1978). Spread and effects of *Armillaria luteobubalina* sp. nov. in an Australian *Eucalyptus regnans* plantation. *Transactions of the British Mycological Society* **71**, 77–87.

POPOOLA, T.O.S. (1991). *Role of stress in Armillaria infections.* Ph.D. Thesis. U.K.: University of Reading.

POPOOLA, T.O.S. AND FOX, R.T.V. (1996). Effects of root damage on honey fungus.

Arboricultural Journal **20**, 329–337.

POWELL, K.A. AND RAYNER, A.D.M. (1983). Ultrastructure of the rhizomorph apex in *Armillaria bulbosa* in relation to mucilage production. *Transactions of the British Mycological Society* **81**, 529–534.

POWELL, K.A. AND RAYNER, A.D.M. (1984). Occurrence of bundles of microfilaments in circum-medullary cells underlying the apical dome of *Armillaria bulbosa* rhizomorphs. *Transactions of the British Mycological Society* **83**, 217–221.

PRIESTLEY, R., MOHAMMED, C. AND DEWEY, F.M. (1994). The development of monoclonal antibody-based ELISA and dipstick assays for the detection of *Armillaria* species in infected wood. In: *Modern assays for plant pathogenic fungi: Indentification, detection and quantification.* Eds. A. Schots, F.M. Dewey and R. Oliver, pp 149–156. Wallingford: CAB International.

PRONOS, J. AND PATTON, R.F. (1979). The effect of chlorophenoxy acid herbicides on growth and rhizomorph production of *Armillaria mellea. Phytopathology* **69**, 136–141.

RAABE, R.D. (1962). Host list of the root rot fungus, *Armillaria mellea. Hilgardia* **33**, 25–88.

RAABE, R.D. (1966). Variation of *Armillaria mellea* in culture. *Phytopathology* **56**, 1241–1244.

RAABE, R.D. (1967). Variation in pathogenicity and virulence in *Armillaria mellea. Phytopathology* **57**, 73–75. Abstract.

RAABE, R.D. (1979). Some previously unreported hosts of *Armillaria mellea* in California (U.S.A.). *Plant Disease Reporter* **63**, 494–495.

RADZIEVSKAYA, M.G. AND BOBKO, I.N. (1985a). [Dark zonal lines in wood. I. Formation of lines by *Armillaria mellea* (Vahl:Fr.) P. Karst]. (In Russian). *Mikologiya i Fitopatologiya* **19** (3), 214–220.

RADZIEVSKAYA, M.G. AND BOBKO, I.N. (1985b). [Dark zonal lines in wood. II. Using the lines in studying the population structure of wood-destroying hymenomycetes]. (In Russian). *Mikologiya i Fitopatologiya* **19** (5), 394–405.

RAYLE, D.L. (1973). Auxin-induced hydrogen ion secretion in *Avena* coleoptiles and its implications. *Planta* **114**, 63–73.

RAYNER, A.D.M. (1976). Dematiaceous hyphomycetes and narrow dark zones in decaying wood. *Transactions of the British Mycological Society* **67**, 546–549.

RAYNER, A.D.M. AND TODD, N.K. (1979). Population and community structure and dynamics of fungi in decaying wood. *Advances in Botanical Research* **7**, 333–420.

RAYNER, A.D.M., POWELL, K.A. AND THOMPSON, W. (1985). Morphogenesis of vegetative organs. In: *Developmental biology of higher fungi*. Eds. D. Moore, L.A. Casselton and D.A. Wood, pp 249–280. Cambridge: Cambridge University Press.

REAVES, J.L., ALLEN, T.C., SHAW, C.G., III, DASHEK, W.V. AND MAYFIELD, J.E. (1988). Occurrence of viruslike particles in isolates of *Armillaria. Journal of Ultrastructure and Molecular Structure Research* **98** (2), 217–221.

REDFERN, D.B. (1970). The ecology of *Armillaria mellea*: rhizomorph growth through soil. In: *Root diseases and soil-borne pathogens: Proceedings of the symposium*: July, 1968, London: Imperial College. Eds. T.A. Toussoun, R.V. Bega and P.E. Nelson, pp 147–149. Berkeley: University of California Press.

REDFERN, D.B. (1973). The growth and behaviour of *Armillaria mellea* rhizomorphs in soil. *Transactions of the British Mycological Society* **61**, 569–581.

REDFERN, D.B. (1975). The influence of food base on the rhizomorph growth and pathogenicity of *Armillaria mellea* isolates. In: *Biology and Control of soil-borne plant pathogens*. Ed. G.W. Bruehl (The Am. Phytopath. Soc.), pp 69–73.

REDFERN, D.B. (1978). Infection by *Armillaria* and some factors affecting host resistance and severity of disease. *Forestry* **51** (2), 121–135.

REHMAN, A.U. AND THURSTON, C.F. (1992). Purification of laccase I from *Armillaria mellea*. *Journal of General Microbiology* **138**, 1251–1257.

REIJNDERS, A.F.M., ed. (1963). *Les Problèmes du Dévelopment des Carpophores des Agaricales et des quelques groupes voisins*. 412pp. Den Haag: Junk.

REITSMA, J. (1932). Studien über *Armillaria mellea* (Vahl) Quél. *Phytopathologische Zeitschrift* **4**, 461–522.

RHOADS, A.S. (1956). The occurrence and destructiveness of Clitocybe root rot of woody plants in Florida. *Lloydia* **19**, 193–240.

RICHARD, C. (1971). Sur l'activité antibiotique de l'*Armillaria mellea*. [Concerning the antibiotic activity of *Armillaria mellea*.] *Canadian Journal of Microbiology* **17**, 1395–1399.

RISHBETH, J. (1968). The growth rate of *Armillaria mellea*. *Transactions of the British Mycological Society* **51**, 575–586.

RISHBETH, J. (1970). The role of basidiospores in stump infection by *Armillaria mellea*. In: *Root diseases and soil-borne pathogens: Proceedings of the symposium*: July, 1968, London: Imperial College. Eds. T.A. Toussoun, R.V. Bega and P.E. Nelson, pp 141–146. Berkeley: University of California Press.

RISHBETH, J. (1972). The production of rhizomorphs by *Armillaria mellea* from stumps. *European Journal of Forest Pathology* **2**, 193–205.

RISHBETH, J. (1976). Chemical treatment and inoculation of hardwood stumps for control of *Armillaria mellea*. *Annals of Applied Biology* **82** (1), 57–70.

RISHBETH, J. (1978). Effects of soil temperature and atmosphere on growth of *Armillaria* rhizomorphs. *Transactions of the British Mycological Society* **70**, 213–220.

RISHBETH, J. (1980). *Armillaria* on cacao in São Tomé. *Tropical Agriculture, Trinidad* **57**, 155–165.

RISHBETH, J. (1982). Species of *Armillaria* in southern England. *Plant Pathology* **31**, 9–17.

RISHBETH, J. (1983). The importance of the honey fungus (*Armillaria*) in urban forestry. *Aboricultural Journal* **7**, 217–225.

RISHBETH, J. (1985a). *Armillaria*: resources and hosts. In: *Developmental biology of higher fungi*. Eds. D. Moore, L.A. Casselton and D.A. Wood, pp 87–101. Cambridge: Cambridge University Press.

RISHBETH, J. (1985b). Infection cycle of *Armillaria* and host response. *European Journal of Forest Pathology* **15** (5–6), 332–341.

RISHBETH, J. (1986). Some characteristics of English *Armillaria* species in culture. *Transactions of the British Mycological Society* **85**, 213–218.

RITCHIE, J.H. (1932). Some observations on the Honey Agaric (*Armillaria mellea* syn. *Agaricus melleus*). *Scottish Forestry Journal* **46**, 132–142.

RITTER, G. AND PONTOR, G. (1969). Root development of young Scots pine as a factor in resistance to attack by *Armillaria mellea*. *Archiv für Forstwesen* **18**, 1037–1042.

RIZZO, D.M., BLANCHETTE, R.A. AND PALMER, M.A. (1992). Biosorption of metal ions by *Armillaria* rhizomorphs. *Canadian Journal of Botany* **70** (8), 1515–1520.

ROBENE-SOUSTRADE, I., LUNG-ESCARMANT, B., BONO, J. AND TARIS, B. (1992). Identification and partial characterization of an extracellular manganese-dependent peroxidase in *Armillaria ostoyae* and *Armillaria mellea*. *European Journal of Forest Pathology* **22** (4), 227–236.

ROBENE-SOUSTRADE, I. AND LUNG-ESCARMANT, B. (1997). Laccase isoenzyme patterns of European *Armillaria* species from culture filtrates and infected woody plant tissues. *European Journal of Forest Pathology* **27** (2), 105–114.

ROSS, E.W. (1970). Sand pine root rot – pathogen: *Clitocybe tabescens*. *Journal of Forestry* **68**, 156–158.

RYKOWSKI, K. (1974). Obserwacje nad owocowaniem opienki miodowej *Armillaria mellea* (Vahl) Karst. w czystych kulturach. [Observations on the fructification of the honey fungus *Armillaria mellea* (Vahl) Karst in pure cultures.] (In Polish.) *Prace Intytutu Badawczego Lesnictwa* **431**, 146–167.

RYKOWSKI, K. (1976). Recherche sur la nutrition azotée de plusieurs souches de l'*Armillaria mellea*, I. L'influence de différentes concentrations du carbone et de l'azote (C:N). [Studies on the nitrogen nutrition of several strains of *Armillaria mellea*, I. The influence of different concentrations of carbon and nitrogen (C:N).] *European Journal of Forest Pathology* **6**, 264–274.

RYKOWSKI, K. (1984). Niektore troficzne uwarunkowania patogenicznosci *Armillaria mellea* (Vahl) Quél. w uprawach sosnowych. [Some trophic factors in the pathogenicity of *Armillaria mellea* in Scots pine plantations.] (In Polish.) *Prace Instytutu Badawczego Lesnictwa* **640**, 1–140.

SANCHEZ-HERNANDEZ, E., GARCIA-MENDOZA, C. AND NOVAES-LEDIEU, M. (1990). Chemical characterization of the hyphal walls of the basidiomycete *Armillaria mellea*. *Experimental Mycology* **14** (2), 178–183.

SCHINNER, F. AND CONCIN, R. (1981). Carbon dioxide fixation by wood rotting fungi. *European Journal of Forest Pathology* **11**, 120–123.

SCHMID, R. AND LIESE, W. (1970). Feinstruktur der Rhizomorphen von *Armillaria mellea*. *Phytopathologische Zeitschrift* **68**, 221–231.

SCHMITZ, J. (1848). Beitrage zur Anatomie und Physiologie der Schwamme, III. Uber den Bau, das Wachstum and einige besondere behenserschienungen der *Rhizomorpha fragilis* Roth. *Linnaea* **17**, 487–535.

SCHÜTTE, K.H. (1956). Translocation in the fungi. *New Phytologist* **55**, 164–182.

SHAW, C.G., III. (1980). Characteristics of *Armillaria mellea* on pine root systems in expanding centers of root rot. *Northwest Science* **54**, 137–145.

SHAW, C.G., III. (1985). *In vitro* responses of different *Armillaria* taxa to gallic acid, tannic acid, and ethanol. *Plant Pathology* **34**, 594–602.

SHAW, C.G., III, MACKENZIE, M. AND TOES, E.H.A. (1981). Cultural characteristics and pathogenicity to *Pinus radiata* of *Armillaria novae-zelanidae* and *A. limonea*. *New Zealand Journal of Forestry Science* **11**, 65–70.

SHAW, C.G., III AND KILE, G.A. (1991). *Armillaria Root Disease*. *Agriculture Handbook (Washington) No. 691*, xi+239pp. Washington, U.S.A.: USDA Forest Service.

SHEARER, B.L. AND TIPPETT, J.T. (1988). Distribution and impact of *Armillaria luteobubalina* in the *Eucalyptus marginata* forest of South-western Australia. *Australian Journal of Botany* **36**, 433–445.

SHIELDS, W.J., JR. AND HOBBS, S.D. (1979). Soil nutrient levels and pH associated with *Armillaria mellea* on conifers in northern Idaho. *Canadian Journal of Forest Research* **9**, 45–48.

SHIGO, A.L. AND TIPPETT, J.T. (1981). Compartmentalization of decayed wood associated with *Armillaria mellea* in several tree species. *Res. Pap. NE488*. 20pp. U.S. Department of Agriculture, Forest Service.

SIETSMA, J.H. AND WESSELS, J.G.H. (1977). Chemical analysis of the hyphal wall of *Schizophyllum commune*. *Biochimica et Biophysica Acta* **496**, 225–239.

SINGH, P. (1981). *Armillaria mellea*: growth and distribution of rhizomorphs in the forest soils of Newfoundland. *European Journal of Forest Pathology* **11**, 208–220.

SINGH, P. AND BAL, A.K. (1973). Ultrastructure of the fructification of *Armillaria mellea*. In: *Abstracts of papers: 2nd international congress of plant pathology,* September 5–12, 1973. Abstract 0157. Minneapolis: University of Minnesota.

SINGH, P. AND CAREW, G.C. (1983). *Armillaria* root rot of urban trees: another perspective to the root problem in Newfoundland. *Canadian Plant Disease Survey* **63**, 3–6.

SMITH, A.M. AND GRIFFIN, D.M. (1971). Oxygen and the ecology of *Armillariella elegans* Heim. *Australian Journal of Biological Sciences* **24**, 231–262.

SMITH, J.E. AND BERRY, D.R., eds. (1978). The filamentous fungi, III. *Developmental mycology*. 464pp. London: Edward Arnold.

SMITH, M.L., BRUHN, J.N. AND ANDERSON, J.B. (1992). The fungus *Armillaria bulbosa* is among the largest and oldest living organisms. *Nature (London)* **256**, 428–431.

SONNENBICHLER, J., DIETRICH, J. AND PIEPP, H. (1994). Secondary metabolites and their biological activities V. Investigations concerning the induction of the biosynthesis of toxic secondary metabolites in basidiomycetes. *Biological Chemistry Hoppe-Seyler* **375**, 71–79.

SONNENBICHLER, J., GUILLAUMIN, J.J., PEIPP, H. AND SCHWARZ, D. (1997). Secondary metabolites from dual cultures of genetically different *Armillaria* isolates. *European Journal of Forest Pathology* **27** (4), 241–249.

SORTKJAER, O. AND ALLERMANN, K. (1972). Rhizomorph formation in fungi, I. Stimulation by ethanol and acetate and inhibition by disulfiram of growth and rhizomorph formation in *Armillaria mellea*. *Physiologia Plantarum* **26**, 376–380.

SORTKJAER, O. AND ALLERMANN, K. (1973). Rhizomorph formation in fungi, III. The effect of ethanol on the synthesis of DNA and RNA and uptake of asparagine and phosphate in *Armillaria mellea*. *Physiologia Plantarum* **29**, 129–133.

STEWART, P.R. AND ROGERS, P.J. (1978). Fungal dimorphism: a particular expression of cell wall morphogenesis. In: *The filamentous fungi, III. Developmental mycology.* Eds. J.E. Smith and D.R. Berry, pp 164–196. London: Edward Arnold.

SWIFT, M.J. (1968). Inhibition of rhizomorph development by *Armillaria mellea* in Rhodesian forest soils. *Transactions of the British Mycological Society* **51**, 241–247.

TERASHITA, T. (1996). Biological species of *Armillaria* symbiotic with *Galeola septentrionalis.* (In Japanese.) *Nippon Kingakukai Kaiho* **37** (2), 45–49.

THOMAS, H.E. (1934). Studies on *Armillaria mellea* (Vahl) Quél., infection, parasitism and host resistance. *Journal of Agricultural Research* **48**, 187–218.

THOMPSON, W. (1984). Distribution, development and functioning of mycelial cord systems of decomposer basidiomycetes of the deciduous woodland floor. In: *The ecology and physiology of the fungal mycelium.* Eds. D.H. Jennings and A.D.M. Rayner, pp 185–214. Cambridge: Cambridge University Press.

TOGASHI, I. (1995). Effects of using *Armillaria* species cultural waste as a substrate in the bottle cultivation of Hiratake mushrooms, *Pleurotus ostreatus.* (In Japanese.) *Mokuzai Gakkaishi* (*Journal of the Japan Wood Research Society*) **41** (10), 956–962.

TOGASHI, I. (1996). Effects of using *Armillaria* species cultural waste as a substrate in the bottle cultivation of Hiratake mushrooms, *Pleurotus ostreatus.* (In Japanese with English figures and tables.) *Journal of the Hokkaido Forest Products Research Institute* **10** (2), 9–13.

TOGASHI, I. (1996). Effects of substrates and seeding method on fruiting body production in the bottle cultivation of Armillaria species. (In Japanese with English figures and tables.) *Mokuzai Gakkaishi* (*Journal of the Japan Wood Research Society*) **42** (2), 186–193.

TOGASHI, I. AND TAKIZAWA, N. (1994). Effect of adding carrot, *Daucus carota* variety *sativa*, to media in the rhizomorph production of *Armillaria* species. (In Japanese.) *Mokuzai Gakkaishi* (*Journal of the Japan Wood Research Society*) **40** (2), 213–219.

TOGASHI, I. AND TAKIZAWA, N. (1995). Effect of primordium formation method and moisture content of medium on fruiting body production of *Armillaria* species. (In Japanese.) *Mokuzai Gakkaishi* (*Journal of the Japan Wood Research Society*) **41** (2), 211–217.

TOGASHI, I. AND TAKIZAWA, N. (1996). Effects of primordium formation method and moisture content of medium on fruiting body production of Armillaria species. *Journal of the Hokkaido Forest Products Research Institute* **10** (1), 11–16.

TURCSANYI, G., SILLER, I., FUHRER, E., KOVACS, M., PENKSZA, K., BUTTNER, S., FIGECZKY, G., DUDAS, J. AND MIHALY, B. (1996). Amount and chemical element content of rhizomorphs in the stemflow and throughfall areas of beech stands on different soil types. *Zeitschrift-fur-Pflanzenernahrung-und-Bodenkunde* **159** (5), 513–518.

TURNER, J.A. (1991). *Biology and control of Armillaria.* Ph.D. Thesis. U.K.: University of Reading.

ULLRICH, R.C. (1977). Isozyme patterns and cellular differentiation in *Schizophyllum. Molecular and General Genetics* **156**, 157–162.

ULLRICH, R.C. AND ANDERSON, J.B. (1978). Sex and diploidy in *Armillaria mellea. Experimental Mycology* **2**, 119–129.

WANG, C.S., SCHWALB, M.N. AND MILES, P.C. (1968). A relationship between cell wall composition and mutant morphology in the basidiomycete *Schizophyllum commune. Canadian Journal of Microbiology* **14**, 809–811.

WANG, C., GUO, S., CHEN, X., CAO, W., XU, J. AND XIAO, P. (1996). Contents of armillarin and melleolid at different stages of Armillaria mellea (Vahl ex Fr.) Quél. (In Chinese.) *China Journal of Chinese Materia Medica* **21** (318), 274–276.

WARGO, P.M. (1972). Defoliation-induced chemical changes in sugar maple roots stimulate growth of *Armillaria mellea. Phytopathology* **62**, 1278–1283.

WARGO, P.M. (1975). Lysis of the cell wall of *Armillaria mellea* by enzymes from forest trees. *Physiological Plant Pathology* **5**, 99–105.

WARGO, P.M. (1980a). Interaction of ethanol, glucose, phenolics and isolate of *Armillaria mellea. Phytopathology* **70**, 480. Abstract.

WARGO, P.M. (1980b). *Armillaria mellea* : an opportunist. *Journal of Aboriculture* **6**, 276–278.

WARGO, P.M. (1981). Defoliation and Secondary-action organism attack: with emphasis on *Armillaria mellea. Journal of Arboriculture* **7**, 64–69.

WARGO, P.M. (1983). The interaction of *Armillaria mellea* with phenolic compounds in the bark of roots of black oak. *Phytopathology* **73**, 838. Abstract.

WARGO, P.M. (1984). Changes in phenols affected by *Armillaria mellea* in bark tissue of roots of oak, *Quercus* spp. In: *Proceedings of the 6th international conference on root and butt rots of forest trees*, August 25–31, 1983, Victoria and Gympie, Queensland, Australia. Ed. G.A. Kile, pp 198–206. Melbourne: International Union of Forestry Research Organizations.

WARGO, P.M. AND SHAW, C.G., III (1985). *Armillaria* root rot: the puzzle is being solved. *Plant Disease* **69** (10), 826–832.

WATANABE, N., OBUCHI, T., TAMAI, M., ARAKI, H., OMURA, S., YANG, J.S., YU, D.Q., LIANG, X.T. AND HUAN, J.H. (1990). A novel N6-substituted adenosine isolated from Mi Huan Jun (*Armillaria mellea*) as a cerebral-protecting compound. *Planta Medica* **56** (1), 48–52.

WATANABE, T. (1986). Rhizomorph production in *Armillaria mellea in vitro* stimulated by *Macrophoma* sp. and several other fungi. *Transactions of the Mycological Society of Japan* **27** (3), 235–245.

WATANABE, T. (1997). Stimulation of perithecium and ascospore production in *Sordaria fimicola* by *Armillaria* and various fungal species. *Mycological Research* **101** (10), 1190–1194.

WATLING, R. (1985). Developmental characters of agarics. In: *Developmental biology of higher fungi.* Eds. D. Moore, L.A. Casselton and D.A. Wood, pp 281–310. Cambridge: Cambridge University Press.

WEINHOLD, A.R. (1963). Rhizomorph production by *Armillaria mellea* induced by ethanol and related compounds. *Science* **142**, 1065–1066.

WEINHOLD, A.R., HENDRIX, F.F. AND RAABE, R.D. (1962). Stimulation of rhizomorph growth of *Armillaria mellea* by indole-3-acetic acid and figwood extract. *Phytopathology* **52**, 757. Abstract.

WEINHOLD, A.R. AND GARRAWAY, M.O. (1966). Nitrogen and carbon nutrition of *Armillaria mellea* in relation to growth-promoting effects of ethanol. *Phytopathology* **56**, 108–112.

WESSELS, J.G.H. (1966). Control of cell wall glucan degradation in *Schizophyllum commune.* Antonie van Leeuwenhoek. *Journal of Microbiology and Serology* **32**, 341–355.

WHITNEY, R.D. (1997). Relationship between decayed roots and aboveground decay in three conifers in Ontario. *Canadian Journal of Forest Research* **27** (8), 1217–1221.

WOESTE, U. (1956). Anatomische Untersuchungen uber die Infektionswege einiger Wurzelpilze. [Anatomical studies on the methods of infection of several root fungi.] *Phytopathologische Zeitschrift* **26**, 225–272. [*Review of Applied Mycology* **35**, 800.]

WOLKINGER, F., PLANK, S. AND BRUNEGGER, A. (1975). Rasterelektronenmikroskopische Untersuchungen an rhizomorphen von *Armillaria mellea. Phytopathologische Zeitschrift* **84**, 352–359.

WONG, A.L. AND WILLETTS, N.J. (1974). Polyacrylamide-gel electrophoresis of enzymes during morphogenesis of sclerotia of *Sclerotinia sclerotiorium. Journal of General Microbiology* **81**, 101–109.

WORRALL, J.J., CHET, I. AND HÜTTERMANN, A. (1986). Association of rhizomorph formation with laccase activity in *Armillaria* spp. *Journal of General Microbiology* **132**, 2527–2533.

XUE, S.Y. AND XU, X.R. (1991). Effect of yungfujing on the vestibular system. *Acta Oto-Laryngologica – Supplement* **481**, 626–628.

YANG, J.S., CHEN, Y.W., FENG, X.Z., YU, D.Q. AND LIANG, X.T. (1984). Chemical constituents of Armillaria mellea mycelium. I. Isolation and characterization of armillarin and armillaridin. *Planta Medica* **50**, 288–290.

YANG, J.S., SU, Y.L., WANG, Y.L., FENG, X.Z., YU, D.Q., CONG, P.Z., TAMAI, M., OBUCHI, T., KONDOH, H. AND LIANG, X.T. (1989a). Isolation and structures of two new sesquiterpenoid aromatic esters: armillarigin and armillarikin. *Planta Medica* **55** (5), 479–481.

YANG, J.S., CHEN, Y.W., FENG, X.Z., YU, D.Q., HE, C.H., ZHENG, Q.T., YANG, J. AND LIANG, X.T. (1989b). Isolation and structure elucidation of armillaricin. *Planta Medica* **55** (6), 564–565.

YANG, J.S., SU, Y.L., WANG, Y.L., FENG, X.Y., YU, D.Q. AND LIANG, X.T. (1990a). Studies on chemical constituents of Armillaria. V. Isolation and characterization of armillarin and armillaridin. (In Chinese.) *Acta Pharmaceutica Sinica* **25**, 24–28.

YANG, J.S., SU, Y.L., WANG, Y.L., FENG, X.Y., YU, D.Q. HE, C.H., ZHENG, Q.T., YANG, J.J. AND YANG, J. (1990b). Chemical constituents of Armillaria mellea mycelium. VI. Isolation and structure of armillaripin. (In Chinese.) *Acta Pharmaceutica Sinica* **25**, 353–356.

YANG, J.S., SUO, Y.L., WANG, Y.L., FENG, X.Y., YU, D.Q. AND LIANG, X.T. (1991). Chemical constituents of Armillaria. VII. Isolation and characterization of chemical constituents of the acetone extract. (In Chinese.) *Acta Pharmaceutica Sinica* **26**, 117–122.

YI, H., YESHOU, S. AND YEYANG F. (1998). Isolation, purification and some properties of polysaccharides from the rhizomorph of Armillariella mellea. *Chinese Pharmaceutical Journal* **33**, 526–528.

YOSHIDA, H., SUGAHARA, T. AND HAYASHI, J. (1984). Studies on free sugars and free sugar alcohols of mushrooms. (In Japanese.) *Journal of the Japanese Society of Food Science and Technology* **31** (12), 765–771.

ZELLER, S.M. (1926). Observations on infections of apple and prune roots by *Armillaria mellea* Vahl. *Phytopathology* **16**, 479–484.

ZHANG, W.J. AND DONG, Z.B. (1986). Ultrastructure of mature rhizomorph of Armillaria mellea. (In Chinese.) *Acta Mycologica Sinica* **5** (2), 99–103.

ZIMMERMANN, M.H. (1971). Transport in the phloem. In: *Trees, structure and function.* Eds. M.H. Zimmermann and C.L. Brown, pp 221–275. Berlin and New York: Springer.

2
Ecology and epidemiology of *Armillaria*

AAD J. TERMORSHUIZEN

Laboratory of Phytopathology, P.O. Box 8025, 6700 EE Wageningen, The Netherlands

Abstract

Armillaria species are almost always associated with the colonization of dead or living wood, but in a few cases they have been found in connection with herbaceous tissue. Besides this, a rather common relationship exists with several mainly achlorophyllous orchid species that derive carbohydrates from *Armillaria* which, at the same time, derives carbohydrates from wood that it has colonized. Primary colonization of substrates occurs by basidiospores infecting freshly cut stumps. Once a primary infection results in the production of rhizomorphs, shoe-string-like surviving structures, soil infestations of *Armillaria* can be extremely persistent. Disease foci expand 0.2–2.5 m year^{-1}. Rhizomorphs are absent in regions with soil temperatures above *c.* 28°C, and spread of the disease seems only to occur through root-to-root contacts. Rhizomorph production occurs after colonized wood has been exploited to a large extent, which may happen years after entry to the wood substrate. The colonization of wood is, especially for the *Armillaria* species that are plant pathogenic, strongly dependent on the absence of colonizing saprotrophic fungi. Parasitic *Armillaria* species can most efficiently compete with saprotrophs by maintaining quiescent, epiphytic root lesions that become active as soon as the host resistance is broken. It is clear that a given *Armillaria* species thrives better on one soil type than on another, but the mechanisms for this have scarcely been clarified. Several antagonists and predators of rhizomorphs of *Armillaria* and competitors for wood colonization are known, but their ecosystem functioning is hitherto insufficiently understood, except possibly for *Trichoderma* sp.

Introduction

Armillaria species can infect a very wide range of woody hosts or colonize dead woody substrate from which they derive predominantly carbohydrates. The single exception is *A. ectypa*, that grows saprotrophically in peat bogs. A given pathogenic *Armillaria* species has a somewhat limited host range, and hosts differ in their

Armillaria *Root Rot: Biology and Control of Honey Fungus*

susceptibility, but species-specific interactions with a single host are unknown. Among the commonest *Armillaria* species, *A. luteobubalina*, *A. mellea*, and *A. ostoyae* are considered most pathogenic, and, among other species, *A. cepistipes* and *A. lutea* most saprotrophic or weakly pathogenic. Saprotrophic or weakly pathogenic *Armillaria* species only rarely infect trees, and in those cases the trees are usually severely weakened (Luisi *et al.*, 1996), although a notable exception was reported by Blodgett and Worrall (1992a) who found *A. lutea* to cause butt rot. However, the pathogenic species only infect their hosts under environmental conditions that favour development of the fungus or that weaken host resistance. These two factors counterbalance each other: conditions allowing abundant development of soil inoculum of *Armillaria* make hosts more prone to infection, while at constant inoculum density, infection is more likely to occur when trees are weakened. *Armillaria* species can survive for long periods in colonized wood or as rhizomorphs. The opportunistic life style of *Armillaria* may be explained by its ability to persist when there are no food sources, waiting for conditions allowing the weakening of the host by, e.g., insect attack, or a sequence of years of unusual weather.

Host range

Apart from woody hosts, infection of some herbaceous species has been reported (Raabe, 1958; Klein-Gebbinck *et al.*, 1991), including an isolate of *A. mellea* pathogenic to strawberry and potato tubers, and a problem in carrot storage (Harada, 1980). Under experimental conditions, Hughes *et al.* (1996) showed that *A. mellea* and *A. ostoyae* were able to kill *Rumex obtusifolius*, and they suggested the use of *Armillaria* as a mycoherbicide. Garrett (1956) used *Armillaria*-infected potatoes as an inoculum source in several experiments, and potatoes have been used to screen *Armillaria* isolates for pathogenicity (Gregory, 1985). Some other unusual hosts include herbs belonging to *Hyssopus*, *Lavendula*, *Origanum*, *Salvia*, and *Sideritis*, (Thanassoulopoulos and Artopeadis, 1991). While in nature *Armillaria* has never been reported in the absence of woody substrate, the ecological significance of infection of herbs in the presence of woody substrate has rarely been considered. Shaw *et al.* (1976) observed that pine trees in close proximity to *Armillaria*-infected *Cortaderia fulvia* (Gramineae) had a significantly higher disease incidence than pines with no infected grass within 60 cm from their stem base. Perhaps herbs are used as a source for nutrients such as N and P, which are clearly in shortage in woody substrate.

Armillaria is also known to form mycorrhizas with achlorophyllous plants, especially members of the Orchidaceae and Monotropaceae and, in some cases, chlorophyllous plants which are poor in photosynthesis (e.g., leafless or subterrestrial orchids). Rhizomorphs of *A. mellea* have been observed growing from roots of some *Gastrodia* spp. (Orchidaceae) (Kusano, 1911) and the fungus was isolated from orchid roots (Campell, 1962). In this symbiosis, the orchid parasitizes the fungus, while at the same time the fungus is deriving carbons from colonized wood. For this reason, the phytosymbiont has been declared an epiparasite (Smith and Read, 1997). The symbiosis seems to be a parasitic one, although clearly negative effects on the mycobiont have not been reported. The profit is especially clear for achlorophyllous phytobionts, but infection occurs also in autotrophic orchids. Similar symbioses occur for several other basidiomycete species that obtain carbons by saprotrophic or parasitic activity, including, e.g., the common soil-borne parasite *Rhizoctonia solani* (Masuhara *et al.*, 1993). The formation of orchid mycorrhiza appears to occur for many *Armillaria* species. Terashita (1996) isolated *A. cepistipes*, *A. lutea*, *A. mellea*, and *A. tabescens* from roots of *Gastrodia septentrionalis*, and Cha and Igarashi (1995) *A. jezoensis*, *A. lutea*, *A. ostoyae*, *A. sinapina* and *A. singula* from *G. elata*.

An unusual symbiosis between *Armillaria* and the basidiomycete *Entoloma abortivum* was reported by Watling (1974). In the absence of *Armillaria*, *E. abortivum* fruits with normal basidiocarps, but in the presence of *Armillaria* it produces fertile gasteroid fruiting bodies. In these closed, globose fruiting bodies hyphae of *Armillaria* and *Entoloma* are present, but only the latter forms basidiospores. Cha and Igarashi (1996) identified the *Armillaria* species connected to *E. abortivum* on several decayed stumps in Japan as *A. jezoensis* and *A. lutea*. Until now, the nutritional status between the two species is not clear. Remarkably, Cha and Igarashi (1996) showed that, in dual culture, *Entoloma abortivum* inhibited the *Armillaria* isolates more than *vice versa*.

Primary colonization

The now generally accepted role of basidiospores as primary inoculum of *Armillaria* has been underestimated for a long time. The prolonged doubtful status of the function of basidiospores may have been caused by the conspicuous persistence of *Armillaria* once it has infested the soil. It is remarkable that only very few controlled experiments have been carried out to study the function of basidiospores. Among these, Leach (1939) was unsuccessful in inoculating basidiospores, and Rishbeth (1970) reported some rare cases where he obtained stump infection using basidiospores as inoculum. Hood and Sandberg (1987) grew young plants of *Pinus radiata* and *Beilschmiedia tawa* protected from rhizomorphs by plastic sheaths and showed that they became colonized by *Armillaria* within 20–22 months. They concluded that infection originated from nearby fruiting-bodies and that stumps are not a prerequisite for establishment of basidiospores, but the authors did not provide evidence for the total absence of *Armillaria* in the soil that surrounded the plants. Large numbers of small, genetically identical groups, so-called genets, are indicative for basidiospore infection (Hood and Sandberg, 1987; Rishbeth, 1978b, 1988; Worrall, 1994; Legrand *et al.*, 1996). Rishbeth (1978b) found infection foci in first-rotation plantings of oak and beech to be associated with thinning stumps. The fact that the infection foci were scattered and contained different genets led Rishbeth (1978b) to conclude that the foci had arisen as a result of stump infection by basidiospores. In five first-rotation plantations that were still unthinned, *Armillaria* was absent, whereas in four out of six 20–40 year-old plantations that had been thinned, *Armillaria* was found (Rishbeth, 1978b). Rizzo *et al.* (1995) found resinosis or other host responses in less than 10% of stumps that were colonized by *A. ostoyae* in a clear-cut plot of *Pinus resinosa* and *P. banksiana*, indicating that *Armillaria* infected most stumps after cutting. On the other hand, Legrand *et al.* (1996) reported many small genets, often consisting of only a single isolate, in a 7 year-old plantation of *Pinus sylvestris* and 12 years after uprooting. Again, the high frequency of small genets is indicative for infestation by basidiospores, and if this is the case, they have infested the site in the absence of stumps, since these had been removed.

The methodology of genet determination has been subject to some debate. The most widely-used method to delineate genets is by pairing isolates. Isolates, when confronted on an agar medium, are said not to belong to the same genet when a melanized line is produced where the two colonies meet each other (Adams and Roth, 1969; Adams, 1974; Anderson *et al.*, 1979). This technique has been criticized in that sib-related genets may not be distinguished by these pairings (Kay and Vilgalys, 1992). However, Rizzo *et al.* (1995) found no differences between isolate pairings and both nuclear and mitochondrial markers, while the high number of shared mitochondrial types of isolates occurred within a plantation (Smith *et al.*, 1994; Rizzo *et al.*, 1995). Thus, sib-related isolates are characterized well by the compatibility test. In general, somatic incompatibility testing of *Armillaria* isolates has led to only minor differences with other genetic markers (Kile, 1983; Rizzo and Harrington, 1993; Guillaumin *et al.*, 1995), if any (Korhonen, 1978; Guillaumin and Berthelay, 1990; Smith *et al.*, 1990, 1992). In addition, even if sib-related genets cannot be distinguished, this would thwart the conclusion of Rishbeth (1978b) and others only if they would have found similar, spatially remote, genets, which is not the case.

The incidence of stump infection by basidiospores is probably quite low. For heterothallic species, two basidiospores need to germinate close to each other and about the same time, since monokaryotic mycelium degenerates quickly (Guillaumin *et al.*, 1991). Except for the rare *A. ectypa*, all temperate *Armillaria* species are heterothallic (Hintikka, 1973; Korhonen, 1978; Anderson and Ullrich, 1979). In contrast, the tropical *A. heimii* and basidiocarps of *A. mellea* that occur in the tropics are both homothallic, perhaps an adaptation to the hot climate where fruiting of *Armillaria* is less common (Swift, 1972), thus enabling the successful colonization by single basidiospores. In a large survey in first-rotations of oak, the incidence of *Armillaria* foci on tree stumps varied between 0 and 1.2% (Rishbeth, 1978b). Kile (1983) estimated the number of successful basidiospore infections in an Australian Eucalypt forest of *c.* 80,000 ha to be one per year. Given the often enormous abundance of fruiting bodies of *Armillaria* in woodlands, the success rate of basidiospores is extremely low indeed. Rizzo *et al.* (1995) reported a close genetic relationship of isolates occurring within a plantation as compared to the general variability of *Armillaria*. This may indicate that basidiospores are dispersed mainly locally. Perhaps their survival during transport through the air is very low, which is generally the case. In addition, high moisture levels may be critical to successful establishment of basidiospores (Rishbeth, 1970). This may explain, as suggested by Worrall (1994), the relative high frequency of small-sized genets in humid forests (Kile, 1986; Hood and Sandberg, 1987) as opposed to dry forests, where large genets predominate (Adams, 1974; Shaw and Roth, 1976). Probably, stumps are receptive to *Armillaria* basidiospores only during a very short time after cutting. Given the low competitive saprotrophic ability of parasitic *Armillaria* species (see below), basidiospores are likely to infect stumps only if other organisms are absent. So, perhaps the receptivity of stumps may be limited to a few seconds or minutes after cutting. The low incidence of stump infection by basidiospores makes biocontrol using saprotrophs that quickly colonize stumps rather costly, but, alternatively, foresters could postpone felling activities when fruiting bodies of *Armillaria* are present.

Primary colonization may also occur when plants that are infected, or that contain adhering soil infested with *Armillaria*, are introduced.

Substrate colonization

Plant pathogenic *Armillaria* species can be designated as necrotrophs, i.e. organisms that kill hosts and continue to decompose the colonized material after its death. *Armillaria* species differ widely in their parasitic and competitive saprotrophic ability but it seems that, to a greater or lesser extent, all exhibit both abilities. In general, the stronger the parasitic ability, the lower the competitive saprotrophic ability, and *vice versa*. The energy obtained by colonizing substrates is then partly used for rhizomorphs to explore the soil for new hosts. By precolonizing living tissue, *Armillaria* species are, to some extent, able to outcompete other saprotrophs which may colonize the substrate after death of the substrate. However, *Armillaria* is usually outcompeted if saprotrophs arrive at the same time. Li and Hood (1992) showed that *Rigidoporus catervatus* and a *Trichoderma* species suppressed *A. novae-zelandiae* that was pre-inoculated in wood segments of *Pinus radiata*. *Armillaria* is clearly favoured compared to saprotrophs when substrate has a weakened, but not totally disfunctioning resistance reaction. Entry *et al.* (1991a) hypothesized that the probability of successful colonization of substrate depends on energy needed to decompose phenolic and lignin compounds and the amount of energy available to the fungus in the form of simple sugars. Entry *et al.* (1991a) tested this hypothesis in an experiment where the effect of light and nitrogen nutrition on the susceptibility of seedlings of several coniferous hosts was studied by placing wood blocks colonized by *A. ostoyae* against the primary root. It appeared that the ratio of sugar concentration in secondary root tissue to that of lignin or of phenolic compounds (as affected by light and nitrogen condition) was inversely related with disease rating. High disease ratings were observed only if the lignin:sugar ratio was <30 or if the phenolics:sugar ratio was <2.5.

All *Armillaria* species can de designated as opportunists; the fitness of the host is decisive in determining whether pathogenic infection occurs or not. There are many references in the literature to greater vulnerability of hosts if they are weakened. In fact, predisposition to infection may be essential because soil infestations by high densities of rhizomorphs of pathogenic *Armillaria* spp. have been reported in locations where infected trees could not be found (Stanosz and Patton, 1991). On the other hand, if conditions allow for host infection, disease is more severe at high rhizomorph densities in soil (MacKenzie and Shaw, 1977; Podger *et al.*, 1978; Kile, 1981). In general, care should be taken to discuss predisposition in relation to *Armillaria* disease since it may be easy to find predisposing factors for such long-lived organisms as trees. Nevertheless, convincing evidence exists that predisposition stimulates infection by *Armillaria*. Thus, planting density (Gerlach *et al.*, 1997), nutrient deficiency (Singh, 1983; Entry *et al.*, 1986), reduced light conditions (Redfern, 1978; Entry *et al.*, 1986, Entry *et al.*, 1991a), defoliation (Wargo, 1977), air pollution (Grzywacz and Wazny, 1973), use of bare-rooted stock material (Singh and Richardson, 1973; Livingston, 1990), changes of the nutrient status of the host caused by pruning (Popoola and Fox, 1996), and attack by insects (Hudak and Wells, 1974) are documented examples where predisposition of the host has led to *Armillaria* infection. Predisposition apparently weakens the host, enabling *Armillaria* to infect the host. *Armillaria* can also strongly develop in herbicide-killed plantations (Pronos and Patton, 1978; Schutt *et al.*, 1978). This has been partly explained by the growth-stimulating effect of some herbicides, notably 2,4–D, or absence of a direct effect on

Armillaria (Pronos and Patton, 1979). The opportunism of *Armillaria* is further exemplified by the observation of MacKenzie (1987) that 31% of 9 year-old *Pinus radiata* trees recorded as being *Armillaria*-infected were rated uninfected 9 years later. Apparently, *Armillaria* may even disappear if a tree recovers from a temporal stress factor. In some cases, however, predisposition is not clearly present resulting in so-called primary attacks. Rosso and Hansen (1998) found that Douglas firs were attacked irrespective of their actual or previous vigorousness. High rhizomorph densities, such as those developed after formation of rhizomorphs on stumps, may result in infection and decline of apparently healthy trees (MacKenzie and Shaw, 1977; Podger *et al.*, 1978; Kile, 1981).

The reason for the often observed severe attack by *Armillaria* after weakening of the host is probably due to the presence of non-pathogenic, quiescent root lesions of *Armillaria* which can bc present for years prior to host colonization (Leach, 1939; Marsh, 1952; Swift, 1972; Reaves *et al.*, 1993). These epiphytic lesions enable *Armillaria* to rapidly colonize weakened hosts that are unable to respond with a defense reaction. The ability of *Armillaria* to persist many years in a quiescent stage on a healthy host is an important strategy to remain ahead of other weak pathogens and saprotrophs. Rizzo *et al.* (1995) observed epiphytic lesions of *A. ostoyae* measuring only 1–2 cm on the major lateral roots of apparently healthy trees. The primary pathogen *A. ostoyae* was frequently isolated from root lesions from healthy *Populus tremuloides* trees (Banik *et al.*, 1995), while no lesions were found yielding the weak pathogen *A. lutea* (Banik *et al.*, 1995).

Armillaria can become especially very active after the cutting of trees in an infested field. The ability of *Armillaria* to colonize stumps can be influenced by ring-barking trees at the stem base the year prior to cutting (Leach, 1937, 1939; Lanier, 1971). Ring-barking of trees prior to felling resulted in only 1.8% infection 8 years after the planting of *Aleurites* sp., while 19% were infected in a part of the plantation that was not ring-barked (Wiehe, 1952). Apparently, loss of resistance against saprotrophs is so fast, and colonization by *Armillaria* so slow, that saprotrophs effectively limit the expansion of *Armillaria* (Garrett, 1960). It is, however, difficult to explain why herbicide treatment or insect attack would be factors stimulating attack by *Armillaria*, whilst ring-barking inhibits colonization by *Armillaria*. Perhaps rates of loss of resistance are decisive. The situation has become complicated by the results of Redfern (1968), who reported that ring-barking of a *Quercus* stand two years prior to felling did not affect, or even stimulate, stump colonization by *Armillaria*. Garrett (1970) suggested that high rhizomorph densities may have caused this unexpected result.

Rhizomorph growth

Rhizomorphs constitute the most important inoculum for secondary spread in soil of many *Armillaria* species in the temperate regions of the world. The vernicular name 'shoe-string fungus' for *Armillaria* species is because the rhizomorphs resemble shoe-strings. Rhizomorphs of *Armillaria* are black, 1–5 mm wide, complex structures that are unique among the fungi. Somewhat similar structures have been described only for the dry rot fungus *Serpula lacrymans*. They consist of a black outer layer, the cortex, which contains melanin and a white inner layer, the medulla, which consists of fine tubes (Hartig, 1874; Stanosz *et al.*, 1987; Cairney *et al.*, 1988). The fine structure of rhizomorphs has been described in detail by Garraway *et al.* (1991). The melanin layer is highly persistent and can be found with the medulla already decomposed. Rhizomorphs are highly persistent and survival has been reported for up to 12 years (Reaves *et al.*, 1993). Rhizomorphs are often also produced on woody substrates, where they usually grow lengthways in high abundance, often giving rise to large sheets by lengthways fusion of rhizomorphs. *Armillaria* species differ in their ability to form rhizomorphs in soil. For example, *A. lutea* and *A. cepistipes* produce most extensive rhizomorph systems in soil, *A. ostoyae* less so, *A. mellea* only sparsely, and *A. tabescens* not at all (Gregory, 1989; Rishbeth, 1982). Hyphae of *Armillaria* have never been reported to occur in soil. This may be due to difficulties in detecting these structures. The low competitive saprotrophic abilities of most *Armillaria* species (Garrett, 1960; Holmer and Stenlid, 1996) suggest that only rhizomorphs occur in soil as large transporting organs and that hyphae are lacking from soil. On the other hand, fruiting bodies of *Armillaria* can often be found not connected to either rhizomorphs or roots, in my view indicating that hyphae must be present in soil. Clearly, this aspect warrants more attention in future research.

Rhizomorphs are specialized organs that are generally assumed to function as, other than their role in survival, a means of efficient transport of nutrients. However, there is still uncertainty about the nutrient economy of rhizomorphs. The observation that rhizomorphs usually develop years after substrate colonization indicates the need for energy reserves for rhizomorph initiation and growth, and thus a translocation of nutrients from the food base to the rhizomorph tips. Anderson and Ullrich (1982) showed that transport, but not uptake, of water and nutrients by rhizomorphs required aerobic respiration, indicating that translocation is not a passive, capillary, process. Morrison (1975) and Anderson and Ullrich (1982) showed that nutrients applied basipetally were transferred to an actively growing rhizomorph tip, but not *vice versa*. However, Granlund *et al.* (1985) and Gray *et al.* (1996) demonstrated that translocation of nutrients in rhizomorphs can be bidirectional. Possibly, nutrient translocation rate and direction depends on the concentration difference between the rhizomorph parts, very much the same as the electric current depends on the potential difference between two ends of a wire. The need for nutrients during rhizomorph growth has not been quantified. If *Armillaria* withdraws nutrients only from wood, a shortage of nitrogen and phosphorus would likely occur. Possibly, nutrient uptake occurs also with small, short hyphae extending from the medulla through the cortex into the soil, as observed by Cairney *et al.* (1988), but the function of these hyphae have not been studied in detail. That the cortex layer consists of short hyphae extending into the soil was already observed as early as 1874 by Hartig. Perhaps nitrogen and phosphorus compounds are derived from root exudates of herbaceous hosts. When rhizomorphs are broken due to soil cultivation, every rhizomorph piece may develop several new branches, thus increasing inoculum density dramatically (Redfern, 1973).

Generally, most rhizomorphs are found in the top soil layers (Morrison, 1976). Growth and survival of rhizomorphs is negatively affected by high soil moisture levels (Pearce and Malajczuk, 1990; Redfern, 1970). This may be related to reduced rhizomorph growth under lowered oxygen or raised carbon dioxide levels (Rishbeth, 1978a). Morrison (1976) reported that the direction of rhizomorph growth is towards increasing oxygen levels and decreasing carbon dioxide levels, which may explain the poor growth of rhizomorphs on heavier soil types with impeded drainage (Singh, 1981; Blodgett and Worrall, 1992b; Whitney, 1984). This may also be the cause of the small foci and limited rhizomorph formation by *A. mellea* in a study of *Armillaria* species occurring in southern England (Rishbeth, 1982). Larger genets of *A. mellea* were observed at sites on lighter soils. For rhizomorph growth, the apex needs to be covered by a film of water (Smith and Griffin, 1971). Dry conditions would suppress especially the saprotrophic species which grow through soil more than parasitic species which may also be able to spread through root systems (Cruickshank *et al.*, 1997). Temperature has a profound effect on rhizomorph initiation, growth, and branching. The lack of any rhizomorphs in tropical Africa at low elevations (Swift, 1968) has been explained by prevailing soil temperatures which are too high for rhizomorph growth (Rishbeth, 1978a). Pearce and Malajczuk (1990) ascribed the paucity of rhizomorphs of *A. luteobubalina* in *Eucalyptus* forests in SW-Australia to unfavourably high temperatures and low soil moisture contents. Cultures of *Armillaria luteobubalina* failed to form rhizomorphs at 30°C, but they did so abundantly after the temperature was lowered to 20°C, suggesting an enzyme inactivation process at 30°C (Pearce and Malajczuk, 1990). *Armillaria* isolates from tropical lowlands do produce rhizomorphs *in vitro* at 25°C (Swift, 1968). Onsando *et al.* (1997) report the presence of rhizomorphs in tea plantations at low elevations on residual roots from trees that were present before the tea was planted and ascribed this to the prevailing moderate temperatures at deeper soil layers. The temperature range for rhizomorph growth varies for the different species (Rishbeth, 1978a), but, in general, rhizomorphs are not formed above 28°C. Rhizomorph initiation has a narrower temperature range than rhizomorph growth (Risbeth, 1968). Temperature affects rhizomorph growth also under more moderate conditions. Pearce and Malajczuk (1990) found that rhizomorphs of *A. luteobubalina* produced *in vitro* different amounts of rhizomorphs at fluctuating temperatures than expected based on rhizomorph growth data at constant temperatures. Redfern (1973) reported that rhizomorph production and intensity of branching were higher at 25°C than at 15°C, while dry weight production was not influenced. Rhizomorph growth and branching pattern is also influenced by the C/N ratio of the organic material. Substances rich in lignin and cellulose (peat, pine bark), added to unsterile soil, stimulated the growth of rhizomorphs but reduced their degree of branching, and substances high in N (lucerne, guano) reduced growth and branching of rhizomorphs (Guillaumin and Leprince, 1979).

Rhizomorphs are usually formed most prominently after the host has died. The importance of colonized stumps as a source for rhizomorph formation was shown by removing stumps (Roth *et al.*, 1980). The rate of appearance of new rhizomorphs from dead stumps depends on the rate of colonization. Slowly dying hosts are supposed to be the best substrate for *Armillaria* since they are thought to maintain some resistance against the entrance of saprotrophic fungi (Leach, 1939). Stumps showing some regrowth are quite slowly colonized, and rhizomorph production may be much delayed (Twery *et al.*, 1990). Maximum rhizomorph production occurred 10 years after a herbicide treatment to kill an unproductive oak stand (Pronos and Patton, 1978). Rishbeth (1972) found that, 14 years after felling, rhizomorph formation was greater on stumps of *Quercus robur* and *Acer pseudoplatanus* in contrast to stumps of *Pinus sylvestris*, which generally produced fewer rhizomorphs. Guillaumin and Lung (1985) reported that both *A. mellea* and *A. ostoyae* decompose wood of *Fagus sylvatica* faster and form more rhizomorphs than on wood of *Pinus sylvestris*, although the last host is known to be strongly favoured by *A. ostoyae*. Significant effects of substrate on rhizomorph formation has also been observed for *A. luteobubalina* (Pearce and Malajczuk, 1990). It has been suggested that the competitive ability of *Armillaria* with other wood-decomposing fungi is stronger in softwoods than in hardwoods (Rishbeth, 1972). Alternatively, hardwoods may contain compounds that stimulate rhizomorph initiation and formation, or softwoods contain inhibitory compounds. Major episodes of tree death caused by *A. mellea* and *A. ostoyae* have often been related to situations where broad-leaved trees were present (Hartig, 1874; Rishbeth, 1982). In his monumental standard work, Peace (1962) characterized *Armillaria* as a fungus of areas with a hardwood forest history. Redfern (1975) reported that killing by *Armillaria* may continue even during successive rotations of conifers. Statements on plantation history and *Armillaria* infection clearly have been biased by the replacement of much hardwood by softwood (Gregory, 1989), resulting in limited observations for cases where softwood is followed by hardwood. MacKenzie and Self (1988) and Gregory (1989) reported *A. ostoyae* from sites that have always been planted to conifers. Furthermore, the presence of susceptible species in the understorey of plantation needs to be taken into consideration when studying the behaviour of rhizomorph growth. For example, the presence of hardwood shrubs may very well influence the dynamics of *Armillaria* in a softwood plantation.

Formation of disease foci

In temperate areas, rhizomorphs are the most important means of dispersal by *Armillaria*. In addition, parasitic species may spread by root-to-root contact (Shearer and Tippett, 1988), and in (sub)tropical areas where rhizomorph formation is impeded, root-to-root contact is the only mechanism of focus formation. Thus, root-to-root contact is the main cause of *Armillaria* root disease in tea (Wiehe, 1952; Onsando *et al.*, 1997), and *Armillaria* disease was absent on tea plantations established on former arable land (Onsando *et al.*, 1997). In a *Eucalyptus marginata* forest, Shearer and Tippett (1988) were unable to find any rhizomorphs of *A. luteobubalina*, and ascribed formation of disease foci to root-to-root contact. On the other hand, in peach orchards, Kable (1974) explained focus formation by rhizomorph growth rate, and not by root-to-root contact.

Reports on the size of genets vary from very large (600 ha in a *Pinus ponderosa* forest: Shaw and Roth, 1976) to small (single stumps: Rishbeth, 1978b; single sample points–0.07 ha in forests in NE-USA: Worrall, 1994) or variable (0.02–6.5 ha in coastal dune vegetations in SW-Australia: Shearer *et al.*, 1997b; Legrand *et al.*, 1996; up to 1.5 ha: Rizzo *et al.*, 1995). The clonal size depends on clone age, growth rate, environmental conditions such as soil moisture, and on the species' ability to form rhizomorphs. Thus, *A. lutea*, which produces relative extensive rhizomorph systems (Rishbeth, 1982), was found to produce, together with *A. cepistipes*, much larger foci in an ancient coppice broad-leaved woodland than *A. mellea* and *A. ostoyae*. *Armillaria* typically forms disease foci that slowly expand. The rate of extension has been assessed to range from 0.2 to 2.5 m year^{-1} (Durrieu *et al.*, 1981; Kile, 1981, 1983; Peet *et al.*, 1996; Podger *et al.*, 1978; Shaw and Roth, 1976; Shearer *et al.*, 1997a; van der

Kamp, 1993). Large disease foci have therefore been estimated to be very old, up to 1,500 years (Smith *et al.*, 1992). On the other hand, foci of considerable size may also be formed by root contacts with *Armillaria*-infected roots alone, as has been found in high-temperature regions that do not allow rhizomorph growth (Wiehe, 1952; Onsando *et al.*, 1997; Shearer and Tippett, 1988) and for *A. tabescens*, which is unable to form any rhizomorphs, Rishbeth (1991) reported disease foci up to 160 m in length. Genets of *A. cepistipes* and *A. lutea* have found to be never overlapping, indicating occupation of similar ecological niches; while they did overlap with *A. ostoyae* (Legrand *et al.*, 1996).

Within a disease focus, dieback of trees occurs, and, especially in natural ecosystems, the vegetation community may shift towards more resistant plant species as a combined result of presence of *Armillaria* but also of changes in light and nutrient conditions (Shearer *et al.*, 1997b). *Armillaria* occurs frequently in natural ecosystems (Kile *et al.*, 1991). Hood and Sandberg (1987) reported a frequency of 13–89% of rhizomorphs occurring in 2,200 cm^3 soil samples in indigenous forests in New Zealand. It may be interesting to study in more detail the significance of *Armillaria* as a selection pressure in the composition of the vegetation.

Interaction with micro-organisms and disease suppressiveness

It is remarkable that only few hyperparasites or predators of *Armillaria* have received scientific attention, although many fungicolous fungi have been isolated from rhizomorphs. In exhaustive reviews on fungicolous fungi, only very few fungicolous fungi occurring on *Armillaria* are mentioned (Jeffries and Young, 1994). The most notable hyperparasite is *Trichoderma* sp. which frequently parasitizes rhizomorphs. The relative sensitivity of *Armillaria*, relative to the hyperparasite, to carbon disulphide (Bliss, 1951; Garrett, 1957; Munnecke *et al.*, 1973), methyl bromide (Ohr and Munnecke, 1974), ammonium sulphamate (Nelson *et al.*, 1995) and heat (Munnecke *et al.*, 1976; Filip and Yang-Erve, 1997) has been exploited to improve control of *Armillaria* by combining *Trichoderma* with reduced dosages of chemicals or heat treatment, respectively. Mycophagous nematodes (*Aphelenchoides cibolensis* and *A. composticola*) were found to reduce seedling mortality of *Pinus ponderosa* caused by *Armillaria*, but they did not kill the fungus (Riffle, 1973). *Aphelenchus avenae* inhibits *Armillaria* growth but not that of the mycoparasite *Trichoderma viride* and *T. polyspermum* (Cayrol *et al.*, 1978). Several *Erwinia* and *Pseudomonas* species were isolated from rhizomorphs (Samyn *et al.*, 1980). Virus-like particles were recognized in several isolates of *A. ostoyae* (Reaves *et al.*, 1988). Some bacteria have been observed to inhibit rhizomorph growth *in vitro* (Dumas, 1992). As stated above, antagonists that suppress *Armillaria* significantly have hitherto been reported only sporadically. This may not be so unexpected, since the estimated high age of some disease foci (Smith *et al.*, 1992) indicates that antagonistic populations fail to suppress *Armillaria* successfully. On the other hand, it is remarkable that many competent fungicolous fungi have been isolated from sclerotia of species such as *Sclerotinia sclerotiorum* and *Sclerotium cepivorum*, that contain a similar melanin-containing sheath as the rhizomorphs of *Armillaria* (Jeffries and Young, 1994). Perhaps there is scope for finding effective antagonists in vegetation communities where *Armillaria* is absent. For example, it is intriguing that *Armillaria* spp. can destroy *Rumex obtusifolius* (Hughes *et al.*, 1996), whereas *Armillaria* is never found at localities that lack wood. Olembo (1972) reported that non-sterile leachates from soils free from *Armillaria* inhibited its penetration and colonization of wood cylinders compared with soils from areas with moderate or high infection.

Disease suppressiveness caused by the constitutive presence of biotic or abiotic factors, as reported for several soil-borne plant pathogens (Cook and Baker, 1983), has only rarely been reported for *Armillaria*. Within a natural *Pinus uncinata*/*P. sylvestris* forest, localities that were destroyed by *Armillaria* showed abundant regeneration that was not attacked by *Armillaria* (Durrieu *et al.*, 1985). It was postulated that the regeneration was due to an induced disease suppressiveness at these sites, but the precise nature of this suppressiveness was not characterized. Blenis *et al.* (1989) found large differences in disease incidence, severity and rhizomorph production of *A. ostoyae* in seedlings of *P. contorta* that were inoculated following planting on four soil types. Seedling infection and mortality varied from 0 to 96 and 0 to 84%, respectively. Again, these observations do not prove that soils were disease suppressive, since plant vigour may also have been affected by soil type. Shields and Hobbs (1979) found that *Armillaria* infection in Douglas fir occurred significantly more at sites low in N and pH, while in *Abies grandis* at sites low in soil Ca and P and high soil K. In southern England, Rishbeth (1982) did not find *A. ostoyae* attacking pines planted on alkaline soils, while at comparable sites on acid soils it did so. *A. ostoyae* was found on slightly but significantly more acid soils than *A. lutea* (Blodgett and Worrall, 1992b). Termorshuizen and Arnolds (1994) showed that the distribution of *A. ostoyae*, *A. lutea* and *A. mellea* in the Netherlands was primarily determined by soil type, and to a lesser extent by host species. The former species occurred on acid sandy soils, while the latter two species occurred on alkaline soils, including calcareous dune sands. *Armillaria* root rot in coniferous forests in North-America has been reported more from low pH soils (Shields and Hobbs, 1979; Singh, 1981). It is likely that these studies dealt with *A. ostoyae*, since in North-America this is the commonest pathogenic *Armillaria* species on conifers (Hood *et al.*, 1991). The mechanism of the pH-effect has not been completely clarified, but Browning and Edmonds (1993) showed that mycelial growth of *A. ostoyae* on buffered media was two times higher at pH 3 or 4 than at pH 5 or 6. Swift (1968) reported the presence of an unidentified autoclavable compound water-extracted from an English and an African soil that strongly inhibited rhizomorph growth in sterile soil from an African *Armillaria* isolate. However, the results of Swift (1968) have never been repeated, and his experiments were performed under quite unnatural conditions. In addition, as Swift (1968) admits, the absence of subcortical rhizomorphs is not explained by inhibitive factors present in soil. Thus, it seems much more likely that prevailing high temperatures were the dominant factor suppressing rhizomorph growth.

Only a few other interactions between *Armillaria* and other organisms have been described. The interaction with *Entoloma abortivum* and the mycorrhizal interaction with orchids were described above. Tryptophol, a secondary metabolite of *Zygorrhynchus moelleri*, a common micromycete on decaying stumps (Kwasna, 1996), was found to enhance the production of rhizomorphs of *A. ostoyae in vitro* (Kwasna and Llakomy, 1998). Other fungi have been reported to stimulate rhizomorph production *in vitro* (Watanabe, 1986). Madziara-Borusiewicz and Strzelecka (1977) observed that the engraver beetle (*Ips typographicus*) preferred to attack *Armillaria*-

infected trees of *Picea excelsa* and related this to an increased production of volatile oils in the infected trees. Four chloroform-soluble antibiotic substances with antifungal and anti-Gram+ bacteria activity were isolated from rhizomorphs growing in pure culture (Oduro *et al.*, 1976). Stimulatory growth of rhizomorphs may also be caused by auxin (Garraway, 1975). In addition, many studies have shown that ethanol induces and stimulates rhizomorph formation and growth (e.g., Sortkjaer and Allermann, 1972; Vance and Garraway, 1973).

Conclusions

Genet analysis by the widely-used incompatibility test has been confirmed using molecular markers and has contributed to our understanding of the importance of basidiospores as a source of primary colonization. However, experimental evidence is scarce and warrants more research. Usually, *Armillaria* species are found in connection with wood, but for obtaining sufficient nutrients other than C, the importance of herbaceous infections or direct uptake from the soil may have been underestimated hitherto. While there seems to be general acceptance that rhizomorph production depends on the *Armillaria* species involved, the effect of previous croppings remains obscure. A confounding factor in studying this aspect is the tendency that many conifer plantations were preceded by hardwoods, but not conversely. Also, effects of the composition of the understorey vegetation on rhizomorph behaviour have not been studied. Rhizomorphs are generally thought to be absent in tropical areas, but the persistence of *Armillaria* in these regions may very well be caused by the presence of rhizomorphs at deep soil layers, where temperatures are more conducive to their formation. Except for *Trichoderma* sp., surprisingly little information exists about the interaction of antagonists or predators with *Armillaria*. This may be no surprise given the high persistence of *Armillaria*, but more attention should be paid to the possibility of biotic causes of the absence of *Armillaria* in some soil types.

Acknowledgements

I am indebted to Dr. P.J. Keizer and Prof. Dr. M.J. Jeger for their comments on an earlier version of this paper.

References

ADAMS, D.H. (1974). Identification of clones of *Armillaria mellea* in young-growth ponderosa pine. *Northwest Science* **48**, 21–28.

ADAMS, D.H. AND ROTH, L.F. (1969). Intraspecific competition among genotypes of *Fomes cajanderi* decaying young-growth Douglas-fir. *Forest Science* **15**, 327–331.

ANDERSON, J.B. AND ULLRICH, R.C. (1979). Biological species of *Armillaria mellea* in North America. *Mycologia* **71**, 402–414.

ANDERSON, J.B., ULLRICH, R.C., ROTH, L.F. AND FILIP, G.M. (1979). Genetic identification of clones of *Armillaria mellea* in coniferous forests in Washington. *Phytopathology* **69**, 1109–1111.

ANDERSON, J.B. AND ULLRICH, R.C. (1982). Translocation in rhizomorphs of *Armillaria mellea*. *Experimental Mycology* **6**, 31–40.

BANIK, M.T., PAUL, J.A. AND BURDSALL, H.H. (1995). Identification of *Armillaria* species from Wisconsin and adjacent areas. *Mycologia* **87**, 707–712.

BLENIS, P.V., MUGALA, M.S. AND HIRATSUKA, Y. (1989). Soil affects *Armillaria* root rot of lodgepole pine. *Canadian Journal of Forest Research* **19**, 1638–1641.

BLISS, D.E. (1951). The destruction of *Armillaria mellea* in citrus soils. *Phytopathology* **41**, 665–683.

BLODGETT, J.T. AND WORRALL, J.J. (1992a). Distributions and hosts of *Armillaria* species in New York. *Plant Disease* **76**, 166–170.

BLODGETT, J.T. AND WORRALL, J.J. (1992b). Site relationships of *Armillaria* species in New York. *Plant Disease* **76**, 170–174.

BROWNING, J.E. AND EDMONDS, R.L. (1993). Influence of soil aluminium and pH on Armillaria root rot in Douglas-fir in Western Washington. *Northwest Science* **67**, 37–43.

CAIRNEY, J.W.G., JENNINGS, D.H. AND VELTKAMP, C.J. (1988). Structural differentiation in maturing rhizomorphs of *Armillaria mellea* (Tricholomatales). *Nova Hedwiga* **46**, 1–25.

CAMPBELL, E.O. (1962). The mycorrhiza of *Gastrodia cunninghamii* Hook. *Transactions of the Royal Society of New Zealand. Botany* **10**, 63–67.

CAYROL, J.C., DUBOS, B. AND GUILLAUMIN, J.J. (1978). Étude preliminaire in vitro de l'aggressivité de quelques nematodes mycophages vis-a-vis de *Trichoderma viride* Pers., *T. polyspermum* (Link. ex Pers.) Rifai et *Armillariella mellea* (Vahl) Karst. *Annales de Phytopathologie* **10**, 177–185.

CHA, J.Y. AND IGARASHI, T. (1995). *Armillaria* species associated with *Gastrodia elata* in Japan. *European Journal of Forest Pathology* **25**, 319–326.

CHA, J.Y. AND IGARASHI, T. (1996). Biological species of *Armillaria* and their mycoparasitic associations with *Rhodophyllus abortivus* in Hokkaido. *Mycoscience* **37**, 25–30.

COOK, R.J. AND BAKER, K.F. (1983). *The nature and practice of biological control of plant pathogens*. St. Paul, Minnesota, U.S.A.: APS Press.

CRUICKSHANK, M.G., MORRISON, D.J. AND PUNJA, Z.K. (1997). Incidence of *Armillaria* species in precommercial thinning stumps and spread of *Armillaria ostoyae* to adjacent Douglas-fir trees. *Canadian Journal of Forest Research* **27**, 481–490.

DUMAS, M.T. (1992). Inhibition of *Armillaria* by bacteria isolated from soils of the boreal mixedwood forest of Ontario. *European Journal of Forest Pathology* **22**, 11–18.

DURRIEU, G., LISBONA, F. AND BITEAU, X. (1981). L'Armillaire en Forêt D'Osséja Premieres Observations. *106° Congrés national des Sociétés savantes, Perpignan sciences II*, pp 175–185.

DURRIEU, G., BENETEAU, A. AND NIOCEL, S. (1985). *Armillaria obscura* dans l'ecosysteme forestier de Cerdagne. *European Journal of Forest Pathology* **15**, 350–355.

ENTRY, J.A., MARTIN, N.E., CROMACK, K. AND STAFFORD, S.G. (1986). Light and nutrient limitation in *Pinus monticola*: seedling susceptibility to *Armillaria* infection. *Forest Ecology and Management* **17**, 189–198.

ENTRY, J.A., CROMACK, K., HANSEN, E. AND WARING, R. (1991a). Response of western coniferous seedlings to infection by *Armillaria ostoyae* under limited light and nitrogen. *Phytopathology* **81**, 89–94.

ENTRY, J.A., CROMACK, K., KELSEY, R.G. AND MARTIN, N.E. (1991b). Response of Douglas-fir to infection by *Armillaria ostoyae* after thinning or thinning plus fertilization. *Phytopathology* **81**, 682–689.

FILIP, G.M. AND YANG-ERVE, L. (1997). Effects of prescribed burning on the viability of *Armillaria ostoyae* in mixed-conifer forest soils in the blue mountains of Oregon. *Northwest Science* **71**, 137–144.

GARRAWAY, M.O. (1975). Stimulation of *Armillaria mellea* growth by plant hormones in relation to the concentration and type of carbohydrate. *European Journal of Forest Pathology* **5**, 35–43.

GARRAWAY, M.O., HÜTTERMANN, A. AND WARGO, P.M. (1991). Ontogeny and physiology. In: *Armillaria Root Disease*, pp 21–47. *Forest Service Handbook No. 691*. Washington, D.C.: USDA.

GARRETT, S.D. (1956). Rhizomorph behaviour in *Armillaria mellea* (Vahl) Quél. II. Logistics of infection. *Annals of Botany* **20**, 193–209.

GARRETT, S.D. (1957). Effect of a soil microflora selected by carbon disulphide fumigation on survival of *Armillaria mellea* in woody host tissues. *Canadian Journal of Microbiology* **3**, 135–149.

GARRETT, S.D. (1960). Rhizomorph behaviour in *Armillaria mellea* (Fr.) Quél. III. Saprophytic colonization of woody substrates in soil. *Annals of Botany* **24**, 275–285.

GARRETT, S.D. (1970). *Pathogenic root-infecting fungi.* Cambridge: Cambridge University Press.

GERLACH, J.P., REICH, P.B., PUETTMANN, K. AND BAKER, T. (1997). Species, diversity, and density affect tree seedling mortality from *Armillaria* root rot. *Canadian Journal of Forest Research* **27**, 1509–1512.

GRANLUND, H.I., JENNINGS, D.H. AND THOMPSON, W. (1985). Translocation of solutes along rhizomorphs of *Armillaria mellea. Transactions of the British Mycological Society* **84**, 111–119.

GRAY, S.N., DIGHTON, J. AND JENNINGS, D.H. (1996). The physiology of basidiomycete linear organs. *New Phytologist* **132**, 471–482.

GREGORY, S.C. (1985). The use of potato tubers in pathogenicity studies of *Armillaria* isolates. *Plant Pathology* **34**, 41–48.

GREGORY, S.C. (1989). *Armillaria* species in northern Britain. *Plant Pathology* **38**, 93–97.

GRZYWACZ, A. AND WAZNY, J. (1973). The impact of industrial air pollutants on the occurrence of several pathogenic fungi of forest trees in Poland. *European Journal of Forest Pathology* **3**, 129–141.

GUILLAUMIN, J.J. AND LEPRINCE, S. (1979). Influence de divers types de matière organique sur l'initiation et la croissance des rhizomorphes d'*Armillariella mellea* (Vahl) Karst. dans le sol. *European Journal of Forest Pathology* **9**, 355–366.

GUILLAUMIN, J.J. AND LUNG, B. (1985). Etude de la spécialisation d'*Armillaria mellea* (Vahl) Kumm. et *Armillaria obscura* (Secr.) Herink en phase saprophytique et en phase parasitaire. *European Journal of Forest Pathology* **15**, 342–349.

GUILLAUMIN, J.J., LUNG, B., ROMAGNESI, H., MARXMÜLLER, H., LAMOURE, D., DURRIEU, G., BERTHELAY, S. AND MOHAMMED, C. (1985). Systématique des Armillaires du groupe Mellea. Conséquences phytopathologique. *European Journal of Forest Pathology* **15**, 268–277.

GUILLAUMIN, J.J. AND BERTHELAY, S. (1990). Comparaison de deux méthodes d'identification des clones chez le Basidiomycète parasite *Armillaria obscura* (syn.: *A. ostoyae*). *European Journal of Forest Pathology* **20**, 257–268.

GUILLAUMIN, J.J., ANDERSON, J.B. AND KORHONEN, K. (1991). Life cycle, interfertility, and biological species. In: *Armillaria Root Disease*, pp 10–20. *Forest Service Handbook No. 691*. Washington, D.C.: USDA.

GUILLAUMIN, J.J., ANDERSON, J.B., LEGRAND, P., GHAHARI, S. AND BERTHELAY, S. (1995). A comparison of different methods for the identification of genets of *Armillaria* spp. *New Phytologist* **133**, 333–343.

HARADA, Y. (1980). Rhizomorphs of *Armillariella mellea* attacking taproot of carrot in storage. (In Japanese.) *Transactions of the Mycological Society of Japan* **21**, 518.

HARTIG, R. (1874). Wichtige Krankheiten der Waldbäume. *Beiträge zur Mykologie und Phytopathologie für Botaniker und Forstmänner*, 127pp. Berlin: Springer.

HINTIKKA, V. (1973). A note on the polarity of *Armillariella mellea. Karstenia* **13**, 32–39.

HOLMER, L. AND STENLID, J. (1996). Diffuse competition for heterogeneous substrate in soil among six species of wood-decomposing basidiomycetes. *Oecologia* **106**, 531–538.

HOOD, I.A. AND SANDBERG, C.J. (1987). Occurrence of *Armillaria* rhizomorph populations in the soil beneath indigenous forests in the Bay of Plenty, New Zealand. *New Zealand Journal of Forestry Science* **17**, 83–99.

HOOD, I.A., REDFERN, D.B. AND KILE, G.A. (1991). *Armillaria* in planted hosts. In: *Armillaria Root Disease*, pp 122–149. *Forest Service Handbook No. 691*. Washington, D.C.: USDA.

HUDAK, J. AND WELLS, R.E. (1974). *Armillaria* root rot in aphid-damaged Balsam Fir in Newfoundland. *Forestry Chronicle* **50**, 74–76.

HUGHES, C.N.G., WEST, J.S. AND FOX, R.T.V. (1996). Control of broad-leaved docks by *Armillaria mellea*. In: *Proceedings of the 9th International Symposium on the Biological Control of Weeds, University of Cape Town, Stellenbosch, South Africa*, January 19–26, 1996. Eds. V.C. Moran and J.H. Hoffmann, pp 531–534. South Africa: University of Cape Town.

JEFFRIES, P. AND YOUNG, T.W.K. (1994). *Interfungal parasitic relationships*. Wallingford, U.K.: CAB International.

KABLE, P.F. (1974). Spread of *Armillariella* sp. in peach orchard. *Transactions of the British Mycological Society* **62**, 89–98.

KAY, E. AND VILGALYS, R. (1992). Spatial distribution and genetic relationships among individuals in a natural population of the oyster mushroom *Pleurotus ostreatus*. *Mycologia* **84**, 173–182.

KILE, G.A. (1981). *Armillaria luteobubalina* a primary cause of decline and death of trees in mixed species eucalypt forests in central Victoria. *Australian Forest Research* **11**, 63–77.

KILE, G.A. (1983). Identification of genotypes and the clonal development of *Armillaria luteobubalina* Watling & Kile in Eucalypt forests. *Australian Journal of Botany* **31**, 657–671.

KILE, G.A. (1986). Genotypes of *Armillaria hinnulea* in wet sclerophyll eucalypt forest in Tasmania. *Transactions of the British Mycological Society* **87**, 312–314.

KILE, G.A., MCDONALD, G.I. AND BYLER, J.W. (1991). Ecology and disease in natural forests. In: *Armillaria Root Disease*, pp 102–121. *Forest Service Handbook No. 691*. Washington, D.C.: USDA.

KLEIN-GEBBINCK, H.W., BLENIS, P.V. AND HIRATSUKA, Y. (1991). Spread of *Armillaria ostoyae* in juvenile lodgepole pine stands in west central Alberta. *Canadian Journal of Forest Research* **21**, 20–24.

KORHONEN, K. (1978). Interfertility and clonal size in the *Armillaria mellea* complex. *Karstenia* **18**, 31–42.

KUSANO, S. (1911). *Gastrodia elata* and its symbiotic association with *Armillaria mellea. Imperial University of Tokyo, Journal of the College of Agriculture* **4**, 1–65.

KWASNA, H. (1996). Mycobiota of birch and birch stump roots and its possible effect on the infection of *Armillaria* spp. I. *Acta Mycologica* **31**, 101–110.

KWASNA, H. AND LLAKOMY, P. (1998). Stimulation of *Armillaria ostoyae* vegetative growth by tryptophol and rhizomorph by *Zygorhynchus moelleri*. *European Journal of Forest Pathology* **28**, 53–61.

LANIER, L. (1971). Application au Pin sylvestre d'un essai de traitement par annélation circulaire contre l'Armillaire. *Annales de Phytopathologie* **3**, 531.

LEACH, R. (1937). Observations on the parasitism and control of *Armillaria mellea. Royal Society of London. Proceedings. Series B* **121**, 561–573.

LEACH, R. (1939). Biological control and ecology of *Armillaria mellea* (Vahl) Fr. *Transactions of the British Mycological Society* **23**, 320–329.

LEGRAND, P., GHAHARI, S. AND GUILLAUMIN, J.J. (1996). Occurrence of genets of *Armillaria* spp. in four mountain forests in central France: the colonization strategy of *Armillaria ostoyae*. *New Phytologist* **133**, 321–332.

LI, Y. AND HOOD, I.A. (1992). A preliminary study into the biological control of *Armillaria-novae-zelandiae* and *A. limonea*. *Australasian Plant Pathology* **21**, 24–28.

LIVINGSTON, W.H. (1990). *Armillaria ostoyae* in young spruce plantations. *Canadian Journal of Forest Research* **20**, 1773–1778.

LUISI, N., SICOLI, G., LARARIO, P., DREYER, E. AND AUSSENAC, G. (1996). Proceedings of the International Symposium on Ecology and Physiology of Oaks in a Changing Environment. (In French.) *Annales des Sciences Forestieres* **53**, 389–394.

MACKENZIE, M. (1987). Infection changes and volume loss in a 19-year-old *Pinus radiata* stand affected by *Armillaria* root-rot. *New Zealand Journal of Forestry Science* **17**, 100–108.

MACKENZIE, M. AND SHAW, C.G. (1977). Spatial relationships between *Armillaria* root-rot of *Pinus radiata* and the stumps of indigenous trees. *New Zealand Journal of Forestry Science* **7**, 374–383.

MACKENZIE, M. AND SELF, N.M. (1988). *Armillaria* in some New Zealand second rotation pine stands. In: *Proceedings of the 36th Annual Western International Forest Disease Work Conference,* 1988, Park City, U.S.A. Ed. B.J. van der Kamp, pp 82–87.

MADZIARA-BORUSIEWICZ, K. AND STRZELECKA, H. (1977). Conditions of spruce (*Picea excelsa*) infestation by the engraver beetle (*Ips typographus* L.) in mountains of Poland. I. Chemical composition of volatile oils from healthy trees and those infected with the honey fungus (*Armillaria mellea* (Vahl) Quél.). *Zeitschrift fuer Angewandte Entomologie* **83**, 409–415.

MARSH, R.W. (1952). Field observations on the spread of *Armillaria mellea* in apple orchards and in a black-currant plantation. *Transactions of the British Mycological Society* **35**, 201–207.

MASUHARA, G., KATSUYA, K. AND YAMAGUNCHI, K. (1993). Potential of symbiosis of *Rhizoctonia solani* and binucleate *Rhizoctonia* with seeds of *Spiranthes sinensis* var. *amoena in vitro*. *Mycological Research* **97**, 746–752.

MORRISON, D.J. (1975). Ion uptake by rhizomorphs of *Armillaria mellea*. *Canadian Journal of Botany* **53**, 48–51.

MORRISON, D.J. (1976). Vertical distribution of *Armillaria mellea* rhizomorphs in soil. *Transactions of the British Mycological Society* **66**, 393–399.

MUNNECKE, D.E., KOLBEZEN, M.J. AND WILBUR, W.D. (1973). Effect of methyl bromide or carbon disulphide on *Armillaria* and *Trichoderma* growing on agar medium and the relation to survival of *Armillaria* in soil following fumigation. *Phytopathology* **63**, 1352–1357.

MUNNECKE, D.E., WILBUR, W. AND DARLEY, E.F. (1976). Effect of heating or drying on *Armillaria mellea* or *Trichoderma viride* and the relation to survival of *A. mellea* in soil. *Phytopathology* **66**, 1363–1368.

NELSON, E.E., PEARCE, M.H. AND MALAJCZUK, N. (1995). Effects of *Trichoderma* spp. and ammonium sulphamate on establishment of *Armillaria luteobubalina* on stumps of *Eucalyptus diversicolor*. *Mycological Research* **99**, 957–962.

ODURU, K.A., MUNNECKE, D.E., SIMS, J.J. AND KEEN, N.T. (1976). Isolation of antibiotics in culture by *Armillaria mellea*. *Transactions of the British Mycological Society* **66**, 195–199.

OHR, H.D. AND MUNNECKE, D.E. (1974). Effects of methyl bromide on antibiotic production by *Armillaria mellea*. *Transactions of the British Mycological Society* **62**, 65–72.

OLEMBO, T.W. (1972). Studies on *Armillaria mellea* in East Africa. Effect of soil leachates on penetration and colonization of *Pinus patula* and *Cupressus lusitanica* wood cylinders by *Armillaria mellea* (Vahl ex Fr.) Kummer. *European Journal of Forest Pathology* **2**, 134–140.

ONSANDO, J.M., WARGO, P.M. AND WAUDO, S.W. (1997). Distribution, severity, and spread of *Armillaria* root disease in Kenya tea plantations. *Plant Disease* **81**, 133–137.

PEACE, T.R. (1962). *Pathology of trees and shrubs*. Oxford: Oxford University Press.

PEARCE, M.H. AND MALAJCZUK, N. (1990). Factors affecting growth of *Armillaria luteobubalina* rhizomorphs in soil. *Mycological Research* **94**, 34–48.

PEET, F.G., MORRISON, D.J. AND PELLOW, K.W. (1996). Rate of spread of *Armillaria ostoyae* in two Douglas-fir plantations in the southern interior of British Columbia. *Canadian Journal of Forest Research* **26**, 148–151.

PODGER, F.D., KILE, G.A., WATLING, R. AND FRYER, J. (1978). Spread and effects of *Armillaria luteobubalina* sp. nov. in an Australian *Eucalyptus regnans plantation*. *Transactions of the British Mycological Society* **71**, 77–87.

POPOOLA, T.O.S. AND FOX, R.T.V. (1996). Effect of root damage on honey fungus. *Arboricultural Journal* **20**, 329–337.

PRONOS, J. AND PATTON, R.F. (1978). Penetration and colonization of oak roots by *Armillaria mellea* in Wisconsin. *European Journal of Forest Pathology* **8**, 259–267.

PRONOS, J. AND PATTON, R.F. (1979). The effect of chlorophenoxy acid herbicides on growth and rhizomorph production of *Armillariella mellea*. *Phytopathology* **69**, 136–141.

RAABE, R.D. (1958). Some previously unreported non-woody hosts of *Armillaria mellea* in California. *Plant Disease Reporter* **42**, 1025.

REAVES, J.L., ALLEN, T.C., SHAW, C.G., DASHEK, W.V. AND MAYFIELD, J.E. (1988). Occurrence of viruslike particles in isolates of *Armillaria*. *Journal of Ultrastructure and Molecular Structure Research* **98**, 217–221.

REAVES, J.L., SHAW, C.G. AND ROTH, L.F. (1993). Infection of ponderosa pine trees by *Armillaria ostoyae*: residual inoculum versus contagion. *Northwest Science* **67**, 156–162.

REDFERN, D.B. (1968). The ecology of *Armillaria mellea* in Britain: biological control. *Annals of Botany* **32**, 293–300.

REDFERN, D.B. (1970). The ecology of *Armillaria mellea*: rhizomorph growth through soil. In: *Root Diseases and Soil-borne Pathogens: Proceedings of the Symposium*, July 1968, Imperial College, London. Eds. T.A. Toussoun, R.V. Bega and P.E. Nelson, pp 147–149. Berkeley: University of California Press.

REDFERN, D.B. (1973). Growth and behaviour of *Armillaria mellea* rhizomorphs in soil. *Transactions of the British Mycological Society* **61**, 569–581.

REDFERN, D.B. (1975). The influence of food base on rhizomorph growth and pathogenicity of *Armillaria mellea* isolates. In: *Biology and Control of Soil-Borne Plant Pathogens*. Ed. G.W. Bruehl, pp 69–73. St. Paul, Minnesota, U.S.A.: The American Phytopathological Society.

REDFERN, D.B. (1978). Infection by *Armillaria mellea* and some factors affecting host resistance. *Forestry* **51**, 120–135.

RIFFLE, J.W. (1973). Effect of two mycophagous nematodes on *Armillaria mellea* root rot of *Pinus ponderosa* seedlings. *Plant Disease Reporter* **57**, 355–357.

RISHBETH, J. (1968). The growth rate of *Armillaria mellea. Transactions of the British Mycological Society* **51**, 575–586.

RISHBETH, J. (1970). The role of basidiospores in stump infection by *Armillaria mellea*. In: *Root diseases and soil borne pathogens*. Eds. T.A. Tousson, R. Bega and P. Nelson, pp 141–146. Berkeley: University of California Press.

RISHBETH, J. (1972). The production of rhizomorphs by *Armillaria mellea* from stumps. *European Journal of Forest Pathology* **2,** 193–205.

RISHBETH, J. (1978a). Effects of soil temperature and atmosphere in growth of *Armillaria* rhizomorphs. *Transactions of the British Mycological Society* **70**, 213–220.

RISHBETH, J. (1978b). Infection foci of *Armillaria mellea* in first-rotation conifers. *Annals of Botany* (London) **42**, 1131–1139.

RISHBETH, J. (1982). Species of *Armillaria* in southern England. *Plant Pathology* **31**, 9–17.

RISHBETH, J. (1988). Stump infection by *Armillaria* in first-rotation conifers. *European Journal of Forest Pathology* **18**, 401–408.

RISHBETH, J. (1991). *Armillaria* in an ancient broadleaved woodland. *European Journal of Forest Pathology* **21**, 239–249.

RIZZO, D.M. AND HARRINGTON, T.C. (1993). Delineation and biology of clones of *Armillaria ostoyae*, *A. gemina* and *A. calvescens*. *Mycologia* **85**, 164–174.

RIZZO, D.M., BLANCHETTE, R.A. AND MAY, G. (1995). Distribution of *Armillaria ostoyae* genets in a *Pinus resinosa – Pinus banksiana* forest. *Canadian Journal of Botany* **73**, 776–787.

ROSSO, P. AND HANSEN, E. (1998). Tree vigour and the susceptibility of Douglas fir to *Armillaria* root disease. *European Journal of Forest Pathology* **28**, 43–52.

ROTH, L.F., ROLPH, L. AND COOLEY, S. (1980). Identifying infected ponderosa pine stumps to reduce costs of controlling *Armillaria* root rot. *Journal of Forestry* **78**, 145–151.

SAMYN, G., VAN VAERENBERG, J. AND WELVAERT, W. (1980). Endogenic bacteria in rhizomorphs of *Armillariella mellea* (Vahl ex Fr.) Karst. *22nd International Symposium Phytopharmacy and Phytiatry*. Mededlingen Faculteit Landbouwwetenschappen Rijksuniversiteit. *Gent* **45**, 411–416.

SCHUTT, P., MASCHNING, E. AND HERMECKE, C. (1978). Entwicklung des Hallimasch in mechanisch und chemisch getoteten Erlen. *Forstwissenschaftliches Centralblatt* **97**, 26–32.

SHAW, C.G. AND ROTH, L.F. (1976). Persistence and distribution of a clone of *Armillaria mellea* in a Ponderosa Pine forest. *Phytopathology* **66**, 1210–1213.

SHAW, C.G., SIJNJA, D. AND MACKENZIE, M. (1976). Toetoe (*Cortaderia fulvida*), a new graminaceous host for Armillaria root rot. *New Zealand Journal of Forestry* **21**, 263–268.

SHEARER, B.L. AND TIPPETT, J.T. (1988). Distribution and impact of *Armillaria luteobubalina* in the *Eucalyptus marginata* forest of south-western Australia. *Australian Journal of Botany* **36**, 443–445.

SHEARER, B.L., BYRNE, A., DILLON, M. AND BUEHRIG, R. (1997a). Distribution of *Armillaria luteobubalina* and its impact on community diversity and structure in *Eucalyptus wandoo* woodland of Southern Western Australia. *Australian Journal of Botany* **44**, 151–165.

SHEARER, B.L., CRANE, C.E., FAIRMAN, R.G. AND GRANT, M.J. (1997b). Occurrence of *Armillaria luteobubalina* and pathogen-mediated changes in coastal dune vegetation of south-western Australia. *Australian Journal of Botany* **45**, 905–917.

SHIELDS, W.J. AND HOBBS, S.D. (1979). Soil nutrient levels and pH associated with *Armillariella mellea* on conifers in northern Idaho. *Canadian Journal of Forest Research* **9**, 45–48.

SIEPMANN, R. AND LEIBIGER, M. (1989). Über die Wirtsspezialisierung von *Armillaria*-Arten. *European Journal of Forest Pathology* **19**, 334–342.

SINGH, P. (1981). *Armillaria mellea*: growth and distribution of rhizomorphs in the forest soil of Newfoundland. *European Journal of Forest Pathology* **11**, 208–220.

SINGH, P. (1983). *Armillaria* root rot: influence of soil nutrients and pH on the susceptibility of conifer species to the disease. *European Journal of Forest Pathology* **13**, 92–101.

SINGH, P. AND RICHARDSON, J. (1973). *Armillaria* root rot in seeded and planted areas in Newfoundland. *Forestry Chronicle* **49**, 180–182.

SMITH, A.M. AND GRIFFIN, D.M. (1971). Oxygen and the ecology of *Armillaria elegans* Heim. *Australian Journal of Biological Sciences* **24**, 231–262.

SMITH, M.L., DUCHESNE, L.C., BRUHN, J.N. AND ANDERSON, J.B. (1990). Mitochondrial genetics in a natural population of the plant pathogen *Armillaria*. *Genetics* **126**, 575–582.

SMITH, M.L., BRUHN, J.N. AND ANDERSON, J.B. (1992). The fungus *Armillaria bulbosa* is among the largest and oldest living organisms. *Nature* **356**, 428–431.

SMITH, M.L., BRUHN, J.N. AND ANDERSON, J.B. (1994). Relatedness and spatial distribution of *Armillaria* genets infecting red pine seedlings. *Phytopathology* **84**, 822–829.

SMITH, S.E. AND READ, D.J. (1997). *Mycorrhizal symbiosis*. San Diego: Academic Press.

SORTKJAER, O. AND ALLERMANN, K. (1972). Rhizomorph formation in fungi. I. Stimulation by ethanol and acetate and inhibition by disulfiram of growth and rhizomorph formation in *Armillaria mellea*. *Physiologia Plantarum* **26**, 376–380.

STANOSZ, G.R., PATTON, R.F. AND SPEAR, R.N. (1987). Structure of *Armillaria* rhizomorphs from Wisconsin aspen stands. *Canadian Journal of Botany* **65**, 2124–2127.

STANOSZ, G.R. AND PATTON, R.F. (1991). Quantification of *Armillaria* rhizomorphs in Wisconsin aspen sucker stands. *European Journal of Forest Pathology* **21**, 5–16.

SWIFT, M.J. (1968). Inhibition of rhizomorph development by *Armillaria mellea* in Rhodesian forest soils. *Transactions of the British Mycological Society* **51**, 241–247.

SWIFT, M.J. (1972). The ecology of *Armillaria mellea* Vahl (ex Fries) in the indigenous and exotic woodlands of Rhodesia. *Forestry* **45**, 67–86.

TERASHITA, T. (1996). Biological species of *Armillaria* symbiotic with *Galeola septentrionalis*. (In Japanese with English summary.) *Nippon Kingakukai Kaiho* **37**, 45–49.

TERMORSHUIZEN, A.J. AND ARNOLDS, E.J.M. (1994). Geographical distribution of the *Armillaria* species in the Netherlands in relation to soil type and hosts. *European Journal of Forest Pathology* **24,** 129–136.

THANASSOULOPOULOS, C.C. AND ARTOPEADIS, M.C. (1991). Some previously unreported hosts of *Armillaria* root rot. *Plant Disease* **75**, 101.

TWERY, M.J., MASON, G.N., WARGO, P.M. AND GOTTSCHALK, K.W. (1990). Abundance and distribution of rhizomorphs of Armillaria spp. in defoliated mixed oak stands in western Maryland. *Canadian Journal of Forest Research* **20**, 674–678.

VANCE, C.P. AND GARRAWAY, M.O. (1973). Growth stimulation of *Armillaria mellea* by ethanol and other alcohols in relation to phenol concentration. *Phytopathology* **63**, 743–748.

VAN DER KAMP, B.J. (1993). Rate of spread of *Armillaria ostoyae* in the central interior of British Columbia. *Canadian Journal of Forest Research* **23**, 1239–1241.

VAN DER KAMP, B.J. (1995). The spatial distribution of *Armillaria* root disease in an uneven-aged, spatially clumped Douglas-fir stand. *Canadian Journal of Forest Research* **25**, 1008–1016.

WARGO, P.M. (1977). *Armillariella mellea* and *Agrillus bileneatus* and mortality of defoliated oak trees. *Forest Science* **23**, 485–492.

WATANABE, T. (1986). Rhizomorph production in *Armillaria mellea* in vitro stimulated by *Macrophoma* sp. and several other fungi. *Mycological Society of Japan. Transactions* **27**, 235–245.

WATLING, R. (1974). Dimorphism in *Entoloma abortivum. Société Linneenne de Lyon. Bulletin* **43**, 449–470.

WHITNEY, R.D. (1984). Site variation of *Armillaria mellea* in three Ontario conifers. In: *Proceedings of the 6th International Conference on root and butt rots of forest trees*, 1983, Melbourne, Australia. Ed. G.A. Kile, pp 122–130.

WIEHE, P.O. (1952). The spread of *Armillaria mellea* (Fr.) Quél. in Tung orchards. *East African Agricultural Journal* **18**, 67–72.

WORRALL, J.J. (1994). Population structure of *Armillaria* species in several forest types. *Mycologia* **86**, 401–407.

3
Quantitative aspects of the epidemiology of *Armillaria* in the field

ANGELIQUE LAMOUR[1] AND MICHAEL J. JEGER[2]

[1]*Laboratory of Phytopathology, Wageningen Agricultural University, P.O. Box 8025, 6700 EE Wageningen, The Netherlands and* [2]*Wye College, University of London, Wye, Ashford, Kent TN25 5AH, U.K.*

Abstract

Armillaria root rot is a serious disease in many forests and horticultural tree crops worldwide. Consequently, there is much interest in determining how different silvicultural practices influence disease incidence and options for avoiding or restricting the spread of disease. However, published information on the biology and ecology of *Armillaria* frequently is not available in a form that can directly assist decision-making. A model to forecast disease development and severity would be invaluable in forest management to facilitate choices between different silvicultural practices. It would also serve as a quantitative statement of hypotheses about root disease dynamics, behaviour, and impact. Serious data gaps can be identified and thus research needs can be defined and prioritized. Studies are reviewed in which the number and dry weight of *Armillaria* rhizomorphs are determined and the distributions of rhizomorphs, both vertically and horizontally, are quantified. Studies on disease progress, both temporal and spatial, in permanent plots are also reviewed and models developed to simulate disease progress are described, notably the Western Root Disease Model.

Introduction

Armillaria root rot is a cause of continuing concern in forest management. Even-aged plantations and silvicultural measures such as soil cultivation favour infection and spread of the disease. Consequently, there is considerable interest in knowing how silvicultural practices influence quantitatively disease incidence and how to avoid or restrict spread of the disease. The wealth of information available on the biology and ecology of *Armillaria* and on root rot disease is frequently not in a form that can directly assist decision-making by forest managers. Besides, a cost-benefit analysis of control options is often lacking (Pawsey and Rahman, 1976). *Armillaria* species can

Armillaria *Root Rot: Biology and Control of Honey Fungus*

Table 3.1. Summary of literature references on the rhizomorph distribution of various *Armillaria* species. Data are from the field, unless indicated.

Reference	Species
Ono, 1965	*Armillaria* sp.
Redfern, 1973	*A. mellea* (lab experiment in soil)
Morrison, 1976	*A. mellea*
Singh, 1981	*A. mellea*
Stanosz & Patton, 1991	*Armillaria* sp.
Rishbeth, 1972	*A. mellea*
Twery *et al.*, 1990	*Armillaria* spp.
Mihail *et al.*, 1995	*A. lutea, A. ostoyae* (lab experiment on agar)
Mihail & Bruhn, 1995	*A. lutea, A. mellea* (lab experiment on agar)
Termorshuizen *et al.*, 1998	*A. ostoyae*

be both plant pathogenic and saprotrophic and have a wide host range (Termorshuizen, this edition). Colonization of substrate provides the energy for production of rhizomorphs to explore the soil for new substrates, often giving rise to an extensive network by means of branching and anastomosis. Various studies have focussed on *Armillaria* rhizomorphs, which are the major means of spread in the temperate regions. Infection spreads from plant to plant when roots encounter rhizomorphs growing from stump roots or when crop roots directly contact infected roots.

Quantification of disease processes can be made with respect to the pathogen and the disease, or ideally both. Quantitative data should then be placed in an appropriate framework so that the consequences for management can be determined. We summarize here what biological data have been quantified, what techniques have been used, and how they have been integrated in predictive models of disease progress. Data on weight and distribution of rhizomorphs in soil are presented (*Table 3.1*). Concepts of fractal geometry (Mandelbrot, 1982) have been used to describe branching patterns and have been applied to a variety of organisms ranging from micro-organisms to plant roots (for example Eghball *et al.*, 1993; Mihail *et al.*, 1994). For quantification of *Armillaria*, the fractal dimension is a measure of branching density (Mihail *et al.*, 1995). Studies on disease incidence, both temporal and spatial, in permanent long-term forest plots are presented (*Table 3.2*). A predictive model to forecast disease

Table 3.2. Summary of literature references on fungal species causing root and butt rot disease in the field.

Reference	Species
Marsh, 1951	*A. mellea*
Vollbrecht & Agestam, 1995a	*Armillaria* spp., *Heterobasidion annosum*
Vollbrecht & Jørgensen, 1995b	*Armillaria* spp., *Heterobasidion annosum*
Williams & Leaphart, 1978	*A. mellea, Phellinus weirii*
Williams & Marsden, 1982	*A. mellea, Phellinus weirii*
Lundquist, 1993	*Armillaria* sp.
van der Kamp, 1995	*A. ostoyae*
Klein-Gebbinck *et al.*, 1990	*A. ostoyae*
Bruhn *et al.*, 1996	*A. ostoyae*
Hughes & Madden, 1998	*A. ostoyae*
Chadoeuf *et al.*, 1993	*Rigidoporus lignosus, Phellinus noxius*
Wiensczyk *et al.*, 1997	*A. ostoyae*
Stage *et al.*, 1990	*Armillaria* spp., *Phellinus weirii*

incidence may prove invaluable in forest management planning to facilitate choices between different silvicultural practices. Such a model can also serve as a quantitative statement of hypotheses about root disease dynamics, behaviour, and impact and can aid scientists in identifying serious data gaps and thus help to define and prioritize research needs. A specific example concerns the Western Root Disease Model (Stage *et al.*, 1990).

Distribution of *Armillaria* rhizomorphs

In temperate forests, rhizomorphs provide a major means of spread and subsequent infection of many tree species by most *Armillaria* species (Morrison, 1976). In general, rhizomorphs are produced most abundantly from wood that was colonized years previously. Studies indicate that the vertical distribution of *Armillaria* rhizomorphs is restricted to a particular part of the soil layer. The pattern of spread from a food base is usually radial (Ono, 1970) and, as a result of branching and anastomosis of individual rhizomorphs, a network is formed (Redfern, 1973). Clearly, on an infested site the chances of contact between rhizomorph growing tips and a root are related to the number of tips present in a unit volume of soil at any given time.

VERTICAL DISTRIBUTION

In Japan, Ono (1965) found more rhizomorphs in the upper 0–10 cm of soil than at 10–20 cm soil depth. Redfern (1973) reported that, at four sites in East Anglia, the distribution of rhizomorphs in soil varied with depth, but the highest densities occurred at 2.5–20 cm soil depth, and they were only rarely found below 30 cm. Morrison (1976) found that the distribution patterns of mature (black) rhizomorphs of *A. mellea* varied from site to site. On moist sites, rhizomorphs were concentrated in the upper 10 cm of soil, whereas on dry sites they were found deeper in the soil profile. In contrast, distribution patterns of immature (red) rhizomorphs were independent of soil moisture conditions. It was suggested that a minimum soil moisture content determines the upper limit of rhizomorph growth. On the other hand, the lower limit of rhizomorph growth, 30–35 cm, seems to be controlled by oxygen or carbon dioxide concentrations. Rhizomorphs initiated on inoculum segments buried at 30 or 60 cm depth frequently grew toward the soil surface. Laboratory experiments showed that the direction of rhizomorph growth was along gradients of increasing oxygen and decreasing carbon dioxide concentrations (Morrison, 1976). Singh (1981) showed that the vertical distribution of rhizomorphs in Newfoundland forests depended not only on soil moisture conditions but also on the type of site (cutover or burned-cutover from a mixed softwood stand, and pastureland), and host species.

HORIZONTAL DISTRIBUTION

Stanosz and Patton (1991) studied the capacity of aspen stumps following clear-cutting to support the production and growth of rhizomorphs over time by sampling rhizomorphs using ring-trench and core soil sampling methods. Weights of rhizomorphs of undetermined *Armillaria* sp. obtained using the two sampling methods around aspen stumps at different intervals after harvest were highly correlated. The

Figure 3.1a–c. Distribution of rhizomorphs in the soil from a source of inoculum.

mean quantities obtained generally increased as a function of the interval after harvest. In laboratory experiments (Redfern, 1973), the number and dry weight of rhizomorphs produced from woody inocula containing *A. mellea* and buried in soil varied with soil type and incubation temperature. In the same soil, fewer rhizomorphs were initiated at 15°C than at 25°C, but the total dry weight of rhizomorphs was approximately the same. The ability of *A. mellea* to produce rhizomorphs from stump samples was tested by a standard laboratory method by Rishbeth (1972). The amount produced from stumps of broad-leaved trees at first increased with the length of period after felling. Later, the yield decreased, although some rhizomorphs were still produced after 40 years. The rate of decay varies between tree species and between individuals within a species so that, in a given area, food bases are available for *A. mellea* over a long period. Wood from stems or roots of broad-leaved trees generally provides a better substrate for rhizomorph production than similar material from pines. Rishbeth (1972) suggests that other fungi compete more effectively with *A. mellea* in pine stumps than in stumps of broad-leaved trees, although clear proof of this does not exist in the literature. Twery *et al.* (1990) quantified the abundance and distribution of rhizomorphs of *Armillaria* spp. in the soil in undisturbed stands and in stands defoliated 1 and 5 years previously by insects. Trees weakened by biotic stress are often colonized and killed by *Armillaria* and other secondary pathogens. Rhizomorph distribution within the 0.04 ha study plots was uniform in the undisturbed stands, but was significantly greater near dead trees in the defoliated stands. Total rhizomorph abundance was greater on plots defoliated 5 years before sampling than on more recently defoliated plots, and it was least on undefoliated plots. Rhizomorph density near dead trees was highly correlated with overall rhizomorph density.

QUANTIFICATION OF HORIZONTAL DISTRIBUTION

Historically, fungal branching patterns have been described graphically or in qualitative terms. With the development of fractal geometry (Mandelbrot, 1982), a quantitative tool is available for the investigation of branching phenomena in a systematic fashion. For any object showing self-similarity at different scales (Pfeifer and Obert, 1989), the fractal dimension can be calculated. Most applications have been made to fungal branching patterns *in vitro*, not in the field. However, fractal geometry has been applied to *Armillaria* rhizomorphs in growth cultures (Mihail *et al.*, 1995) and in the field (Termorshuizen *et al.*, 1998). A model that can be adapted to quantify rhizomorph growth (Brown *et al.*, 1997) has been described.

Description of fractal dimension To conceptualize the applications of fractal geometry to fungal branching (Mihail *et al.*, 1994), first consider a germinating propagule from which a single, unbranched hyphen extends across the surface of an agar medium. Geometrically, this system is represented as a line, which is a one-dimensional object (dimension=1) measured in units of length, e.g. mm. Next, consider the case of a fungal thallus growing on a nutrient-rich agar medium, such that the *entire* surface of the medium is completely covered. Geometrically, this system can be represented as a two-dimensional object (dimension=2), measured in units of area, e.g. mm^2. For both of these systems, the dimension of the object is an integer. Finally, consider the more typical (and more interesting) case of a less densely branched fungal thallus growing on an agar surface. Geometrically, the branched thallus should have a dimension greater than 1.0 (that of a line), but smaller than 2.0 (that of a plane), for example 1.6. Mandelbrot (1982) proposed the term 'fractal' to describe objects with 'fractional' dimensions. The fractal dimension (D) describes the space-filling characteristics of an object. Thus, D would be close to 1.0 for a sparsely branched thallus, and close to 2.0 for a highly branched thallus.

Description of self-similarity Fractal objects are, by definition, self-similar (Pfeifer and Obert, 1989). This means that if a small portion of a fractal object is expanded to the same size as the entire object, the expanded part is indistinguishable from the total object. An idealized fractal object is identically self-similar over an infinite range of scales. However, real-world fractal objects are self-similar only over a finite range of scales. For fungal thalli, the lower scaling limit is set by the hyphal diameter, and the upper scaling limit may be no greater than the largest diameter of the thallus, which increases as the thallus grows (Obert *et al.*, 1990). Also, real-world fractal objects are self-similar only in the statistical sense that the objects have the same branching density within a defined range of scales.

Calculation of fractal dimension Evaluation of the fractal dimension using a box-counting method can be carried out as follows. The mycelium is covered with a grid of cells (boxes) of side length ε and the number of boxes, $N_{box}(\varepsilon)$, intersected by the mycelium is counted. $N_{box}(\varepsilon)$ is then calculated over a range of different side lengths ε. The mycelium has a well-defined fractal dimension if $\log\{N_{box}(\varepsilon)\}=|D|\log\{\varepsilon\}+C$, where C is a proportionality constant. In the case where N_{box} is plotted versus ε on logarithmic scales and a linear relationship is apparent, then the absolute value of the slope of the regression line represents the fractal dimension D. There are many examples of successful applications of this procedure (for example Mihail *et al.*, 1994). Rather than counting how many boxes are intersected by the mycelium, the mycelial mass contained within a given radius r (originating from the centre of the colony), M (r), can be calculated over a range of different radii (Ritz and Crawford, 1990). The mathematical representation is then $\log\{M(r)\}=D \log\{r\}+C$. Instead of mass, area of mycelium can also be used (Bolton and Boddy, 1993) where, for example, the mycelial area is calculated from photographs using image analysis (Bolton *et al.*, 1991).

Application of fractal geometry to Armillaria Fractal dimension (D) is a useful descriptor of foraging pattern, described as a balance between exploratory and exploitative growth strategies (Rayner, 1991), as it provides a quantitative summary of two-dimensional space utilization. Although rhizomorph systems occur in a three-dimensional soil matrix, the depth dimension is small in comparison with the two-dimensional area (Morrison, 1976; Rishbeth, 1978). Thus, the reduction of rhizomorph branching from three to two dimensions is quite appropriate. Within a specified environment, D is highly consistent for multiple thalli representing a genet, which implies a strong degree of genetic control over rhizomorph branching pattern (Mihail *et al.*, 1995). In these studies, D for rhizomorph systems ranged from 1.4 to 1.9, depending on the genet and species examined (Mihail *et al.*, 1995; Mihail and

Bruhn, 1995). Indeed, it was possible to distinguish rhizomorph branching patterns of *A. lutea* and *A. mellea* based solely on the magnitude of D (*A. lutea*, D=1.57; *A. mellea*, D=1.68; $P<0.05$) (Mihail and Bruhn, 1995). Further, D was temporally constant for rhizomorph systems, which is consistent with their foraging function. Although consistent in a specified environment, D is, however, sensitive to varying environmental conditions.

In a stand of *Pinus nigra* heavily affected by *Armillaria* root rot, a dense rhizomorph network of *A. ostoyae* was mapped over an area of 25 m^2 (*Figure 3.1*) and described quantitatively (Lamour *et al.*, unpublished data; Termorshuizen *et al.*, 1998). Basic characteristics of the network were: rhizomorph length, number of branches, number of cases in which rhizomorphs crossed each other, both with and without forming connections, and number of interconnected rhizomorph loops. The loops varied in size from very small (lengths 1–5 cm) to fairly large (lengths up to 800 cm). The fractal dimension of the network (D) was 1.30. Quantitative description of branching density using the fractal dimension might become a novel supplement to the morphological criteria traditionally used in fungal taxonomy.

Rhizomorph growth model There are no examples of mathematical models specifically designed to describe rhizomorph growth, although Brown *et al.* (1997) presented a step towards a generic framework for modelling root-architecture and the spatial structure of associated microbial ecosystems. This framework can be adapted for other root system features, such as rhizomorph interactions with root systems. A simulation programme produces and manipulates computer representations of a plant root-microbial ecosystem. The root system is described by a set of nodes. A node records a position in three-dimensional space and may represent the origin of another root, a bend, or some change in the root's microbial status. Disease lesions, closely associated microbial populations such as mycorrhizal fungi, and free soil microbial populations may also be represented. A multi-dimensional matrix structure provides a useful conceptual framework for the numerous stochastic functions required by the model. Procedures for composing the node-based root map, and for simulating root growth and microbial interaction, are presented in terms of the matrix structure, which in conjunction with the necessary manipulations of the node lists represents the root system.

Armillaria disease distribution

To study *Armillaria* root rot, dead and decaying tree species have been mapped in various regions over long periods of time using ground surveys (Vollbrecht and Agestam, 1995a; Vollbrecht and Jørgensen, 1995b) and aerial photography (Williams and Leaphart, 1978; Lundquist, 1993). Studies in which the rate of temporal disease mortality increase was determined or the spatial distribution was evaluated are given below and models developed to simulate disease progress are described.

DISTRIBUTION IN TIME

Records of *A. mellea* infections were made in a blackcurrant plantation from 1932–1941 (Marsh, 1951). The progress of infection could reasonably be explained by assuming that the major factor influencing fungal spread from bush to bush was root contact, and that it could invade the root system of a blackcurrant sufficiently quickly to bring about death of the entire bush in the season of initial infection. Vollbrecht and Agestam (1995a) presented an empirical model to forecast the incidence of butt rot caused by *Heterobasidion annosum* and *Armillaria* spp. at the stand level. The model is based on data from 152 permanent plots of pure Norway spruce plantations in southern Sweden, where the incidence of butt rot at stump height in thinned trees had been recorded after each thinning. According to simulations with the model, areas previously used as fields or for grazing were particularly susceptible to butt rot, while old hardwood sites were less susceptible. Furthermore, the model predicted that the earlier, the harder or more often a stand was thinned, the faster will be development of the disease. Development of butt rot in permanent plots of pure Norway spruce had also been recorded in Denmark (Vollbrecht and Jørgensen, 1995b). Regression analysis was carried out to predict incidence of butt rot using variables describing site, stand, and silvicultural treatments. The model predicted the fastest disease development in stands that had been planted on previous hardwood sites and in thinned stands. This result does not corroborate the data from the study in Sweden (Vollbrecht and Agestam, 1995a), where Norway spruce on old hardwood sites exhibited the slowest disease development. As the epidemiology of the disease is complex, the many simplifications introduced in empirical models makes application to different geographical areas or soils dangerous.

It is useful to identify sites with a high probability of developing root disease centres, making it possible to allocate land usage or manage appropriately on a short-term basis. Perhaps more importantly, the development of a model will help determine factors that contribute to damage and provide specific long-range direction for alleviating or managing these factors. Williams and Leaphart (1978) reported a survey system employing aerial photography with follow-up ground evaluations of root disease centres, indicated by openings in the forest canopy. The most frequently identified root pathogens in the disease centres were *Phellinus weirii* and *A. mellea*. Williams and Marsden (1982) evaluated their data and correlated forest stand and site characteristics with occurrence of root disease centres, among which stand age, average stand diameter, timber type, soil type, aspect, habitat type, and elevation showed significant associations. A logistic regression model was used for predicting the probability of root disease centre occurrence as a function of stand conditions. For wet aspects, the highest probability was found on soils with lowest year-round moisture availability. The reverse was true for dry aspects. Increased slope was associated with increased probability of the development of a root disease centre. Root disease centre frequency was inversely related to elevation and positively related to occurrence of *Pseudotsuga menziesii* and *Abies grandis*. Maximum probability of root disease centre occurrence was at 60–100 years of stand age across all habitat types (Williams and Marsden, 1982). Lundquist (1993) also used aerial photo surveys together with roadside reconnaissance to characterize the frequency and size of canopy gaps caused by *Armillaria* root disease in stands of *Pinus* spp. in the eastern Transvaal, South Africa. Disease severity (proportion of stand area with gaps) increased rapidly to age 3, remained nearly constant between ages 6–17, increased rapidly again between 17–28, and then decreased abruptly between ages 28–40. Gap incidence (proportion of all stands with gaps) rapidly increased to age 10, after which it remained nearly constant. Host species present in previous stands influenced whether disease developed in current stands. Lundquist (1993) also presented empirical models describing the various phases of the disease progress curves.

SPATIAL DISTRIBUTION

The following three questions can be posed: (1) Does *Armillaria* infection occur in clusters, and if so, how large are these clusters? (2) Is *Armillaria* infection associated with old stumps? (3) Is the spatial distribution such that removal of all trees within a set distance of *Armillaria* killed or dying trees has a reasonable chance of preventing further spread? To answer these questions, van der Kamp (1995) developed an analytical procedure to distinguish between spatial clustering of *Armillaria* infection attributable to clustering of the host, and clustering caused by the spatial distribution of *Armillaria* inoculum. The location, species, and infection status of all trees and stumps in nine 40 by 40 m plots located in a single large area infested by *A. ostoyae* in British Columbia were recorded. The area was logged to a diameter limit and then left undisturbed. Spatial analysis using variance-to-mean ratios of number of trees per grid square for a series of grid sizes showed that stumps were randomly distributed, trees were strongly clustered, and infected trees occurred in small clusters (1–29 trees) that were themselves randomly distributed. van der Kamp (1995) concluded that 30 years after the last major spatial disturbance by cutting, *Armillaria* occurs largely on the root systems of trees regenerated since logging. Removing all trees in close vicinity of an infected tree may isolate most of the viable *Armillaria* inoculum from the remainder of the stand.

To determine the inoculum sources, and methods and pattern of spread by *A. ostoyae* in lodgepole pine stands in West Central Alberta, juvenile lodgepole pine was excavated at three sites (Klein-Gebbinck *et al.*, 1991). In all cases in which *A. ostoyae* had become established in the root collar or taproot, it was also able to colonize lateral roots. In cases in which only lateral roots were infected, subsequent colonization was primarily distal to the point of infection. Rhizomorphs were associated with 89% of 121 infected roots, whereas only 19% of 70 roots with no associated rhizomorphs were infected. Stumps, roots, and debris from the previous stand were the major source of inoculum. There was no spatial relationship between stumps and symptomatic trees. Nearest neighbour analysis (Pielou, 1961) was used to determine whether there was a spatial association between stumps and symptomatic trees and to determine if pines near symptomatic trees were at a greater risk of developing symptoms than pines near asymptomatic trees. Nearest neighbour analysis indicated that the likelihood of an individual tree developing symptoms was dependent on whether trees within 0.15 m distance were dead or dying but independent of the apparent health of trees at greater distances. Trees within 0.15 m distance of each other often had arisen from the same cone, therefore it is likely that the same source of inoculum or secondary spread of the fungus caused infection.

In general, there have been few mathematical models developed to describe the spatial dynamics of diseases caused by soil-borne pathogens (Jeger, 1990). Bruhn *et al.* (1996) determined the temporal progress and spatial distribution of *Armillaria* disease mortality in black spruce plantations. Between 1982 and 1989, 22 black spruce seed plantations were established on cleared jack pine forest land in north-west Ontario. These plantations were located on suboptimal sites to hasten seed production. Mortality caused by *A. ostoyae* was observed in most of these plantations within three years of establishment. In four plantations where epidemics developed, temporal disease progress was best described by a monomolecular function rather than a Gompertz or logistic function. Monomolecular rates of disease increase were 0.0062–0.0346 per year. Applying these rates, they estimated that cumulative *Armillaria* root disease mortality would be 9–41% and 25–79% at 20 and 50 years after planting, respectively. *Armillaria* root disease mortality was spatially aggregated in all four plantations. The incidence of disease mortality was significantly higher for neighbouring trees in clusters adjacent to *Armillaria*-killed trees than expected under the null hypothesis of random spatial pattern. Hughes and Madden (1998) reanalysed the data of Bruhn *et al.* (1996) on the basis that the use of thc binomial distribution was inappropriate. They considered that the ß-binomial distribution model should be used to describe aggregated patterns of binary disease incidence data, rather than the binomial distribution. Characterizing heterogeneity in the patterns of diseased or dead trees, and using this information in the formulation of management recommendations relating to forest health, are important developments in forest research.

Site characteristics seem to play an important role in the incidence and spread of *Armillaria* root disease both directly, through their effects on the fungus, and indirectly, through their effects on the host. Redfern and Filip (1991) discuss several environmental factors, including soil temperature, pH, moisture, organic matter content, and nutrient status, that may directly affect the growth of *Armillaria* rhizomorphs through the soil. The ecological aspects of these interactions are described by Termorshuizen (this edition). Damage by *Armillaria* root disease is known to increase in severity when trees are stressed by either abiotic and biotic factors (Wargo and Harrington, 1991), and certain site characteristics such as soil texture and moisture regime relate well to stress susceptibility. Tree vigour and the resistance of the tree to infection may be a function of site. Wiensczyk *et al.* (1997) found a significant relationship between several environmental variables and *A. ostoyae* infection levels. This could predict the potential impacts of *Armillaria* root disease and therefore be helpful in forest management.

Spatial models have been developed for tree root rot diseases other than *Armillaria*. Chadoeuf *et al.* (1993) presented a Markov model for spread of a root rot epidemic in Côte d'Ivoire from 1977 to 1984 due to *Rigidoporus lignosus* and *Phellinus noxius*, where these root pathogens spread along planting lines without interline spread. This reduced the study of the stand to that of a single line. The health status of the trees on a planting line was considered as a Markov process. Statistical analysis showed that the two pathogens represented two independently developing processes. The pathogenicity of the fungi did not decrease with time and a healthy tree, whatever its age, always remained very susceptible to attack. Control methods aiming at reducing initial infection, secondary infection, or both, could be evaluated using the model.

A SPECIFIC EXAMPLE: THE WESTERN ROOT DISEASE MODEL

The Western Root Disease Model (Stage *et al.*, 1990) was developed to simulate the spread and impact of *Armillaria* spp. and *Phellinus weirii* in western coniferous forests in the U.S. The simulations are sensitive to information on initial disease conditions supplied to the model, including the number, size, and location of root disease centres. The model has limited capacity to simulate the spatial pattern of root disease centres. When the disease is scattered throughout the stand, it can be simulated as a single, large centre. The model can project up to 40 growth cycles of stand development, normally of 10 years each, and operate in stands up to 100 ha. It consists of three submodels, described by Shaw *et al.* (1991). A root disease submodel provides the status and spread of root disease, where alternative hypotheses concerning root disease dynamics can be explored. A second submodel structures the interactions between root diseases and other mortality agents (for example windthrow or bark beetles). Finally, the stand-interface submodel links the stand-development model, to which the Western Root Disease Model is attached, with the two above-mentioned submodels.

Forest inventory requirements and model limitations were demonstrated by Marsden (1992a). Tests of the Western Root Disease Model have identified the user-supplied parameters for which the model is most sensitive (Marsden, 1992b). Various management options for dealing with *Armillaria*, including stump removal, were also examined (Marsden *et al.*, 1993a). A refinement in the model is documented by Marsden *et al.* (1993b). Modelling the initial conditions of root disease in a stand is more accurate if combined with the use of generated random proportions for the degree of root system colonized by the disease. This is necessary to reflect the variation in the proportion of root systems colonized by the disease. A truncated-binomial distribution is used to generate the sequence of proportions for each sample tree. This technique may have application in the modelling of other root rot diseases.

Concluding remarks

Quantification of the epidemiology of *Armillaria* is essential for the understanding of the disease in the field and for determining optimal and cost-effective control interventions. Routine management procedures carried out in forest and horticultural plantations, for example thinning, cultivation, or control of other pests and diseases, may directly influence disease development and severity, often without affecting the amount of inoculum present. In practice, such management is rarely conducted specifically for disease control because such operations are costly and because reliable information on the expected economic gain is lacking (Pawsey and Rahman, 1976). Indeed, such an analysis is barely possible without appropriate quantification of the biological processes involved in disease development. The procedures used to develop the Western Root Disease Model may serve as a prototype for modelling the dynamics and behaviour of *Armillaria* root disease in other forest ecosystems or plantations. Only when existing biological data are put in a mathematical or quantitative framework can they be helpful to forest management.

References

BOLTON, R.G., MORRIS, C.W. AND BODDY, L. (1991). Non-destructive quantification of growth and regression of mycelial cords using image analysis. *Binary* **3**, 127–132.

BOLTON, R.G. AND BODDY, L. (1993). Characterization of the spatial aspects of foraging mycelial cord systems using fractal geometry. *Mycological Research* **97** (6), 762–768.

BROWN, T.N., KULASIRI, D. AND GAUNT, R.E. (1997). A root-morphology based simulation for plant/soil microbial ecosystem modelling. *Ecological modelling* **99**, 275–287.

BRUHN, J.N., MIHAIL, J.D. AND MEYER, T.R. (1996). Using spatial and temporal patterns of *Armillaria* root disease to formulate management recommendations for Ontario's black spruce (*Picea mariana*) seed orchards. *Canadian Journal of Forest Research* **26**, 298–305.

CHADOEUF, J., PIERRAT, J.C., NANDRIS, D., GEIGER, J.P. AND NICOLE, M. (1993). Modeling rubber tree root disease epidemics with a Markov spatial process. *Forest Science* **39** (1), 41–54.

EGHBALL, B., SETTIMI, J.R., MARANVILLE, J.W. AND PARKHURST, A.M. (1993). Fractal analysis for morphological description of corn roots under nitrogen stress. *Agronomy Journal* **85** (2), 287–289.

HUGHES, G. AND MADDEN, L.V. (1998). Comment – Using spatial and temporal patterns of *Armillaria* root disease to formulate management recommendations for Ontario's black spruce (*Picea mariana*) seed orchards. *Canadian Journal of Forest Research* **28**, 154–158.

JEGER, M.J. (1990). Mathematical analysis and modeling of spatial aspects of plant disease epidemics. In: *Epidemics of plant diseases: mathematical analysis and modeling*. Ed. J. Kranz, pp 53–95. Berlin: Springer-Verlag.

KLEIN-GEBBINCK, H.W., BLENIS, P.V. AND HIRATSUKA, Y. (1991). Spread of *Armillaria ostoyae* in juvenile lodgepole pine stands in west central Alberta. *Canadian Journal of Forest Research* **21**, 20–24.

LUNDQUIST, J.E. (1993). Spatial and temporal characteristics of canopy gaps caused by *Armillaria* root disease and their management implications in lowveld forests of South Africa. *European Journal of Forest Pathology* **23**, 362–371.

MANDELBROT, B.B. (1982). *The fractal geometry of nature*. San Francisco, California: W.H. Freeman and Co.

MARSDEN, M.A. (1992a). Sensitivity of the Western Root Disease Model: inventory of root disease. *USDA Forest Service* Research Paper RM–303.

MARSDEN, M.A. (1992b), Sensitivity analyses of the Western Root Disease Model to user-specified starting parameters. *USDA Forest Service* Research Paper RM–306.

MARSDEN, M.A., SHAW, C.G. AND MORRISON, M. (1993a). Simulation of management options for stands of southwestern ponderosa pine attacked by *Armillaria* root disease and dwarf mistletoe. *USDA Forest Service* Research Paper RM–308.

MARSDEN, M.A., EAV, B.B. AND THOMPSON, M.K. (1993b). Modeling initial conditions for root rot in forest stands: random proportions. *USDA Forest Service* Research Note RM–524.

MARSH, R.W. (1951). Field observations on the spread of *Armillaria mellea* in apple orchards and in a blackcurrant plantation. *Transactions of the British Mycological Society* **35**, 201–207.

MIHAIL, J.D., OBERT, M., TAYLOR, S.J. AND BRUHN, J.N. (1994). The fractal dimension of young colonies of *Macrophomina phaseolina* produced from microsclerotia. *Mycologia* **86** (3), 350–356.

MIHAIL, J.D. AND BRUHN, J.N. (1995). Using fractal geometry to compare rhizomorph foraging strategies among six *Armillaria* species. (Abstract). *Phytopathology* **85**, 1127.

MIHAIL, J.D., OBERT, M., BRUHN, J.N. AND TAYLOR, S.J. (1995). Fractal geometry of diffuse mycelia and rhizomorphs of *Armillaria* species. *Mycological Research* **99**, 81–88.

MORRISON, D.J. (1976). Vertical distribution of *Armillaria mellea* rhizomorphs in soil. *Transactions of the British Mycological Society* **66**, 393–399.

OBERT, M., PFEIFER, P. AND SERNETZ, M. (1990). Microbial growth patterns described by fractal geometry. *Journal of Bacteriology* **172**, 1180–1185.

ONO, K. (1965). *Armillaria* root rot in plantations of Hokkaido. Effects of topography and soil conditions on its occurrence. *Bulletin of the Government Forest Experiment Station, Meguro* **179**, 1–62.

ONO, K. (1970). Effect of soil conditions on the occurrence of *Armillaria* root rot of the Japanese Larch. *Bulletin of the Government Forest Experiment Station, Meguro* **229**, 1–219.

PAWSEY, R.G. AND RAHMAN, M.A. (1976). Chemical control of infection by honey fungus, *Armillaria mellea*: a review. *Arboricultural Journal* **2**, 468–479.

PFEIFER, P. AND OBERT, M. (1989). Fractals: basic concepts and terminology. In: *The fractal approach to heterogeneous chemistry*. Ed. D. Avnir, pp 11–43. New York: John Wiley and Sons.

PIELOU, E.C. (1961). Segregation and symmetry in two species populations as studied by nearest neighbor relationships. *Journal of Ecology* **49**, 255–269.

RAYNER, A.D.M. (1991). The challenge of the individualistic mycelium. *Mycologia* **83**, 48–71.

REDFERN, D.B. (1973). Growth and behaviour of *Armillaria mellea* rhizomorphs in soil. *Transactions of the British Mycological Society* **61** (3), 569–581.

REDFERN, D.B. AND FILIP, G.M. (1991). Inoculum and infection. In: *Armillaria Root Disease*. Eds. C.G. Shaw and G.A. Kile, pp 48–61. *Forest Service Handbook No. 691*. Washington, D.C.: USDA.

RISHBETH, J. (1972). The production of rhizomorphs by *Armillaria mellea* from stumps. *European Journal of Forest Pathology* **2**, 193–205.

RISHBETH, J. (1978). Effect of soil temperature and atmosphere on growth of *Armillaria* rhizomorphs. *Transactions of the British Mycological Society* **70**, 213–220.

RITZ, K. AND CRAWFORD, J. (1990). Quantification of the fractal nature of colonies of *Trichoderma viride*. *Mycological Research* **94**, 1138–1152.

SHAW, C.G., STAGE, A.R. AND MCNAMEE, P. (1991). Modeling the dynamics, behavior, and impact of *Armillaria* root disease. In: *Armillaria Root Disease*. Eds. C.G. Shaw and G.A. Kile, pp 150–156. *Forest Service Handbook No. 691*. Washington, D.C.: USDA.

SIMS, R.A., TOWILL, W.D., BALDWIN, K.A. AND WICKWARE, G.M. (1989). *Field guide to the forest ecosystem classification for northwestern Ontario*. Toronto: Ontario Ministry of Natural Resources.

SINGH, P. (1981). *Armillaria mellea*: growth and distribution of rhizomorphs in the forest soils of Newfoundland. *European Journal of Forest Pathology* **11**, 208–220.

STAGE, A.R., SHAW, C.G., MARSDEN, M.A., BYLER, J.W., RENNER, D.L., EAV, B.B., MCNAMEE, P.J., SUTHERLAND, G.D. AND WEBB, T.M. (1990). User's manual for Western Root Disease Model. *USDA Forest Service* General Technical Report INT–267.

STANOSZ, G.R. AND PATTON, R.F. (1991). Quantification of *Armillaria* rhizomorphs in Wisconsin aspen sucker stands. *European Journal of Forest Pathology* **21**, 5–16.

TERMORSHUIZEN, A.J., LAMOUR, A. AND JEGER, M.J. (1998). Networks formed by rhizomorphs of *Armillaria ostoyae*. *Proceedings of the Sixth International Mycological Congress*, Jerusalem, Israel, August 23–28.

TWERY, M.J., MASON, G.N., WARGO, P.M. AND GOTTSCHALK, K.W. (1990). Abundance and distribution of rhizomorphs of *Armillaria* spp. in defoliated mixed oak stands in western Maryland. *Canadian Journal of Forest Research* **20**, 674–678.

VAN DER KAMP, B.J. (1995). The spatial distribution of *Armillaria* root disease in an uneven-aged, spatially clumped Douglas-fir stand. *Canadian Journal of Forest Research* **25**, 1008–1016.

VOLLBRECHT, G. AND AGESTAM, E. (1995a). Modelling incidence of root rot in *Picea abies* plantations in southern Sweden. *Scandinavian Journal of Forest Research* **10**, 74–81.

VOLLBRECHT, G. AND JØRGENSEN, B.B. (1995b). Modelling the incidence of butt rot in plantations of *Picea abies* in Denmark. *Canadian Journal of Forest Research* **25**, 1887–1896.

WARGO, P.M. AND HARRINGTON, T.C. (1991). Host stress and susceptibility. In: *Armillaria Root Disease*. Eds. C.G. Shaw and G.A. Kile, pp 88–101. *Forest Service Handbook No. 691*. Washington, D.C.: USDA.

WIENSCZYK, A.M., DUMAS, M.T. AND IRWIN, R.N. (1997). Predicting *Armillaria ostoyae* infection levels in black spruce plantations as a function of environmental factors. *Canadian Journal of Forest Research* **27**, 1630–1634.

WILLIAMS, R.E. AND LEAPHART, C.D. (1978). A system using aerial photography to estimate area of root disease centres in forests. *Canadian Journal of Forest Research* **8**, 214–219.

WILLIAMS, R.E. AND MARSDEN, M.A. (1982). Modelling probability of root disease centre occurrence in northern Idaho forests. *Canadian Journal of Forest Research* **12**, 876–882.

SECTION 2
Diversity

4
Taxonomy, Nomenclature and Description of *Armillaria*

DAVID N. PEGLER

The Herbarium, Royal Botanic Gardens, Kew, Richmond, Surrey TW9 3AB, U.K.

Synopsis

The nomenclatural status of the genus *Armillaria* is provided, together with a generic description, a list of the World species with synonymy and references, a key to the European species, and an illustrated account of the type species, *A. mellea*.

The genus *Armillaria*

The taxonomic and nomenclatural confusion over the name and identity of the 'Honey Fungus' or 'Bootlace Fungus', *Armillaria mellea* (Vahl: Fr.) P. Kumm., has continued for almost two centuries. There have been difficulties on the delimitation of the genus, the recognition of the type species and choice of the generic name.

In 1766, the Danish botanist, Martin Vahl, published a beautiful painting, with a short description, under the name of *Agaricus melleus* in *Flora Danica* (Vahl, 1766: pl. 1013). The Swedish mycologist, Elias Fries (1819) included this species when introducing the name *Armillaria* for a new tribe in the genus *Agaricus*, and this was expanded in his sanctioning work, *Systema Mycologicum* (1821), to include twelve species. The tribe was defined as follows: *Velum partiales, simplex, annuli-forme, persistens, stipiti innatum margineque pilei in statu juniori adnexum. Stipes solidus!, firmus, subfibrillosus, inaequalis. Pileus carnosus, convexus, expansus, obtusus!, epidermide semper contigua (etiam in squamosis), a velo plane discreta. Caro alba, firma. Lamellae latae, inaequales, postice subacutae. Color lamell. albus l. pallescens. Obs. Species autumnales, diu persistentes, firmae, esculentae, habitu inter se sat diverso. Annulus nunc superus h. e. ab apice stipitis reflexus, nunc infernus h. e. ad medium cum epiderm. stipitis contiguus: vel etiam proprius, supra medium insertus.* The name is based on the Latin word *armilla*, or ring, referring to the annular veil present on the stipe. In later years, Fries (1825, 1854, 1857) abandoned the tribe *Armillaria*, transferring most species to the genus *Lepiota* (Pers.: Fr.) Gray. Fries included twelve species in the *Systema*, namely: *Agaricus robustus* (= *Tricholoma robustum* (Alb. & Schw.: Fr.) Ricken), *A. persoonii* (= ?*Lepiota* sp.), *A. guttatus*

(= *Limacella guttata* (Pers.: Fr.) Konr. & Maubl.), *A. bulbiger* (= *Leucocortinarius bulbiger* (Alb. & Schw.: Fr.) Sing.), *A. constrictus* (= *Calocybe constricta* (Fr.: Fr.) Kühn.), *A. subcavus* (?), *A. mucidus* (= *Oudemansiella mucida* (Schrad.: Fr.) Höhn.), *A. vagans* (?), *A. griseofuscus* (?), *A. denigratus* (= *Agrocybe erebia* (Fr.) Kühn.), *A. rhagadiosus* (?), and *A. melleus*.

TYPIFICATION OF *ARMILLARIA*

Seven species have been selected by different authorities. Four of these species were not included in the original list of Fries, namely *A. ramentaceus* Bull.: Fr. by Earle (1909), *A. luteovirens* Alb. & Schwein.: Fr. by Singer (1936), Konrad and Maublanc (1948) and Locquin (1952), *A. focalis* Fr. by Romagnesi (1950), and *A. imperialis* Quél. by Gilbert (1950). Imai (1938) chose *A. robustus* Alb. & Schwein.: Fr., being the first of the species listed by Fries, but this can be rejected as Fries stated that he had no personal knowledge of the species. This leaves *A. mucidus* Schrad.: Fr. and *A. melleus* as the only species common to both the original circumscription of Fries (1821), and also in the subsequent generic recognitions by Staude (1857), Kummer (1871) and Quélet (1872). *Agaricus mucidus* was transferred by Patouillard (1887) to his own genus *Mucidula* Pat., a genus since referred to *Oudemansiella* Speg.

The tribe *Armillaria* of Fries was raised to generic rank by Staude (1857), who included four species, *A. mucidus*, *A. melleus, A. aurantius* and *A. robustus*. Singer (1956) rejected Staude's genus, contending that Staude had merely raised the status of the Friesian taxon and was not proposing a new genus. Singer regarded *Armillaria* proposed by Kummer (1871) as the sole authority for the genus. This view has been amply disputed by Donk (1962), Watling, Kile and Gregory (1982), Volk and Burdsall (1995) and Bon (1997). Unfortunately, Singer used the genus *Armillariella* (P. Karst.) P. Karst. (Karsten, 1881), with *A. mellea* as the type species, whilst utilizing *Armillaria* Kumm. for *A. luteovirens* Alb. & Schw.: Fr. The latter species is not related to *A. mellea*, and Pouzar (1957) has subsequently proposed *Floccularia* as a new generic name for this taxon. *Armillariella* is now regarded as an obligate synonym of earlier *Armillaria* under the International Code of Botanical Nomenclature, Article 14.4, because both names are based upon the same type species.

In the past, many authors have used *Armillaria* (Fr.) Quél. as the correct generic name, for Quélet (1872) had also raised the Fries tribe to generic status, unaware that this had already been done by both Kummer and Staude. *Polymyces* was proposed by Earle (1909), but this is merely a typonym of *Armillaria,* and *Rhizomorpha* Roth, based upon the sterile rhizomorphs, is not acceptable as a teleomorph name. Further discussion on the nomenclatural problems of *Armillaria* may be found in Burdsall and Volk (1993), Dennis, Wakefield and Bisby (1954), Jaques-Felix (1977), Watling (1976), and Watling, Kile and Burdsall (1991).

As Fries (1821) had based *A. mellea* on a published illustration, no type specimen was available for study. Watling, Kile and Gregory (1982) have recently selected a neotype collection, gathered at Klampenburg, near Copenhagen in Denmark, a locality which is likely to have been visited by Vahl.

CLASSIFICATION OF *ARMILLARIA*

Singer (1986) has placed *Armillaria* within the Subtribe *Omphalinae* Singer of the Tribe *Tricholomateae* Singer within *Tricholoma* (Fr.) Kumm., a genus of the family *Tricholomataceae* Hein ex Pouzar. However, ultrastructural examination of the basidiospore wall teguments reveal longitudinally raised ridges and thickenings of the corium, overlaid by a continuous tectum, which raises the possibility of a relationship with *Clitopilus* (Fr.) Kumm. and *Rhodocybe* R. Maire, genera currently placed in *Entolomataceae* Kotl. & Pouzar.

Generic description

ARMILLARIA (Fr.: Fr.) Staude, *Schwämme Mitteld.*: xxvii, 180 (1857).

Armillaria Tribe Armillaria Fr.: Fr., *Syst. Mycol.* **1**: 9, 26 (1821); *Fr. Spec. Syst. Mycol.* **8** (1819).

Armillaria Sect. *Armillaria* P. Karst. in *Ryssl., Finl. Skand. Halføns Hattsvamp*: xii (1979).

Armillariella (P. Karst.) P. Karst. in *Acta Soc. Fauna. Fl. fenn.* **2** (1): 4 (1881).

Basidioma clitocyboid to tricholomatoid, generally caespitose. *Pileus* conico-convex to applanate or depressed, fleshy; surface generally expallent, somewhat viscid when moist, brownish to ochraceous or yellowish, with darker fibrillose squamules which remain at disk at maturity, often arranged concentrically; otherwise squamules and fibrils detersile; margin thin, finally striate. *Lamellae* arcuate, adnato-decurrent to deeply decurrent or with a decurrent tooth, pale coloured, moderately thick, subdistant to crowded. *Stipe* central, slender often clavate to subbulbous, fibrous-fleshy, fistulose to hollow; surface brownish to yellowish, with white, brown or yellowish floccules or smooth; annulate or not. *Partial veil* present, annulate, floccoso-membranous to arachnoid or absent. *Context* thin to fleshy, inamyloid, consisting of a monomitic hyphal system of inflated generative hyphae; *clamp-connexions* absent. *Rhizomorphs* present at stipe base, and often bioluminescent, or absent, but frequent in culture. *Spore deposit* pure white to cream, drying deep cream colour. *Basidiospores* ovo-ellipsoid, oblong to ellipsoid, hyaline, inamyloid, weakly cyanophilic, with a thickened wall, smooth or with vague longitudinal ridges.Ultrastructurally, the ridges are seen to be formed by thickenings of the corium wall-tegument beneath a smooth tectum. *Basidia* clavate, tetrasporic, at times bisporic; occasional sclerobasidia present, with or without a basal clamp-connexion. *Lamella-edge* sterile; *cheilocystidia* present, not conspicuous, variable, at times catenulate-septate, hyaline, thin-walled. *Pleurocystidia* none or occasional. *Hymenophoral trama* initially bilateral with divergent lateral strata, becoming more regular at maturity. *Pileipellis* a repent cutis or epicutis, disrupting to become subtrichodermial and forming fibrillose squamules; hypodermium of radially parallel hyphae. On living or dead wood of trees and shrubs, rarely on herbs or terrestrial in marshland; parasitic, saprotrophic, forming endotrophic mycorrhizas in tropics; cosmopolitan. *Development*: bivelangiocarpic, metavelangiocarpic or monovelangiocarpic. *Agaricales* Clements, *Tricholomataceae* Heim ex Pouzar tribe *Tricholomateae* subtr. *Omphalinae* Singer. *Lectotype species basionym* (Watling, Kile and Gregory, 1982: 275): *Agaricus melleus* Vahl: Fr.

Speciation in *Armillaria*

Once the generic definition and range was finally determined, many species were eliminated but others were proposed for the genus. Problems have arisen in correlating morphological characteristics with life-patterns and degree of pathogenicity (Rishbeth, 1982). It was already realized that *A. mellea* was not a single, morphologically variable species of widespread distribution and host range, when incompatibility tests based on monosporous (haploid) isolates resulted in the recognition of 'biological species'. Korhonen (1978) distinguished five European 'biological species', whilst Anderson and Ullrich (1979) recognized ten 'biological species' from North America. The acceptance of 'biological species' made it difficult to apply the older names (Termorshuizen and Arnolds, 1987). However, most of the European and North American 'biological species' have now been given names (Marxmüller, 1992; Volk and Burdsall, 1995).

The accepted World species of *Armillaria*

1. EUROPE

Section 1. Armillaria: provided with a membranous annulus or a distinct annular zone.
borealis Marxm. & Korhonen in *Bull. Soc. Mycol. Fr.* **98**: 122 (1982).
Ref: Bon (1997: 93, pl. 4B); Gregory and Watling (1985: 47); Marxmüller and Printz (1982: 3 as species A); Termorshuizen (1995: 36).
cepistipes Vel., *Ceské Houby*, Dil. **II**: 283 (1920).
Ref: Bon (1997: 94); Marxmüller (1982: 87), Romagnesi and Marxmüller (1983: 309, pl. 231–232); Termorshuizen (1995: 38).
lutea Gillet, *Les Hymenom.*: 83 (1874).
Ref: Bon (1997: 93, pl. 4A); Marxmüller (1980: pl. 221 as *A. bulbosa*; 1987: 272; 1992: 272); Termorshuizen (1995: 37 as *A. lutea)*; Watling (1987: 460).
= *A. mellea* (Vahl: Fr.) P. Kumm. var. *bulbosa* Barla in *Bull. Soc. Mycol. Fr.* **3**: 143 (1887).
= *inflata* Vel., *Ceské Houby*, Dil. **II**: 283 (1920).
= *gallica* Marxm. & Romagn. in *Bull. Soc. Mycol. Fr.* **103**: 152 (1987).
mellea (Vahl: Fr.) P. Kumm., *Führ. Pilzk.*: 135 (1871).
Ref: Bon (1997: 91); Jahn (1979: pl. 166); Marxmüller (1982: 87–124); Romagnesi (1973: 195);Termorshuizen (1995: 35); Watling, Kile and Gregory (1978: 271).
= *cerasi* Vel., *Ceské Houby*, Dil. **II**: 282 (1920).
= *nigritula* P.D. Orton in *Notes Roy. Bot. Gard., Edinb.* **38**: 316 (1980).
ostoyae (Romagn.) Herink in *Vysok. Skola Zemed. V Brné. Vyzn. Rádem Prace BRNO*: 42 (1973).
Ref: Bon (1997: 92); Romagnesi (1970: 265); Marxmüller and Printz (1982: 3 *as A. obscura*); Termorshuizen (1995: 36).
= *laricina* (Bolton: Fr.) Sacc., *Syll. Fung.* **5**: 81 (1887).
= *obscura* (Schaeff.) Herink in *Vysok. Skola Zemed. V Brné. Vyzn. Rádem Prace BRNO*: 42 (1973).
= *polymyces* (Pers.) Sing. & Clémençon in *Nova Hedw.* **23**: 311 (1972).
Ref: Marxmüller (1992: 269); Watling (1987: 474); Watling, Kile and Burdsall (1991: 5).

Section 2. Desarmillaria Herink in *Vysok. Skola Zemed. V Brné. Vyzn. Rádem Prace BRNO*: 44 (1973): annulus and annular zone absent or indistinct.

ectypa (Fr.: Fr.) D. Lam. in *C. R. hebd. Seanc. Acad. Sci.., Paris* **260**: 4562 (1965).
Ref: Bon (1997: 95); Emel (1921: 59); Favre (1939: 196); Marchand (1986: pl. 814); Moreau (1929: 93); Termorshuizen (1995: 38).

nigropunctata (Fr.) Herink in *Vysok. Skola Zemed. V Brné. Vyzn. Rádem Prace BRNO*: 44 (1973).
Ref: Bon (1997: 96).

tabescens (Scop.) Emel, *Genre Armillaria*: 50 (1921).
Ref: Bon (1997: 95); Intini (1988: 49–72); Malençon and Bertault (1975: 132); Marchand (1986: pl. 820); Phillips (1981: 32); Termorshuizen (1995: 38).
= *socialis* (DC: Fr.) Herink in *Vysok. Skola Zemed. V Brné. Vyzn. Rádem Prace BRNO*: 44 (1973).

2. NORTH AMERICA

calvescens Bérubé & Dessur. in *Mycologia* **81**: 220 (1989).
cepistipes

gemina Bérubé & Dessur. in *Mycologia* **81**: 217 (1989).
mellea
ostoyae

sinapina Bérubé & Dessur. in *Canad. Journ. Bot.* **66**: 2030 (1988).
tabescens

3. CENTRAL & SOUTH AMERICA

affinis (Sing.) Volk & Burds. in *Synops. Fung.* **8**: 24 (1995).
Ref: Singer (1989: 12).

griseomellea (Sing.) Kile & Watl. in *Trans. Brit. Mycol. Soc.* **81**: 131 (1983).
Ref: Singer (1969: 40).

melleorubens (Berk. & M.A. Curtis) Sacc., *Syll. Fung.* **5**: 81 (1887).
Ref: Berkeley and Curtis (1869: 283); Chandhra and Watling (1981: 78, fig.1l).

montagnei (Sing.) Herink in *Vysok. Skola Zemed. V Brné. Vyzn. Rádem Prace BRNO*: 41 (1973).
Ref: Singer (1956: 182).

procera Speg. in *Bol. Acad. Nac. Córdoba* **11**: 385 (1889).
Ref: Pegler (1997: 9); Singer (1970: 11).

puiggarii Speg. in *Bol. Acad. Nac. Córdoba* **11**: 384 (1889).
Ref: Pegler (1983: 101, pl. 23g–h; 1997: 9).

sparrei (Sing.) Herink in *Vysok. Skola Zemed. V Brné. Vyzn. Rádem Prace BRNO*: 43 (1973).
Ref: Singer (1956: 183).

tigrensis (Sing.) Raithelh. in *Metrodiana* Sonderh. **2**: 3 (1983).
Ref: Singer (1970: 8).

viridiflava (Sing.) Volk & Burds. in *Synops. Fung.* **8**: 119 (1995).
Ref: Singer (1989: 12).

yungensis (Singer) Herink in *Vysok. Skola Zemed. V Brné. Vyzn. Rádem Prace BRNO*: 43 (1973).
Ref: Singer (1970: 12).

4. AFRICA

camerunensis (Henn.) Volk & Burds. in *Synops. Fung*. **8**: 39 (1995).
Ref: Hennings (1895: 107); Watling (1992: 488).
heimii Pegler in *Kew Bull. Addit. Ser*. **6**: 92 (1977).
= *Clitocybe elegans* R. Heim in *Rev. Mycol*. **28**: 94 (1963), non *Armillaria elegans* Beeli. Possibly identical with *A. fuscipes*.
Ref: Kile and Watling (1988: 311 as *A. fuscipes*); Watling (1992: 488 as *A. fuscipes*).
.***pelliculata*** Beeli in *Bull. Soc. Roy. Belg*. **59**: 111 (1927).

5. ASIA

adelpha (Berk.) Sacc., *Syll. Fung*. **5**: 84 (1887).
Ref: Berkeley (1850: 47); Chandhra and Watling (1981: 64, figs. 1d–e, 2b–c, 3f); Manjula (1983: 83).
apalosclerus (Berk.) Chandhra & Watl. in *Kavaka* **10**: 65 (1981).
Ref: Berkeley (1850: 82); Chandhra and Watling (1981: 65, figs. 1h, 3b).
dicupella (Berk.) Sacc., *Syll. Fung*. **5**: 83 (1887 as *A. discupella*).
Ref: Berkeley (1850: 45); Chandhra and Watling (1981: 66, figs. 1k,3e).
duplicata (Berk.) Sacc., *Syll. Fung*. **5**: 83 (1887).
Ref: Berkeley (1850: 45); Chandhra and Watling (1981: 67, fig.1j).
fuscipes Petch in *Ann. Roy. Bot. Gard., Peradeniya* **4**: 299 (1909).
Ref: Chandhra and Watling (1981: 68, fig.1f–g, 2d–e); Pegler (1986: 81, fig. 16a–c).
horrens (Berk.) Sacc., *Syll. Fung*. **5**: 82 (1887).
Ref: Berkeley (1850: 44); Chandhra and Watling (1981: 69, fig.1a, 3c). Possibly identical with *A. ostoyae*.
jezoenis Cha & Igarashi in *Mycoscience* **35**: 42 (1994).
mellea
multicolor (Berk.) Sacc., *Syll. Fung*. **5**: 84 (1887).
Ref: Berkeley (1850: 46); Chandhra and Watling (1981: 70, fig.1i, 3a).
omnituens (Berk.) Sacc., *Syll. Fung*. **5**: 84 (1887).
Ref: Berkeley (1850: 46); Chandhra and Watling (1981: 74, fig.1b, 3d). Manjula (1983: 104).
ostoyae
Ref: Chandhra and Watling (1981: 78 as *A. obscura*).
singula Cha & Igarashi in *Mycoscience* **35**: 45 (1994).
vara (Berk.) Sacc., *Syll. Fung*. **5**: 83 (1887).
Ref: Berkeley (1850: 45); Chandhra and Watling (1981: 75, fig.1c, 2a, 3g).

6. AUSTRALASIA

fellea (Hongo) Kile & Watl. in *Trans. Br. Mycol. Soc*. **81**: 131 (1983).
Ref: Hongo (1976: 97).
fumosa Kile & Watl. in *Trans. Br. Mycol. Soc*. **81**: 129 (1983).
hinnulea Kile & Watl. in *Trans. Br. Mycol. Soc*. **81**: 131 (1983).
limonea (G. Stev.) Boesew. in *N. Z. Journ. Agric. Res*. **20**: 585 (1977).
Ref: Stevenson (1964: 13).

luteobubalina Kile & Watl. in *Trans. Br. Mycol. Soc.* **71**: 79 (1978).
novae-zelandiae (G. Stev.) Herink in *Vysok. Skola Zemed. V Brné. Vyzn. Rádem Prace BRNO*: 43 (1973).
Ref: Stevenson (1964: 14).
pallidula Kile & Watl. in *Trans. Br. Mycol. Soc.* **91**: 131 (1988).

Key to the European species of *Armillaria*

1. Annulus membranous, submembranous or with a well delimited annular zone on stipe .... **Sect. Armillaria, 2**
– Stipe with neither a membranous annulus nor an annular zone, rarely with a few, scattered, silky-fibrillose squamules; basidia with a basal clamp-connexion .... **Sect. Desarmillaria, 6**
2. Annulus membranous and well-formed; basidiomata typically in large clusters, with wide host range **3**
– Annulus cortinate to submembranous but fragile; basidiomata solitary or in small clusters; stipe subclavate; basidia with a basal clamp-connexion **5**
3. Stipe elongate, tapering below, with scattered, white to yellowish floccules; annulus persistent, with yellowish floccules on underside; pileus yellowish to orange-brown or olive-brown, with a few squamules restricted to disk; basidia lacking a basal clamp-connexion; October–November (*Figure 4.1a–e*) .... **A. mellea**
– Stipe base subclavate; basidia with a basal clamp-connexion **4**
4. Pileus dark brown to ochraceous brown, with numerous, concentric, dark brown, pyramidal squamules especially towards disk; stipe with numerous and persistent, white to brown floccules, base up to 2.8 cm in diameter; annulus with blackish brown, marginal squamules; host range variable but especially on *Picea*; widespread and common; autumn (*Figure 4.2a*) .... **A. ostoyae**
– Pileus ochraceous brown to orange-brown, or with an olive tint, with dark brown, fibrillose squamules towards disk; stipe with scattered, yellowish brown floccules, base up to 1.5 cm in diameter; annulus submembranous, white with a brown or yellowish margin; wide host range but more common in Nordic and east European countries; early summer onwards (*Figure 4.2b*) .... **A. borealis**
5. Basidioma robust, stipe with base up to 2.7 cm in diameter; pileus greenish yellow to ochraceous brown, with erect or ascending, grey-brown fibrils towards the disk; elements of squamules often more than 60 µm long (*Figure 4.2c*) .... **A. lutea**
– Basidioma slender, stipe with base up to 1.5 cm in diameter, with scattered, yellow, grey or white floccules; pileus hazel brown, smooth except for a few, dark brown, fibrillose squamules at disk; elements of squamules 25–50 µm long (*Figure 4.2d*) .... **A. cepistipes**
6. Terrestrial, in marshland amongst *Sphagnum*; solitary or in small clusters; pileus ochraceous brown, translucent striate towards margin, faintly squamulose at disk; lamellae subdistant; stipe cylindrical; August–September, rare (*Figure 4.2e*) .... **A. ectypa**
– Lignicolous, basidiomata forming caespitose clusters **7**
7. Pileus yellowish brown to tawny brown, with small, rusty brown , fibrillose squamules towards disk; stipe tapering below, whitish, bruising brown; densely

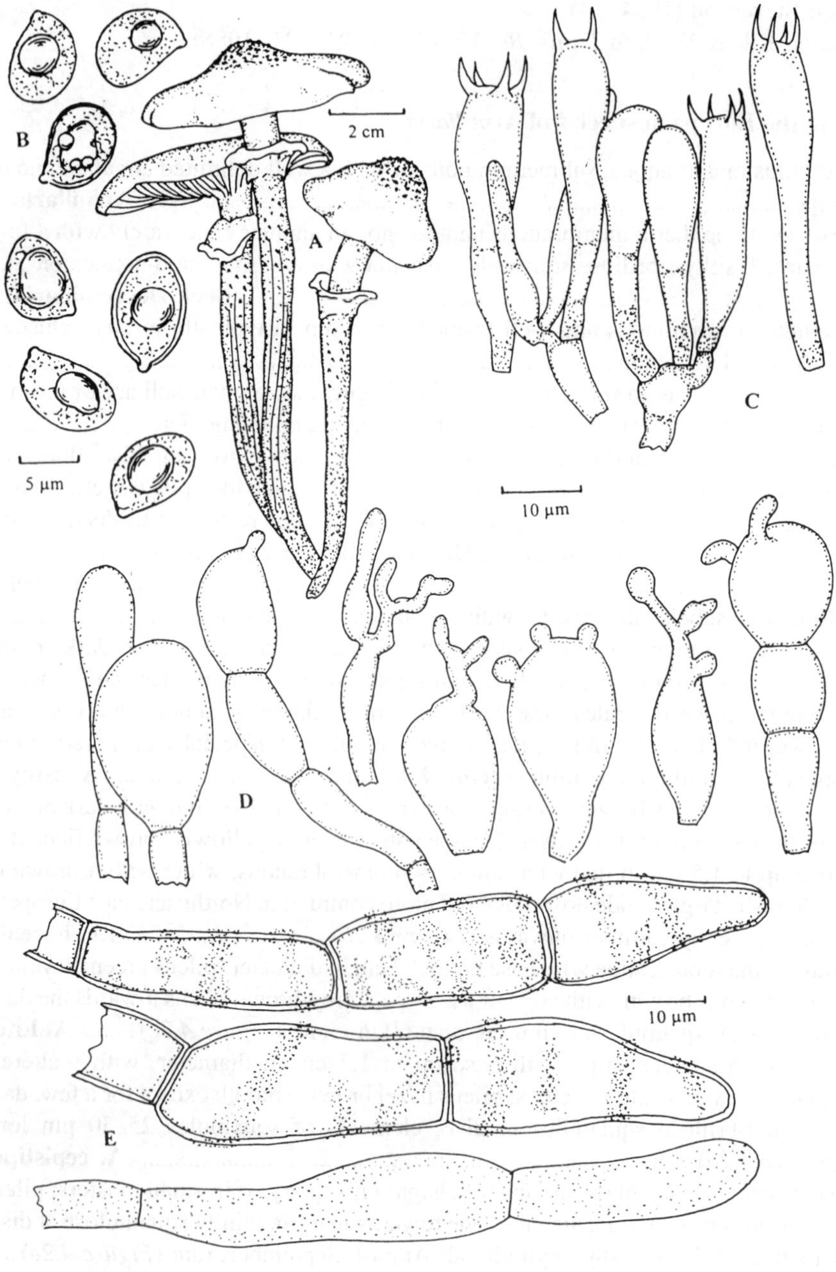

Figure 4.1a–e. *Armillaria mellea*: (a) Habit, showing the membranous annulus, tapering stipe and limited pileal squamules; (b) basidiospores; (c) basidia; (d) cheilocystidia; (e) epicuticular elements.

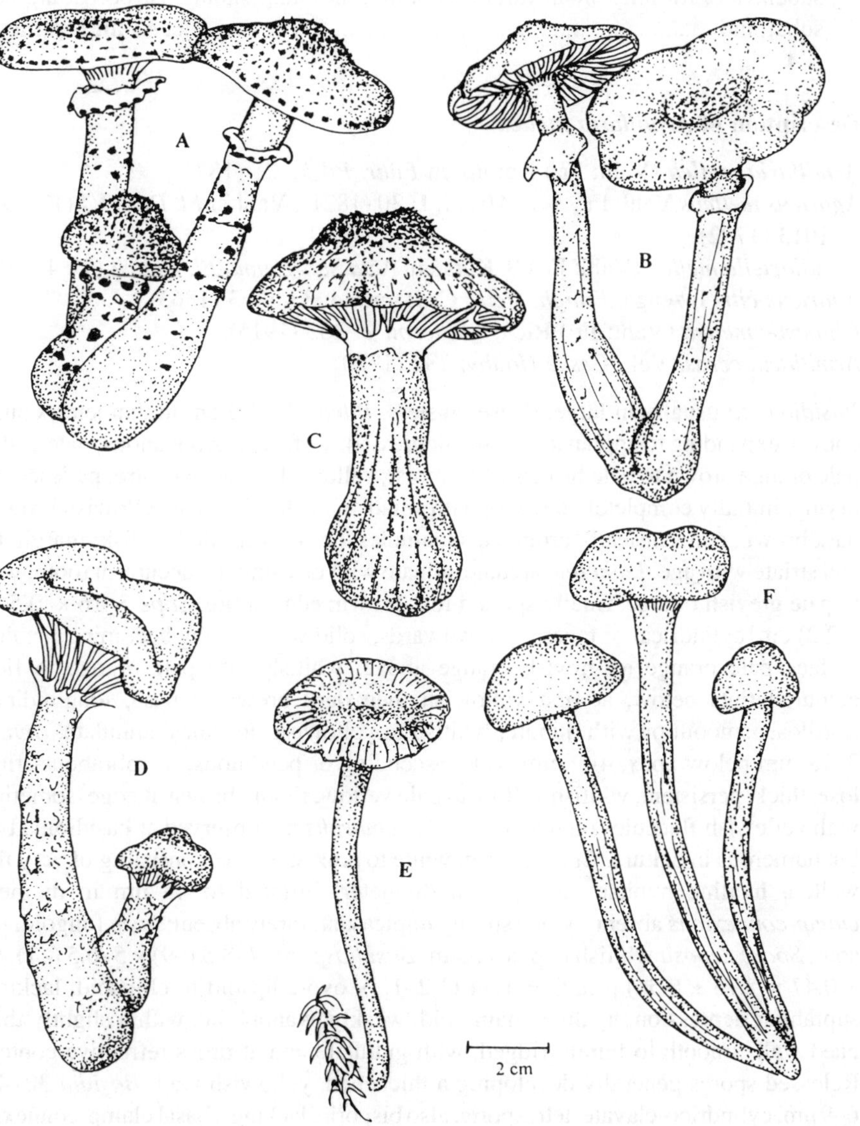

Figure 4.2a–f. European species of *Armillaria*: (a) *A. ostoyae*, showing a membranous annulus with brown squamules, pileus with pyramidal squamules, and squamulose, clavate stipe; (b) *A. broealis*, showing clustered habit, membranous annulus, and subclavate stipe; (c) *A. lutea*, showing a solitary, robust habit, with clavate stipe and cortinoid veil; (d) *A. cepistipes*, showing a slender habit with subclavate stipe, cortinoid veil and almost smooth pileus; (e) *A. ectypa*, showing habitat amongst *Sphagnum*, pileus with a striate margin and absence of a veil; (f) *A. tabescens*, showing a densely caespitose habit, tapering stipe and absence of a veil.

caespitose, usually on wood of deciduous trees; western Europe, not occurring in Nordic countries (*Figure 4.2f*) .... **A. tabescens**

– Pileus brown, with dense, blackish, verrucose squamules towards disk; stipe subclavate; forming small tufts on conifers in damp situations; boreal alpine to subalpine .... **A. nigropunctata**

Description of *Armillaria mellea*

Armillaria mellea (Vahl: Fr.) Kumm. in *Führ. Pilzk.*: 25 (1871).
Agaricus melleus Vahl: Fr., *Syst. Mycol.* **1**: 30 (1821); Vahl in *Fl. Dan.* **VI** (17): 9, pl. 1013 (1792).
Armillariella mellea (Vahl: Fr.) P. Karst. in *Acta Soc. Fauna Fl. fenn.* **2** (1): 4 (1881).
Agaricus citri Inzenga, *Funghi sicil., Cent. prima*: 33, pl. 3 (1865).
Clitocybe mellea (Vahl: Fr.) Rick, *Blätterpilze*: 362 (1915).
Armillaria cerasi Vel., *Ceské Houby*: 282 (1920).

Basidiomata usually in large, dense clusters. *Pileus* 2–11.5 cm in diameter, conico-convex expanding to applanate, finally depressed; surface hygrophanous, pale yellow, pale orange brown to pale brown when moist, yellowish green at centre; pallescent on drying, initially completely covered with small and detersile, pale yellowish brown to dark brown, concentrically arranged squamules, but later limited to disk; margin thin and striate with age. *Lamellae* arcuate, adnate with or without a decurrent tooth, white to pale greyish orange, finally spotted rust-brown; edge entire. *Stipe* 5–16 × 0.5–1.5 (–2.2) cm, cylindrical or tapering downwards, solid when young becoming fistulose; surface pale orange to greyish orange above; whitish with pale orange to flesh-coloured tints below, at first yellowish then dark brown at base, longitudinally fibrillose, smooth or with minute, whitish to yellowish floccules; annulate. *Annulus* 3–18 mm below apex, 4–9 mm wide, ascending or pendulous, membranous- fibrillose, thick, persistent, whitish yellow to pale yellow, finally brown at edge, sometimes with yellowish floccules on underside. *Rhizomorphs* not observed at basidioma base, but numerous in culture. *Context* firm, white to pinkish white, consisting of very thin-walled, hyaline hyphae, 2–5 μm in diameter, inflated to 45 μm in diameter; *clamp-connexions* absent; *odour* strong, unpleasant, rarely absent; *taste* fungoid, later sour. *Spore deposit* whitish to pale cream. *Basidiospores* 7–8.5 (–9) × 5–6.5 (–7), (7.7 ± 0.47 × 5.75 ± 0.56) μm, Q = 1.34 (1.2–1.5), ovo-ellipsoid to ellipsoid, lacking a suprahilar depression, hyaline, inamyloid, weakly cyanophilic, with a slightly thickened wall, smooth to faintly ridged, with guttulate and at times refractive contents. Released spores generally developing a thickened, yellowish wall. *Basidia* 30–45 × 6–9 μm, cylindrico-clavate, tetrasporic, also bisporic, lacking a basal clamp-connexion. *Lamella-edge* sterile, with densely crowded cheilocystidia. *Cheilocystidia* polymorphic, ranging from subglobose, broadly clavate to sinuoso-cylindrical, 15–35 × 7–16 μm, hyaline, thin-walled, often catenulate-septate, of 1–3 elements, with terminal elements bearing irregular outgrowths, nodulose to digitate, up to 35 μm long. *Hymenophoral trama* initially bilateral, with diverging hyphae, soon becoming subregular, with very thin-walled hyphae, 5–7 μm in diameter, inflated to 25 μm in diameter *Subhymenial layer* loosely woven, 8–15 μm wide. *Pileipellis* a cutis of repent, radial hyphae, 2–16 μm in diameter, thin-walled, hyaline or a with a brown vacuolar pigment, lacking clamp-connexions, disrupting towards the pileal disk to form the squamules. *Hypodermial layer* somewhat gelatinized. *Hyphae of squamules* forming fascicles, each composed of chains of short, inflated elements, 20–60 × 6–16 µm, hyaline or brown pigmented, with vacuolar, parietal and encrusting pigment. *Caulocystidia* variable, similar to cheilocystidia. *Substrate* on living, less frequently dead, coniferous and deciduous trees and shrubs, pathogenic or saprotrophic. Europe and North America, more common on sandy soils. October–November (*Figures 4.1 a–e*).

References

ANDERSON, J.B. AND ULLRICH, R.C. (1979). Biological species of *Armillaria* in North America. *Mycologia* **71**, 402–414.

BERKELEY, M.J. (1850). Decades of Fungi XXV to XXX. Sikkim Himalayan fungi collected by Dr. J. D. Hooker. *Hooker's Journal of Botany & Kew Miscellany* **2**, 42–51.

BERKELEY, M.J. AND CURTIS, M.A. (1869). Fungi Cubensis (Hymenomycetes). *Journal of the Linnean Society, Botany* **10**, 280–392.

BON, M. (1997). Flore Mycologique d'Europe: Tricholomataceae (2) Clitocybeae. *Document Mycologique Mémoire Hors séries 4.*

BURDSALL, H.H. AND VOLK, T.J. (1993). The state of taxonomy of the genus *Armillaria. McIlvainea* **11**, 4–12.

CHANDHRA, A. AND WATLING, R. (1981). Studies in Indian *Armillaria* (Fr.: Fr.) Staude (Basidiomycotina). *Kavaka* **10**, 63–84.

DENNIS, R.W.G., WAKEFIELD, E.M. AND BISBY, G.R. (1954). The nomenclature of *Armillaria, Hypholoma* and *Entoloma. Transactions of the British Mycological Society* **37**, 33–37.

DONK, M.A. (1962). The generic names proposed for Agaricaceae. *Beihefte von Nova Hedwiga* **5**, 530.

EARLE, F.S. (1909). The genera of North American gill fungi. *Bulletin of the New York Botanical Garden* **5**, 373–451.

EMEL, L. (1921). *Le genre* Armillaria. *Sa suppression de la systématique botanique,* 85pp.

FAVRE, J. (1939). Champignons rares ou peu connus des haut-marais jurassiens. *Bulletin Trimestriel de la Société Mycologique de France* **55**, 196–219.

FRIES, E.M. (1819). *Specimen systematis mycologicae. 8*. Lundae.

FRIES, E.M. (1821). *Systema mycologicum 1*, 520pp. Greifswald.

FRIES, E.M. (1825). *Systema orbis vegetabilis 1*, 374pp. Lundae.

FRIES, E.M. (1854). *Monographia Armillariarum Sueciae*, 16pp.

FRIES, E.M. (1857). *Mongraphia Hymenomycetem Sueciae*.

GILBERT, E.J. (1950). Annexe au procés-verbal de la séance du 26 Avril 1950. *Bulletin Trimestriel de la Société Mycologique de France* **66**, 76.

GREGORY, S.C. AND WATLING, R. (1985). Occurrence of *Armillaria borealis* in Britain. *Transactions of the British Mycological Society* **84**, 47–55.

HENNINGS, P. (1895). Fungi cameruneses I. *Engler's Botanische Jahrbucher* **22**, 72–111.

HONGO, T. (1976). Agarics from Papua-New Guinea. *Report of the Tottori Mycological Institute (Japan)* **14**, 95–104.

IMAI, S. (1938). Studies of the Agaricaceae of Hokkaido. *Journal of the Faculty of Agriculture, Hokkaido University* **43**, 1–378.

INTINI, M.G. (1988). Contributo alla conoscenza della Agaricales Italiana. Guida al riconosciemento delle Armillarie lignicole. *Micologia e vegetazione Mediterranea* **3**, 49–72.

JACQUES-FELIX, M. (1977). *Le complex de l'Armillaire*. Travaux dediés a George Viennot - Bourgin, pp 143–157.

JAHN, H. (1979). *Pilze die an Holz wachsen*. 268pp. Busseche Verlagshandling, Herford.

KARSTEN, P.A. (1881). Hymenomycetes Fennicae enumerati. *Acta Societatis pro Fauna et Flora Fennica* **2** (1), 44pp.

KILE, G. AND WATLING, R. (1988). Identification and occurrence of Australian *Armillaria* species, including *A. pallidula* sp. nov. and comparative studies between them and non-Australian tropical and Indian *Armillaria*. *Transactions of the British Mycological Society* **91**, 305–315.

KONRAD, P. AND MAUBLANC, A. (1948). Les Agaricales. Classification, revision des espèces, iconographie, comestibilité. *Encyclopédia Mycologique* **14**, 5–649.

KORHONEN, K. (1978). Interfertility and clonal size in the *Armillaria mellea* complex. *Karstenia* **18**, 31–42.

KUMMER, P. (1871). *Der Führer in die Pilzkunde*. 146pp. Zerbst.

LOCQUIN, M. (1952). Sur la non validité de quelques genres d'Agaricales. *Bulletin Trimestriel de la Société Mycologique de France* **68**, 165–169.

MALENÇON, G. AND BERTAULT, R. (1975). *Flore Champignons supérieurs de Maroc* **2**, 132–133.

MANJULA, B. (1983). A revised list of the agaricoid and boletoid basidiomycetes from India and Nepal. *Proceedings of the Indian Academy of Science (Plant Science)* **92**, 81–213.

MARCHAND, A. (1986). *Champignons du Nord et du Midi* **9**. 273pp, pl. 801–900.

MARXMÜLLER, H. (1980). *Armillariella bulbosa* (Barla) Romagnesi. *Bulletin Trimestriel de la Société Mycologique de France* **96**, Atlas pl. 221.

MARXMÜLLER, H. (1982). Etude morphologique des *Armillaria* s. str. a anneau. *Bulletin Trimestriel de la Société Mycologique de France* **98**, 87–124.

MARXMÜLLER, H. (1987). Quelques remargues complémentaires sur les Armillaires annelées. *Bulletin Trimestriel de la Société Mycologique de France* **103**, 137–156.

MARXMÜLLER, H. (1992). Some notes on the taxonomy and nomenclature of five European *Armillaria* species. *Mycotaxon* **44**, 267–274.

MARXMÜLLER, H. AND PRINTZ, P. (1982). Honningsvampe. *Svampe* **5**, 1–10.

MOREAU, F. (1929). Note sur le *Clitocybe ectypa* Fr. non Bres. *Bulletin Trimestriel de la Société Mycologique de France* **45**, 93–95.

PATOUILLARD, N. (1887). *Essai taxonomique des Hyménomycètes*. Lons-le-Saunier.

PEGLER, D.N. (1983). Agaric flora of the Lesser Antilles. *Kew Bulletin Additional Series* **9**. 688pp, 27 pl., 127 figs.

PEGLER, D.N. (1986). Agaric flora of Sri Lanka. *Kew Bulletin Additional Series* **12**. 519pp.

PEGLER, D.N. (1997). *The agarics of São Paulo, Brazil*. 68pp. Kew: Royal Botanic Gardens.

PHILLIPS, R. (1981). *Mushrooms and other fungi of Great Britain & Europe*. 288pp. London: Pan Books Ltd.

POUZAR, Z. (1957). Nova genera macromycetum. *Ceská Mykologie* **11**, 48–50.

QUÉLET, L. (1872). Les champignons du Jura et des Vosges. *Mémoires de la Société Emul. Montbéliard* **sér. 2** (5), 1–332.

RISHBETH, J. (1982). Species of *Armillaria* in southern England. *Plant Pathology* **31**, 9–17.

ROMAGNESI, H. (1950). Annexe au procés-verbal de la séance du 26 Avril 1950. *Bulletin Trimestriel de la Société Mycologique de France* **66**, 76.

ROMAGNESI, H. (1970). Observations sur les *Armillariella* I. *Bulletin Trimestriel de la Société Mycologique de France* **86**, 257–266.

ROMAGNESI, H. (1973). Observations sur les *Armillariella* II. *Bulletin Trimestriel de la Société Mycologique de France* **89**, 195–206.

ROMAGNESI, H. AND MARXMÜLLER, H. (1983). Étude complémentaire sur les armillaires annelées. *Bulletin Trimestriel de la Société Mycologique de France* **99**, 301–321.

SINGER, R. (1936). Das System der Agaricales. *Annales Mycologici* **34**, 286–378.

SINGER, R. (1956). The *Armillariella mellea* group. *Lloydia* **19**, 176–187.

SINGER, R. (1969). Mycoflora australis. *Beihefte zur Nova Hedwiga* **29**, 1–405.

SINGER, R. (1970). Omphalineae. *Flora Neotropica* **Monograph 3**.

SINGER, R. (1986). *The Agaricales in modern taxonomy*. Edit. 4. 981pp, 88pl. Koenigstein: Koeltz Scientific Books.

SINGER, R. (1989). New taxa and new combinations of Agaricales (Diagnoses fungorum novorum agaricalium IV). *Fieldiana (Botany)*, new ser. **21**, 133.

STAUDE, F. (1857). *Die Schwämme Mitteldeutschlands*. 148pp. Coburg.

STEVENSON, G. (1964). The Agaricales of New Zealand V. *Kew Bulletin* **19**, 1–48.

TERMORSHUIZEN, A.J. (1995). *Armillaria*. Eds. C. Bas, T.W. Kuyper, M.E. Noordeloos and E.C. Vellinga. *Flora Agaricina Neerlandica* **3**, 34–39.

TERMORSHUIZEN, A.J. AND ARNOLDS, E. (1987). On the nomenclature of the European species of the *Armillaria mellea* group. *Mycotaxon* **30**, 101–116.

VAHL, M. (1766). *Icones planatarum sponte nascentium in regnis Daniae et Norwegiae, et in ducatibus Slesvici, Holsatiae et Oldenburgi: ad Illustrandum opus de iisdem plantis Daniae nomine inscriptum 6*, **Fasc. 17**, pl. 961–1020. Hauniae.

VOLK, T.J. AND BURDSALL, H.H. (1995). A nomenclatural study of *Armillaria* and *Armillariella* species. *Synopsis Fungorum* **8**. 121pp.

WATLING, R. (1976). A pilot scheme. *Bulletin of the British Mycological Society* **10**, 43–44.

WATLING, R. (1987). The occurrence of annulate *Armillaria* species in northern Britain. *Notes of the Royal Botanic Garden, Edinburgh* **44**, 459–484.

WATLING, R. (1992). *Armillaria* Staude in Cameroon Republic. *Persoonia* **14**, 483–491.

WATLING, R., KILE, G.A. AND GREGORY, N.M. (1982). The genus *Armillaria* – nomenclature, typification, the identity of *Armillaria mellea* and species differentiation. *Transactions of the British Mycological Society* **78**, 271–285.

WATLING, R., KILE, G.A. AND BURDSALL, H.H. (1991). *Armillaria*: Nomenclature, Taxonomy, and Identification. In: *Armillaria Root Disease*. Eds. C.C. Shaw and G.A. Kile, pp 1–20. *Forest Service, Agriculture Handbook No. 691*. Washington, D.C.: USDA.

5
Molecular methods used for the detection and identification of *Armillaria*

ANA PÉREZ-SIERRA[1], DEBRA WHITEHEAD[2] AND MICHAEL WHITEHEAD[2]

[1] *Royal Horticultural Society, Wisley, Woking, Surrey GU23 6QB, U.K. and*
[2]*School of Applied Sciences, University of Wolverhampton, Wolverhampton WV1 1SB, U.K.*

Abstract

Traditional methods used for the identification and isolation of *Armillaria* spp. can be lengthy and problematic. This has partially driven the application of molecular based techniques for the diagnosis of *Armillaria* infections. This chapter will aim to review both the protein and DNA based methods which have been used to study *Armillaria*. The various advantages and disadvantages, in relation to the technique's applicability to diagnosis, will be discussed. To date, the most successful investigations are based upon RFLP analysis of the PCR amplified rDNA region. Immunological based methods remain a potentially very useful mode of analysis. It is likely that the traditional methods will remain in common usage and only gradually come to be replaced by molecular based methods of analysis.

Introduction

In recent years, there has been a great proliferation in the molecular techniques available to the mycologist for the study of fungi. A lot of the methods are extremely useful for fundamental aspects of study such as taxonomy, phylogeny and the investigation of plant pathogen interactions. A consequence of these investigations has been the ability to develop molecular based technologies for the identification and detection of fungi. A standard method used for the identification of *Armillaria* is the analysis of characteristics in culture (Rishbeth, 1986). This procedure is problematical as it is time consuming and requires a great deal of experience. Another standard method of identification is based upon mating analysis with known haploid tester strains (Korhonen, 1978). Field samples are usually dikaryotic or diploid, and the interpretation of results between these and haploid strains is particularly difficult (Guillaumin *et al.*, 1991). This technique is also rather time consuming, taking up to 8 weeks to obtain a result from when a sample is received. Due to the problems outlined above, the application of molecular techniques to the study of *Armillaria* has

Armillaria *Root Rot: Biology and Control of Honey Fungus*

been intense, leading to the development of a variety of methods of analysis with varying degrees of usefulness (Schulze *et al.*, 1997b; Schulze and Bahnweg, 1998). From the point of view of a diagnostician, it is vital that a technique is repeatable, easy and safe to perform, rapid and cheap. The molecular methods which have been used for *Armillaria* analysis will be discussed in relation to their advantages and limitations for identification.

Protein based techniques

WHOLE CELL ANALYSIS

It is possible to obtain a protein profile of an organism by isolating and analysing the total protein content. A protein profile is achieved by extracting the protein from a pure culture of an isolate and separating the protein via denaturing SDS-polyacrylamide gel electrophoresis (SDS-PAGE). Protein banding patterns are visualized by staining and compared for differences. This technique has been used to characterize a number of organisms including fungi (Alfenas *et al.*, 1984).

Some limited investigations using whole protein analysis have been performed on *Armillaria* species. Lung-Escarmant *et al.* (1985) demonstrated that in European isolates the protein profiles of *A. lutea* were different from the other *Armillaria* species. Lin *et al.* (1989) performed native protein analysis (without SDS) to analyse the North American species I, III, V and VII and were able to identify distinct patterns for each group.

This technique has little application for the routine identification of *Armillaria* isolates. The procedure is quite lengthy, requires experience, and the profile produced varies with environmental factors and the age of the culture. As a consequence, the database of known profiles of *Armillaria* species is very limited. However, the detection of unique protein bands can be used for the development of immunological methods of identification.

IMMUNOLOGY BASED TECHNIQUES

Immunological based methods are potentially extremely powerful tools for the identification of plant pathogenic fungi. The development of these methods can be very technically challenging and time consuming but, once available, allow for the possibility of routine identification and diagnosis. These techniques rely on the raising of polyclonal or monoclonal antibodies (MAbs) to mycelial fragments, surface washings of cultures or SDS-PAGE isolated proteins. Once antibodies have been produced and purified, they can be used to detect the presence of particular antigens by techniques such as enzyme linked immunosorbent assays (ELISA).

Early investigations by Lung-Escarmant *et al.* (1985) were able to distinguish *A. mellea* and *A. ostoyae* isolates using polyclonal antibodies raised to a partially purified antigen of *A. mellea*. An investigation by Zollfrank *et al.* (1987) met with very limited success, with anti-*Armillaria* antibodies cross reacting with all the species investigated (*A. borealis*, *A. lutea* and *A. ostoyae*). Using MAbs raised to crude mycelial extracts, Fox and Hahne (1989) demonstrated some success in separating British species of *Armillaria*. They found three MAbs that could distinguish one isolate of *A. ostoyae* from one isolate of *A. lutea* and two isolates of *A. mellea*. However, the differentiation obtained was more qualitative than quantitative. Burdsall and Banik (1990) raised polyclonal antibodies to whole cell antigens of *A. mellea*, *A. ostoyae* and *A. tabescens*. Some of the antibody preparations were able to distinguish homologous species from isolates of heterologous species.

The most successful report to date on the use of antibodies for the identification and detection of *Armillaria* was by Priestley *et al.* (1994). They raised MAbs to *A. mellea* and *A. ostoyae* antigens isolated from SDS-PAGE or from protein extracts of lyophilised mycelial powder. Two MAbs raised against *A. mellea* were found to be specific to *A. mellea* when compared to *A. ostoyae*, *A. lutea*, *A. cepistipes* and *A. tabescens*. Two further MAbs were found to be genus specific. When MAbs were raised against *A. ostoyae*, one was found to be specific to *A. ostoyae* when compared to *A. mellea*, *A. borealis* and *A. lutea*, and two specific to *A. ostoyae* when reacted against *A. mellea*, *A. lutea*, *A. cepistipes* and *A. tabescens*. When some of the MAbs were further investigated for activity against extracts from infected wood, one was found to be *A. mellea* specific, one *A. ostoyae* specific and one specific to *A. ostoyae* and *A. lutea*. They further developed their technique to be able to perform dipstick analysis of extracts from infected and non-infected wood samples. In only one case was there a discrepancy between the results obtained with MAbs and those from pairing tests.

The work of Priestley *et al.* (1994) shows the potentially useful nature of this technique. Not only is it accurate and reproducible, it is also extremely quick. This was exemplified by Priestley *et al.* (1994) who estimated the time for identification of a species by pairing test analysis to be six to eight weeks, while for the immunological based assay it was less than 24 hours. However, they did note that if the infected wood dried out, then identifications were unsuccessful. They also commented on the need for extensive testing upon a wide range of wood-infecting fungi and *Armillaria* wood-rotted material before the dip stick assay can be exploited commercially.

ISOZYME ANALYSIS

Isozyme analysis is a relatively easy and cheap technique to perform which has numerous applications in plant pathology. This technique has a long history in the study of biological organisms, and mycologists have adapted this method of analysis for fungi. Uses of isozyme analysis include taxonomy, identification, patenting, analysing genetic variability, epidemiology and investigation of ploidy levels (Bonde *et al.*, 1993).

Isozyme analysis relies upon the fact that there are different molecular forms (isoforms) of a particular enzyme. These different forms have essentially the same enzymatic activity but differ in the amino acid composition. When these differences result in an altered net charge, or result in large differences in the shape of an enzyme, they can be detected by starch gel electrophoresis, polyacrylamide gel electrophoresis or isoelectric focusing. Following electrophoresis, the position of the enzymes can be determined following staining with an appropriate substrate which will change colour on modification by the enzyme. This will produce a characteristic banding pattern or fingerprint.

Isozymes are formed by allelic variation at a single locus (called allozymes), multiple loci coding for a single enzyme or post-translational modifications of an enzyme. This therefore allows a genetic interpretation of data produced by a fingerprint, and an estimate of the genetic variation between genomes of organisms.

A number of investigations into the use of isozyme analysis in *Armillaria* species have been undertaken. One of the first was performed by Morrison *et al.* (1985) who analysed the isozyme patterns of esterases and polyphenoloxidases in British Columbian isolates. The analysis of six intersterile groups allowed them to separate out *A. lutea*, *A. ostoyae* and groups V and IX. However, it should be noted that these enzymes may play a role in pathogenicity and are therefore not neutral markers of genetic variation (Schulze *et al.*, 1997b).

A more extensive study was performed by Lin *et al.* (1989), who analysed 20 different enzymes from 13 isolates of *Armillaria* species NAB I, II, V, VII. Of the enzymes tested, only 3 (esterase, polyphenol oxidase and alcohol dehydrogenase) showed activity, and only esterase was able to distinguish between all four groups. They also noted some intraspecific variation for group I isolates collected from distinct geographical areas.

A disadvantage of the study performed by Lin *et al.* (1989) was the small number of strains investigated. A more comprehensive study was performed by Wahlström *et al.* (1991) who analysed 94 European isolates of *A. ostoyae*, *A. lutea*, *A. borealis*, *A. mellea* and *A. cepistipes*. This investigation analysed the pectinolytic enzymes pectin esterase, pectin lyase and polygalacturonase. The isozyme pattern produced by polygalacturonase was able to clearly distinguish *A. mellea* from the other four species. The overall pectinolytic pattern produced allowed all five species to be distinguished, and showed a very good correlation (81%–100%) with interfertility tests. The pectinolytic pattern of *A. mellea* was found to be the most divergent, while *A. ostoyae* and *A. borealis* were the most closely related. However, it is noteworthy that pectinolytic enzymes may also play a role in pathogenesis and variation in the isozyme patterns produced were found to be dependent upon the media used for growth.

Rizzo and Harrington (1993) further contributed to the database of enzymes which may be useful for isozyme analysis. They found polymorphic patterns for four enzymes (ß-glucosidase, glucosephosphate isomerase, aconitase and isocitrate dehydrogenase) which were able to distinguish between isolates of *A. gemina*, *A. calvescens* and *A. ostoyae*. Additionally, this combination of enzymes was able to distinguish between all the 16 somatic incompatibility groups (SIGs) determined in the isolates with the exception of two *A. ostoyae* isolates. Guillaumin *et al.* (1996) compared different methods for the identification of genotypes of *A. ostoyae*, *A. lutea* and *A. cepistipes*. This study compared the use of SIG testing with RAPD analysis, mating type alleles and isozyme analysis. The enzymes analysed (malate dehydrogenase, esterase, phosphoglucomutase, acid phosphatase, superoxide dismutase and glutamate dehydrogenase) were clearly able to discriminate between the three species. They were able to find intraspecific differences for the enzymes used but concluded that this was not a suitable method for routine work in epidemiology. Their method was determined to be no more precise than somatic incompatibility testing and suffered from problems of reproducibility.

Recent studies have been performed by Bragaloni *et al.* (1997) and Mwenje and Ride (1997) to identify and characterize strains of *Armillaria* from Europe and Africa respectively. Mwenje and Ride (1997) confirmed the usefulness of pectin enzyme analysis, enabling them to group African isolates of *Armillaria*. They also demonstrated that African isolates of *A. mellea* produced distinct patterns from European *A. mellea* isolates, indicating that these species are not identical. Bragaloni *et al.* (1997) investigated the potential of isozyme analysis for the identification of European *Armillaria* species. Of 22 enzyme patterns analysed only esterase, glutamic-oxalacetic transaminase, phospho-gluco-mutase, alcohol dehydrogenase and polygalacturonase demonstrated sufficient polymorphism to allow the identification of different *Armillaria* species. *A. borealis*, *A. cepistipes*, *A. ostoyae* and *A. lutea* were all found to have discriminating profiles with esterase activity. *A. mellea*, *A. lutea* and *A. tabescens* could be distinguished with polygalacturonase isozyme analysis. Additionally, a distinguishing pattern was observed for *A. tabescens* with glutamic-oxalacetic transaminase, alcohol dehydrogenase and phospho-gluco-mutase. The authors concluded that this analysis could be used for the identification of all six European *Armillaria* species. One disadvantage of the analysis performed was that only those haploid isolates of each species, and three diploid isolates of *A. lutea*, *A. mellea* and *A. tabescens* and two diploid isolates of *A. ostoyae* were analysed. A far more comprehensive analysis of field isolates needs to be undertaken to determine the degree, if any, of intraspecific variation that exists before the technique can be widely applied to identification.

The prolonged interest in the use of isozyme analysis proves the value of this technique in investigating *Armillaria*. The technique is simple, rapid and cheap, and has application in mating analysis, identification, clonal detection, heterozygosity and phylogeny. Problems with this technique can be found with a lack of reproducibility dependent upon growth conditions, culture age and electrophoretic conditions. Additionally, the sample must be grown up in pure culture to obtain a reasonable quantity of sample for analysis. Although the technique is described as rapid, it can take weeks to prepare the culture filtrate for analysis, which compares unfavourably with some of the DNA techniques described below.

DNA analysis methods

DNA HOMOLOGY/REASSOCIATION ANALYSIS

One of the crudest methods for the analysis of DNA is to assess the DNA GC content and sequence similarity, over the entire genome, between two organisms. These techniques have been applied to some *Armillaria* species. Jahnke *et al.* (1987) determined that the nuclear GC contents for *A. mellea*, *A. lutea* and *A. ostoyae* ranged from 46.0 to 48.1%. The DNA homologies were assessed between these species, as well as for two isolates within each species, using a simple spectroscopic method. Intraspecific hybridizations showed 90–100% homologies, while interspecific homologies were 44–70%. A more exacting analysis was performed by Miller *et al.* (1994) using DNA reassociation experiments and S1 nuclease protection. Samples of DNA from two organisms were broken up, radioactively labelled and allowed to hybridize to one another. Where the DNA has not associated, it will be susceptible to digestion by the enzyme S1 nuclease. This will enable the percentage reassociation between the DNA of any two organisms to be assessed and hence the gross DNA similarity between those organisms can be assessed (Johnson, 1985). This technique has primarily been used for analysis in bacterial systematic studies (Johnson, 1989). Miller *et al.* (1994) determined four major DNA sequence similarity clusters for the species investigated, these being *A. calvescens*, *A. lutea*, *A. sinapina*, NABS IX and NABS X; *A. mellea*; *A. gemina* and *A. ostoyae*; and *A. tabescens*. Either due to the crude nature of the results or the complex nature of the analysis, these techniques are not suitable for identification purposes but do have some advantages for use in phylogenetic analysis.

RESTRICTION FRAGMENT LENGTH POLYMORPHISM (RFLP) ANALYSIS WITHOUT PCR

Restriction endonuclease (Type II) enzymes recognize specific sequences within a DNA molecule (usually 4 b.p., 6 b.p. or 8 b.p. in length) and cleave the DNA at that point. The DNA fragments produced can be separated by agarose gel electrophoresis. Following denaturation of the fragments, the DNA can be transferred to nitrocellulose or, more commonly, nylon membranes (Southern blotting; Sambrook *et al.*, 1989). The DNA on the membranes can then be probed for the presence or absence of a particular DNA sequence. The DNA used for a probe is labelled (usually with radioactivity) and will hybridize to any complementary sequences on the membrane. Hybridization of the probe DNA to the target DNA can be visualized by autoradiography which will produce characteristic banding patterns.

Variations within the DNA of two organisms can be analysed using this technique. Changes in the DNA which result in the loss or acquisition of a restriction endonuclease site, or in the deletion or insertion of a length of DNA can be observed. Hence, where such a change has occurred, a change in the restriction length (an RFLP) of a piece of DNA will be observed.

If the DNA to be analysed is relatively small, such as mitochondrial DNA (usually about 100 kb), RFLPs can sometimes be detected directly from the electrophoresis gel. Such an analysis was performed by Jahnke *et al.* (1987) when they isolated the mitochondrial DNA from two species of *A. mellea, A. lutea* and *A. ostoyae*. This DNA was cut by six restriction enzymes and the fragments obtained compared. Intraspecies patterns showed 67–100% similarity, while interpecies patterns were rather dissimilar with similarity values of 0–50%. Additionally, RFLPs in genomic DNA can occasionally be detected directly from the electrophoresis gel, when a highly repetitive sequence demonstrates an RFLP. However, this is rarely the case and usually of limited use.

Critical to the success of this RFLP analysis is the availability and choice of which DNA to use as a probe. Anderson *et al.* (1989) cloned the ribosomal DNA (rDNA) from a genomic library of a strain of NAB I using the entire rDNA repeat segment of *Coprinus cinereus*. The rDNA can be a particularly useful region to analyse (see PCR section for detail). They used the *Armillaria* rDNA clone to detect RFLPs in the rDNA region of 30 isolates of five European *Armillaria* species and eight NABS designations. Amongst the isolates, nine restriction site polymorphisms and two length polymorphisms were observed. These polymorphisms enabled the isolates to be placed into six classes which essentially possessed the same rDNA maps, although some variability within classes was observed. The six classes were class 1: NABS I and *A. ostoyae*; class 2: NABS II; class 3: *A. borealis*; class 4: NABS V, IX and X; class 5: NABS III, VII, *A. lutea* and *A. cepistipes*; and class 6: NABS VI and *A. mellea*. They observed that these rDNA classes did not correlate with either geographical distribution or intersterility, indicating that loss of interfertility in *Armillaria* can develop without detectable changes in the rDNA regions occurring. Three of the sets (classes 1, 2 and 3) were determined to be closely related.

Smith and Anderson (1989) analysed 23 isolates of NABS I, II, III, V, VI VII, IX and X using either the entire mitochondrial genome, or subcloned mitochondrial fragments, as the probe. They were able to determine that the patterns produced were similar within biological species and dissimilar between biological species. Additionally, although the RFLP patterns were similar within biological species, each isolate had a unique pattern. Analysis of the mitochondrial patterns allowed correct identification of the isolates. However, due to the variation of the mitochondrial DNA, the technique was unsuitable for routine identification, or to resolve phylogenetic relationships within *Armillaria*. Smith *et al.* (1990) used a very similar method of analysis with the entire mitochondrial genome as a probe to analyse the transmission and propagation of mitochondrial genotypes in field isolates. They were able to conclude that either matings between haploid isolates are uncommon in nature, or when mating does occur, the mitochondria of one partner will predominate during subsequent vegetative growth.

Smith *et al.* (1992) analysed nuclear RFLPs (along with RAPDs, see later section) to identify the location of a clonal individual of *A. lutea*. Using four anonymous nuclear DNA fragments and a portion of the nuclear rDNA repeat of *A. ostoyae* as probes, they were able to identify 27 nuclear fragments which were invariant in all diploid isolates of the clonal individual. From this investigation, they were able to conclude that *Armillaria* are amongst the largest and oldest living organisms.

Smith and Anderson (1994) performed a comprehensive analysis of the mitochondrial genome of *A. ostoyae*. They determined the overall size of three European isolates to vary between 80–117 kb, while 17 North American isolates possessed a mitochondrial genome of between 125–147 kb, with extensive variation between individuals.

Rizzo *et al.* (1995) investigated the population structure of *A. ostoyae* in a *Pinus resinosa – P. banksiana* stand in North Minnesota. They investigated the molecular variation in 96 of 439 isolates, using the mitochondrial analysis of Smith and Anderson (1989). In addition to this, they used oligonucleotide probes of $(TCC)_5$ and $(CAT)_5$ to detect variable dispersed repetitive DNA (microsatellites) within the nucleus. When the genomic DNA was digested with *Hind*III or *Pst*I, they were able to observe 14 fragments in total. When they correlated the banding pattern produced by this analysis to the 16 SIG of the isolates, they observed no band which was common to all 16 SIG. They stated that most SIGs had unique genotypes and all isolates tested from within an individual SIG had identical genotypes. However, they also observed a few exceptions to this as, on two occasions, the fragments could not distinguish between two SIGs, and three different patterns were observed among 19 isolates of one SIG.

The detection of RFLPs in the rDNA of *Armillaria* was more recently investigated by Schulze *et al.* (1995, 1997a). In their first study they used the *Saccharomyces carlsbergensis* rDNA as a probe to distinguish 44 isolates of the five European species *A. borealis*, *A. lutea*, *A. cepistipes*, *A. mellea* and *A. ostoyae*. Using the restriction endonuclease *Bgl*II in combination with either *Eco*RI, *Bam*HI or *Hind*III, they were able to distinguish all five species. However, more conclusive results were obtained upon digestion with *Ava*II. In their later investigation, they analysed genetic variation among 20 *A. ostoyae* isolates using the *Ava*II analysis described above, in addition to the PCR based analysis described later. They were able to detect intraspecific variation with the RFLP analysis such that the 20 isolates could be assigned to five patterns.

Overall, these investigations show the extremely powerful nature of the genetic analysis of biological specimens by RFLP based techniques. These techniques have been demonstrated to have a variety of applications in taxonomy, phylogeny, population studies, host-pathogen interactions and identification. The different techniques discussed have various strengths and weaknesses depending upon their application. The analysis of rDNA polymorphisms is particularly useful for taxonomic, phylogenetic and identification purposes. In fact, the information gained from these investigations serves as a basis for the later PCR based rDNA analysis. The analysis of mitochondrial DNA is particularly useful for phylogenetic and population studies. However, the high degree of variation exhibited by *Armillaria* mitochondrial genomes make this analysis unsuitable for identification. Beyond these two main themes of *Armillaria* RFLP analysis, there is quite a scarcity of information available. Investigations, such as that performed by Rizzo *et al.* (1995) into new RFLP markers, such as microsatellite DNA, may produce probes useful for identification.

There are some additional drawbacks common to all the RFLP analyses performed. Sizeable amounts of DNA (μg levels) must be isolated from pure cultures, causing these investigations to be protracted. These investigations are expensive to perform and the use of radioactivity is a concern for a routine diagnosis service, although non-radioactive chemiluminescence techniques have been used in some of these investigation (Schulze *et al.*, 1997a) without any apparent loss of sensitivity or reproducibility. Overall, the disadvantages demonstrated by these methods means that there will continue to be a concentration on immunological and PCR based methods.

POLYMERASE CHAIN REACTION BASED TECHNIQUES

Polymerase Chain Reaction (PCR) is an extremely powerful technique for the molecular analysis of organisms. PCR involves the exponential amplification of target DNA sequences via thermostable DNA polymerase. Minute quantities (pg levels) can be amplified to μg levels for analysis (e.g. gel electrophoresis, sequencing) in a matter of hours. The target sequence to be amplified must be flanked by short (approximately 10–30 bases in length) primers which are added to the PCR reaction mix. The use of this procedure has found a wide variety of applications in the analysis of plant pathogenic fungi, including gene cloning, sequencing, taxonomic analysis and strain and species identification. *Armillaria* species are amongst the plant pathogenic fungi for which PCR has proved an extremely useful method of analysis (for review see Foster *et al.*, 1993).

RANDOMLY AMPLIFIED POLYMORPHIC DNA ANALYSIS

One drawback of the standard PCR amplification protocol is that the sequence surrounding the target DNA must be known so primers can be made to amplify the target DNA. This particular problem is circumvented by the use of randomly amplified polymorphic DNA analysis (RAPD; Williams *et al.*, 1990). This technique relies upon the use of a randomly chosen primer (usually about 10 bases in length) which anneals to complementary regions of the genome to be analysed. Provided that the primer anneals to both strands of DNA within about 0.5 kb to 5 kb of each other, then a fragment of DNA will be amplified in an exponential manner. Hence, variations in DNA due to single or multiple base changes or sizeable insertions and deletions can be detected by the change in size or presence or absence of an amplified fragment. This allows discrimination of organisms to the species and even isolate level.

The earliest description of the use of RAPDs for the analysis of *Armillaria* is by Smith *et al.* (1992). Along with the RFLP analysis described in the previous section, they used RAPD analysis to supplement the investigation of the *A. lutea* isolate. In addition to the 27 RFLP markers, RAPD analysis generated markers for 20 loci and were used to draw the conclusions discussed previously. As mentioned earlier, Guillaumin *et al.* (1996) used RAPD analysis among the techniques for the identification of genotypes of *Armillaria* species. For both RAPD analysis and mating type analysis, they investigated 73 isolates of 36 genotypes of *A. ostoyae*, *A. lutea* and *A. cepistipes*, as identified by SI reactions. Using three RAPD primers, they were able to distinguish all 36 genotypes (in some cases characterization was possible with just one or two primers). They concluded that RAPD analysis was equal in its ability to SI to distinguish genotypes of *Armillaria.* Additionally, they observed that RAPDs revealed greater variation than did SI as three genotypes could be further divided into subgroups.

In the investigation of genetic variation of *A. ostoyae* by Schulze *et al.* (1997a), in addition to the rDNA-RFLP analysis, RAPD investigations were performed. Using 10 primers, a total of 138 scorable fragments were determined from the 20 *A. ostoyae* isolates. These fragments were able to characterize the individuals investigated. Additionally, they found a good correlation between determined subgroups and the geographical origins of the genetic individuals. With such a large database of genetic loci information, they have stated their intention to use this for the investigation of the relationship between the genotypes of individuals and their phenotypic interaction (compatible, incompatible) with the host *Picea abies*. They also discovered two fragments (790 b.p. and 690 b.p.) produced by one primer, OPA-3, which was consistently observed in all *A. ostoyae* isolates. This may prove a particularly useful marker for the identification and detection of *A. ostoyae*.

These results show some of the powerful uses of RAPD analysis. The technique allows the easy analysis of a large number of genetic loci and hence is a powerful discriminating tool. However, the technique can suffer from a lack of reproducibility (Guillaumin *et al.*, 1996). The interpretation of results can also require the use of sophisticated analytical programmes. Often, a number of complex patterns will be produced for a particular primer/species combination. Before this technique becomes of use for identification and diagnosis, simple unique patterns will need to be described, as performed by Schulze *et al.* (1997a) for the OPA3/*A. ostoyae* combination. Fortunately, due to the potentially unlimited number of loci that can be analysed, given concerted research, this may be possible.

PCR ANALYSIS OF RDNA

The nature of the rDNA cluster results in it being particularly useful for the analysis of genus, species and isolate variability by PCR analysis. The arrangement of the ribosomal repeat of some lower eukaryotic organisms, including *Armillaria* species, is shown in *Figure 5.1*. The 18S, 5.8S and 26S rDNA is transcribed as one unit, along with the two internal transcribed spacer units (ITS1 and ITS2). The 5S rDNA is transcribed as a separate unit and is separated from the 26S and 18S rDNA by two non-transcribed intergenic sequences (IGS-1 and IGS-2).

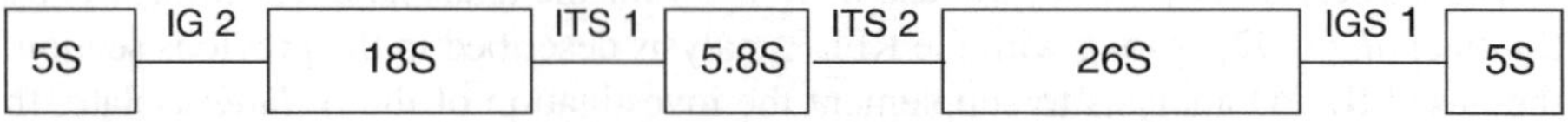

Figure 5.1. Schematic view of the arrangement of rDNA repeat.

Due to evolutionary constraints, the nuclear rDNA sequences evolve slowly, and hence are highly conserved. They are, therefore, particularly useful for studying distantly related organisms. However, there is little evolutionary pressure for the conservation of the non-coding ITS1, ITS2, IGS-1 and IGS-2 regions. These sequences can therefore exhibit high degrees of variability, and are particularly useful for studies at the species and individual level. The fact that the variable ITS1, ITS2, IGS-1 and IGS-2 regions are flanked by highly conserved DNA sequences provides a neat solution for one drawback of PCR, that of the design of primers for the amplification of DNA.

It appears from the recent proliferation in publications that PCR analysis of rDNA is one of the most active areas of research into *Armillaria*. The groundwork for many of the investigations being currently performed was laid by the thorough investigation of Anderson and Stasovski (1992). They amplified and sequenced the IGS region between the 26S and 5S gene using primers CLR12 and O-1 (Duchesne and Anderson, 1990). This was performed on five isolates of *A. ostoyae*, four isolates of *A. lutea*, three isolates of *A. cepistipes* and two isolates of NABS IX and X, *A. calvescens*, *A. gemina*, *A. borealis* and *A. sinapina*. The resulting sequence information was analysed by the parsimony programme PAUP, to determine the phylogeny of this group of *Armillaria*. This produced two distinct groups: *A. gemina*, *A. ostoyae* and *A. borealis*, and *A. sinapina*, NABS X, *A. calvescens*, *A. lutea*, *A. cepistipes* and NABS IX. The ITS regions of the above strains and four of *A. mellea* and one of *A. tabescens* were also amplified using primers ITS1 and ITS4 (White *et al.*, 1990). This region proved recalcitrant to detailed analysis, but they were able to conclude that *A. mellea* and *A. tabescens* demonstrated marked divergence to the other species at the rDNA cluster.

The sequence variation observed by Anderson and Stasovski (1992) suggested that RFLP analysis of this region may be useful to discriminate between *Armillaria* species. This possibility was investigated by Harrington and Wingfield (1995). Using the same primers as Anderson and Stasovski (1992), named as LR12R and O-1, they amplified the IGS-1 region of 74 isolates of 11 European and North American *Armillaria* species. They developed the technique such that the IGS-1 could be amplified directly from mycelial fragments. The resultant PCR products were analysed for RFLPs using the enzymes *Alu*I, *Nde*I and *Bsm*I. The results of the *Alu*I

Table 5.1. Summary of the results obtained by Harrington and Wingfield (1995) following digestion of the PCR amplified IGS-1 region digested with *Alu*I.

Species (RFLP group)	Number of isolates exhibiting pattern	Fragment sizes (b.p.)
A. ostoyae	9	310, 200, 135
A. gemina	7	310, 200, 135
A. borealis (A)	3	310, 200, 135
A. borealis (B)	5	310, 200, 104
A. sinapina	4	399, 200, 135
A. cepistipes (A)	5	399, 200, 183
A. cepistipes (B)	6	310, 20, 135
NABS IX	4	534, 200
NABS X	4	399, 183, 142
A. lutea (European)	2	399, 240, 183
A. lutea (American)	5	582, 240
A. calvescens	6	582, 240
A. mellea (A)	4	490, 180
A. mellea (B)	3	320, 155
A. tabescens (A)	5	430, 240
A. tabescens (B)	1	320, 240, 100

Only fragment sizes in excess of 100 b.p. are given.

analysis are summarized in *Table 5.1*. Only the PCR products of *A. gemina* and *A. borealis* were cleaved by *Nde*I (producing 550 and 370 b.p. fragments). Eight of the nine *A. ostoyae* isolates produced 620 and 300 b.p. fragments upon *Bsm*I digestion, while none of the *A. gemina* or *A. borealis* isolates' PCR products were cleaved. *Hind*II did not cleave the PCR products of the *A. ostoyae*, *A. gemina* and *A. borealis* isolates, but did digest the *A. cepistipes*' products (yielding 580 and 340 b.p. fragments). These results demonstrated the potential for this analysis in the identification of *Armillaria* species and form the basis for a number of investigations described below.

The same primer combination (LR12R and O-1) followed by *Alu*I digestion has been used by a number of researchers (Volk *et al.*, 1996; Banik *et al.*, 1996; White *et al.*, 1998; Frontz *et al.*, 1998; Pérez-Sierra *et al.*, 1999). These investigators have produced a wealth of data confirming the validity of Harrington and Wingfield's (1995) survey. A number of additional RFLP patterns have also been reported for various species. If this technique is going to be valid for identification purposes, it is important to have as complete a list as possible of the patterns produced by a particular primer/restriction endonuclease combination for the different species of *Armillaria*. A typical example of the type of results obtained by this procedure can be seen in *Figure 5.2*.

Volk *et al.* (1996) investigated a new *Armillaria* species, *A. nabsnona*, and revealed three patterns of 563, 200 b.p. (pattern-a); 306, 230, 196 b.p. (pattern-b) and 560, 321, 237, 203 b.p. (pattern-c). This investigation also revealed the first instance of an isolate exhibiting heterozygosity for the rDNA cluster (pattern-c). Such heterozygosity appears a curiously rare phenomenon and has been observed in a small number of isolates of *A. sinapina* (White *et al.*, 1998) and *A. lutea* (Pérez-Sierra *et al.*, 1999). Banik *et al.* (1996) reported a new RFLP pattern for *A. sinapina* of 401, 237 and 184 b.p. This pattern was also observed by White *et al.* (1998); although their measurements recorded the sizes as 400, 235 and 190 b.p., as well as recording three

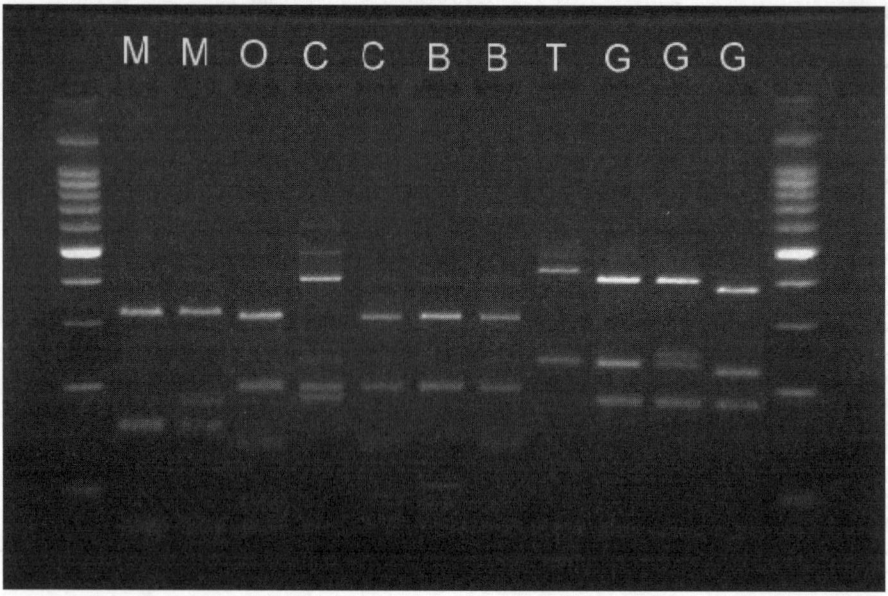

Figure 5.2. *Alu*I digest of PCR amplification products of the IGS-1 region of various *Armillaria* species Lanes 1 and 13, 100 b.p. ladder; lanes 2 and 3, *A. mellea*; lane 4, *A. ostoyae*; lanes 5 and 6, *A. cepistipes*; lanes 7 and 8, *A. borealis*; lane 9, *A. tabescens*; lanes 10, 11 and 12, *A. lutea*.

additional patterns of 400, 235, 200, 190 and 135 b.p.; 400, 200, 190 and 135 b.p.; and 400, 200 and 190 b.p. An altered pattern was also observed for a small number of *A. ostoyae* isolates (two out of a total of 116 isolates from tree stumps) which had a 185 b.p. band. They also observed two patterns for *A. lutea*, these being 400, 245, 190 and 80 b.p. and 400, 235, 190 and 80 b.p., one of which probably corresponds to the Harrington and Wingfield (1995) pattern of 399, 240 and 183 b.p. Pérez-Sierra *et al.* (1999) also observed an additional two patterns for *A. lutea* of 390, 230 and 190 b.p. and 400, 250, 240 and 190 b.p., of which the latter is probably a heterozygote of the two patterns determined by White *et al.* (1998). Pérez-Sierra *et al.* (1999) also observed a new pattern for *A. mellea* isolates of 320, 180 and 155 b.p.

Frontz *et al.* (1998) used the same primer combination to identify 39 *Armillaria* isolates from Pennsylvania. They also found *Alu*I digestion most efficient in separating individual *Armillaria* species, although the pattern produced was identical for both *A. ostoyae* and *A. lutea*. Hence, these two species were discriminated by their *Hae*III digestion patterns.

The IGS-1 region of *Armillaria* species has also been investigated by a further two authors, Chillali *et al.* (1998a) and Terashima *et al.* (1998), using different primer combinations of 5Sa/CNL12 (Henrion *et al.*, 1992) and O-5/O-1 (Anderson and Stasovski, 1992), respectively. Terashima *et al.* (1998) used the sequence determined from the amplified product for a phylogenetic study of two new Japanese species, *A. jezoensis* and *A. singula*, and one new subspecies, *A. mellea* subsp. *nipponica*. They concluded that the two new species belong to the *A. lutea* cluster of species, while the subspecies sequence could not be accurately aligned. Chillali *et al.* (1998a) also found *Alu*I digestion of the amplified IGS-1 region to be most informative in discriminating a few European isolates of *A. borealis*, *A. cepistipes*, *A. lutea*, *A. ostoyae*, *A. mellea*, *A. tabescens* and *A. ectypa*. They were able to identify all isolates (18 in total) of the seven species with *Alu*I digestion. They also analysed the RFLP patterns produced by *Rsa*I, *Taq*I and *Hin*fI, which allowed for some discrimination, but not as comprehensive as that shown by *Alu*I digestion.

A limited investigation of the IGS-2 region was performed by White *et al.* (1998). Two groups of isolates, *A. lutea* (pattern-1) and *A. sinapina* (pattern-1) could not be distinguished by the IGS-1 *Alu*I RFLP pattern. When the IGS-2 region was amplified and subjected to *Alu*I digestion, discriminatory patterns were produced.

The ITS regions of the rDNA genes of *Armillaria* species have also been the subject of recent concerted investigation (Schulze *et al.*, 1997a; Chillali *et al.*, 1998a, 1998b). Schulze *et al.* (1997) designed two *Armillaria* specific primers (ARM1 and ARM2) to amplify the ITS-1 and ITS-2 regions of the rDNA repeat of the 20 *A. ostoyae* isolates analysed by RAPD analysis (see previous section). This DNA product was subjected to restriction endonuclease digestion with various enzymes, of which two, *Alu*I and *Hin*fI, were found to digest the DNA. The resulting digests were of little use for discriminating between isolates. The investigation of Chillali *et al.* (1998a) aimed to discriminate between the species investigated by analysing the IGS-1 sequence (see above), using RFLP analysis of the ITS regions. Using the primers ITS-1 and ITS-4 (White *et al.*, 1990), the ITS regions were amplified and analysed by *Eco*RI, *Cfo*I, *Alu*I, *Hin*fI, *Taq*I and *Nde*I digestion. No single enzyme allowed discrimination between all seven species. Generally, *A. mellea*, *A. tabescens* and *A. ectypa* could be distinguished by most enzymes. In addition, digestion with *Hin*fI and *Taq*I allowed separation of *A. borealis* and *A. ostoyae* from *A. cepistipes* and *A. lutea*. *Nde*II digestion was further demonstrated to distinguish *A. borealis* from *A. ostoyae*. Chillali *et al.* (1998b) sequenced the ITS regions from the species investigated in Chillali *et al.* (1998a). This information may be helpful in revealing potential RFLPs for use in the diagnosis and identification of *Armillaria* species.

This recent proliferation in publications detailing PCR techniques used for the characterization and identification of *Armillaria* species exemplified in this section indicates their potential. A number of these techniques, such as ITS analysis and RAPDs, are still in the relatively early stages of development. The analysis of the IGS-1 (particularly with *Alu*I) has now been performed on a very large selection of isolates from numerous *Armillaria* species throughout the world. These analyses have demonstrated the simplicity, reproducibility and accuracy of the technique. However, they are not without their problems. Reproducibility is not always guaranteed, often without explanation, and can lead to the generation of false negative results. These techniques also merely detect the presence of DNA and may or may not be indicative of the presence of a viable organism. PCR is also particularly susceptible to the problem of contamination. The presence of just one or two contaminating spores could lead to inconclusive or false positive results. Given the realities of a busy diagnosis service, this is a serious concern that is often not adequately addressed. As more patterns are being characterized, it is important that a database of known patterns is available. This may even require the establishment of a strain collection which exhibits the known patterns, especially since the accuracy with which fragment sizes can be assessed varies quite considerably from laboratory to laboratory. A serious concern is that, as more isolates are analysed, more and more patterns will be identified, making the process unwieldy.

Conclusion

Plant pathologists are constantly being informed that molecular based techniques are going to provide the solution to the problems of isolate identification and diagnosis. Yet, despite many years of intensive investigations, it is probably only relatively recently that some of the PCR based techniques have started to be used routinely. No doubt these techniques will continue to be developed and refined to cure some of the problems outlined. This will be helped by the intensive research into human genetic fingerprinting and diagnosis of diseases. The potential for immunological based identification procedures, such as the dipstick analysis described, has yet to be realized. It seems likely that the advantages of such tests will continue to fuel research into these areas, allowing the production of easy-to-use kits. The traditional methods of analysis, such as morphological characterization and mating analysis, will continue to be used. The new technologies described will serve as a powerful addition to these methods, and only gradually come to replace them.

References

ALFENAS, A.C., JENG, R.S. AND HUBBES, M. (1984). Isozyme and protein patterns of isolates of *Cryphonectria cubensis* differing in virulence. *Canadian Journal of Botany* **62**, 1756–1762.

ANDERSON, J.B., BAILEY, S.S. AND PUKKILA, P.J. (1989). Variation in ribosomal DNA among biological species of Armillaria, a genus of root-infecting fungi. *Evolution* **43**, 1652–1662.

ANDERSON, J.B. AND STASOVSKI, E. (1992). Molecular phylogeny of northern hemisphere species of *Armillaria. Mycologia* **84**, 505–516.

BANIK, M.T., VOLK, T.J. AND BURDSALL, H.H. (1996). *Armillaria* species of the Olympic Peninsula of Washington state, including confirmation of North American biological species XI. *Mycologia* **88**, 492–496.

BONDE, M.R., MICALES, J.A. AND PETERSON, G.L. (1993). The use of isozyme analysis for identification of plant-pathogenic fungi. *Plant Disease* **77**, 961–968.

BRAGALONI, M., ANSELMI, N. AND CELLERINO, G.P. (1997). Identification of European *Armillaria* species by analysis of isozyme profiles. *European Journal of Forest Pathology* **27**, 147–157.

BURDSALL, H.H. AND BANIK, M. (1990). Serological differentiation of three species of *Armillaria* and *Lentinula erodes* by enzyme-linked immunosorbent assay using immunized chickens as a source of antibodies. *Mycologia* **82**, 415–423.

CHILLALI, M., IDDER-IGHILI, H., GUILLAUMIN, J.J., MOHAMMED, C., LUNG, B. AND BOTTON, B. (1998a). Variation in the ITS and IGS regions of ribosomal DNA among the biological species of European *Armillaria. Mycological Research* **102**, 533–540.

CHILLALI, M., WIPF, D., GUILLAUMIN, J.J., MOHAMMED, C. AND BOTTON, B. (1998b). Delineation of the European *Armillaria* species based on the sequences of the internal transcribed spacer (ITS) of ribosomal DNA. *New Phytologist* **138**, 553–561.

DUCHESNE, L.C. AND ANDERSON, J.B. (1990). Location and direction of transcription of the 5S rRNA gene in *Armillaria. Mycological Research* **94**, 266–269.

FOSTER, L.M., KOZAK, K.R., LOFTUS, M.G., STEVENS, J.J. AND ROSS, I.K. (1993). The polymerase chain reaction and its application to filamentous fungi. *Mycological Research* **97**, 769–781.

FOX, R.T.V. AND HAHNE, K. (1989). Prospects for the rapid diagnosis of Armillaria by monoclonal antibody ELISA. In: *Proceedings of the 7th international conference on root and butt rots*. Ed. D.J. Morrison, pp 458–468. Victoria, B.C.: International Union of Forestry Research Organizations.

FRONTZ, T.M., DAVIS, D.D., BUNYARD, B.A. AND ROYSE, D.J. (1998). Identification of *Armillaria* species isolated from bigtooth aspen based on rDNA RFLP analysis. *Canadian Journal of Forest Research* **28**, 141–149.

GUILLAUMIN, J.J., ANDERSON, J.B. AND KORHONEN, K. (1991). Life cycle, interfertility and biological species. In: *Armillaria Root Disease*. Eds. C.G. Shaw and G.A. Kile, pp 10–20. *Forest Service Handbook No 691*. Washington, D.C.: USDA.

GUILLAUMIN, J.J., ANDERSON, J.B., LEGRAND, P., GHAHARI, S. AND BERTHELAY, S. (1996). A comparison of different methods for the identification of genets of *Armillaria* spp. *New Phytologist* **133**, 333–343.

HARRINGTON, T.C. AND WINGFIELD, B.D. (1995). A PCR-based identification method for species of *Armillaria*. *Mycologia* **87**, 280–288.

HENRION, B., LE TACON, F. AND MARTIN, F. (1992). Rapid identification of genetic variation of ectomycorrhizal fungi by amplication of ribosomal RNA genes. *New Phytologist* **122**, 289–298.

JAHNKE, E.D., BAHNWEG, G. AND WORRAL, J.J. (1987). Species delimitation in the *Armillaria mellea* complex by analysis of nuclear and mitochondrial DNAs. *Transactions of the British Mycological Society* **88**, 572–575.

JOHNSON, J.L. (1985). DNA reassociation and RNA hybridization of bacterial nucleic acids. In: *Methods in Microbiology*, vol. 18. Ed. G. Gottschalk, pp 33–74. London: Academic Press.

JOHNSON, J.L. (1989). Nucleic acids in bacterial classification. In: *Bergey's Manual of Systematic Bacteriology*, vol. 4. Eds. S.T. Williams, M.E. Sharpe and J.G. Holt, pp 2306–2309. Baltimore: The Williams & Wilkins Co.

KORHONEN, K. (1978). Interfertility and clonal size in the *Armillariella mellea* complex. *Karstenia* **18**, 31–42.

LIN, D., DUMAS, M.T. AND HUBBES, M. (1989). Isozyme and general protein patterns of *Armillaria* spp. collected from the boreal mixedwood forests of Ontario. *Canadian Journal of Botany* **67**, 1143–1147.

LUNG-ESCARMANT, B.C., MOHAMMED, C. AND DUNEZ, J. (1985). Nouvelles méthodes de détermination des Armillaires européens: Immunologie et électrophorése en gel de polyacrylamide. *European Journal of Forest Pathology* **67**, 1143–1147.

MILLER, O.K., JOHNSON, J.L., BURDSALL, H.H. AND FLYNN, T. (1994). Species delimitation in North American species of *Armillaria* as measured by DNA reassociation. *Mycological Research* **98**, 1005–1011.

MORRISON, D.J., THOMSON, A.J., CHU, D., PEET, F.G. AND SAHOTA, T.S. (1985). Isozyme patterns of *Armillaria* intersterility groups occurring in British Columbia. *Canadian Journal of Microbiology* **31**, 651–653.

MWENJE, E. AND RIDE, J.P. (1997). The use of pectic enzymes in the characterization of *Armillaria* isolates from Africa. *Plant Pathology* **46**, 341–354.

PÉREZ-SIERRA, A., WHITEHEAD, D. AND WHITEHEAD, M. (1999). Investigation of a PCR-based method for the routine identification of British *Armillaria* species. *Mycological Research* **103**, 1631–1636.

PRIESTLEY, R., MOHAMMED, C. AND DEWEY, F.M. (1994). The development of monoclonal antibody-based ELISA and dipstick assays for the detection and identification of *Armillaria* species in infected wood. In: *Modern Assays for Plant Pathogenic Fungi*. Eds. A. Schots, F.M. Dewey and R.P. Oliver, pp 149–156. Oxford: CAB International.

RISHBETH, J. (1986). Some characteristics of English *Armillaria* species in culture. *Transactions of the British Mycological Society* **85**, 213–218.

RIZZO, D.M. AND HARRINGTON, T.C. (1993). Delineation and biology of clones of *Armillaria ostoyae*, *A. gemina* and *A. calvescens*. *Mycologia* **85**, 164–174.

RIZZO, D.M., BLANCHETTE, R.A. AND MAY, G. (1995). Distribution of *Armillaria ostoyae* genets in a *Pinus resinosa – Pinus banksiana* forest. *Canadian Journal of Botany* **73**, 776–787.

SAMBROOK, J., FRITSCH, F.E. AND MANIATIS, T. (1989). *Molecular cloning*. Cold Spring Harbor, N.Y.: Cold Spring Harbor Laboratory Press.

SCHULZE, S., BAHNWEG, G., TESCHE, M. AND SANDERMANN, H. (1995). Identification of European *Armillaria* species by restriction-fragment-length polymorphisms of ribosomal DNA. *European Journal of Forest Pathology* **25**, 214–223.

SCHULZE, S., BAHNWEG, G., MÖLLER, E.M. AND SANDERMANN, H. (1997a). Identification of the genus *Armillaria* by specific amplification of an rDNA-ITS fragment and evaluation of genetic variation within *A. ostoyae* by rDNA-RFLP and RAPD analysis. *European Journal of Forest Pathology* **27**, 225–239.

SCHULZE, S., BAHNWEG, G., TESCHE, M. AND SANDERMANN, H. (1997b). Identification techniques for *Armillaria* spp. and *Heterobasidion annosum* root and butt rot diseases. *Journal of Plant Diseases and Protection* **104**, 433–451.

SCHULZE, S. AND BAHNWEG, G. (1998). Critical review of identification techniques for *Armillaria* spp. and *Heterobasidion annosum* root and butt rot diseases. *Journal of Phytopathology* **146**, 61–72.

SMITH, M.L. AND ANDERSON, J.B. (1989). Restriction fragment length polymorphisms in mitochondrial DNAs of *Armillaria*: identification of North American biological species. *Mycological Research* **93**, 247–256.

SMITH, M.L., DUCHESNE, L.C., BRUHN, J.N. AND ANDERSON, J.B. (1990). Mitochondrial genetics in a natural population of the plant pathogen *Armillaria*. *Genetics* **126**, 575–582.

SMITH, M.L., BRUHN, J.N. AND ANDERSON, J.B. (1992). The fungus *Armillaria bulbosa* is among the largest and oldest living organisms. *Nature* **356**, 428–431.

SMITH, M.L. AND ANDERSON, J.B. (1994). Mitochondrial DNAs of the fungus *Armillaria ostoyae*: restriction map and length variation. *Current Genetics* **25**, 545–553.

TERASHIMA, K., CHA, J.Y., YAJIMA, T., IGARASHI, T. AND MIURA, K. (1998). Phylogenetic analysis of Japanese *Armillaria* based on the intergenic spacer (IGS) sequences of their ribosomal DNA. *European Journal of Forest Pathology* **28**, 11–19.

VOLK, T.J., BURDSALL, H.H. AND BANIK, M.T. (1996). *Armillaria nabsnona*, a new species from western North America. *Mycologia* **88**, 484–491.

WAHLSTRÖM, K., KARLSON, J.O., HOLDENRIEDER, O. AND STENLID, J. (1991). Pectinolytic activity and isozymes in European *Armillaria* species. *Canadian Journal of Botany* **69**, 2732–2739.

WHITE, E.E., DUBETZ, C.P., CRUICKSHANK, M.G. AND MORRISON, D.G. (1998). DNA diagnostic for *Armillaria* species in British Columbia: within and between species variation in the IGS-1 and IGS-2 regions. *Mycologia* **90**, 125–131.

WHITE,T.J., BRUNS, T., LEE, S. AND TAYLOR, J. (1990). Amplification and direct sequencing of fungal ribosomal RNA genes for phylogenetics. In: *PCR Protocols: A Guide to Methods and Applications*. Eds. M.A. Innis, D.H. Gelfand, J.J. Sninsky and T.J. White, pp 315–322. San Diego: Academic Press.

WILLIAMS, J.G.K., KUBELIK, A.R., LIVAK, K.J., RAFALSKI, J.A. AND TINGEY, S.V. (1990). DNA polymorphisms amplified by arbitrary primers are useful as genetic markers. *Nucleic Acids Research* **18**, 6531–6535.

ZOLLFRANK, U., SAUTTER, C. AND HOCK, B. (1987). Fluorescence immunohistochemical detection of *Armillaria* and *Heterobasidion* in Norway spruce. *European Journal of Forest Pathology* **17**, 230–237.

SECTION 3
Pathology

6
Pathogenicity

ROLAND T.V. FOX

The University of Reading, School of Plant Sciences, Department of Horticulture and Landscape (Crop Protection), 2 Earley Gate, Reading, Berkshire RG6 6AU, U.K.

Synopsis

Many of the species of *Armillaria* that live in suitable ecological habitats throughout the temperate and tropical parts of the world are non-pathogens. Most are saprophytic on wood or other dead plant material, and a few are symbiotic with orchids. Others are primarily pathogens of a diverse range of woody and some herbaceous plant hosts, which they also decay and recycle. These pathogens are often collectively known as honey fungus. *Armillaria* genets may outlast many generations of trees in some forest soils. Despite this, some mature trees can continue to live with an infection for decades and some hosts can destroy mycelium by lysis. Although basidiospores are produced annually in great abundance, many species of *Armillaria*, such as *A. mellea*, mainly depend upon vegetative infection for their spread. Rhizomorphs penetrate through the bark of the roots, crown or lower trunk, then sprout sheets of hyphae that fan out underneath. This mycelium destroys the active phloem and xylem by separating off the bark from the underlying wood and the phytotoxic substances it secretes block the xylem vessels of the hosts, causing the foliage to change colour and wilt, followed by an abundant but final crop of fruits or cones. The weakened roots of trees often fail to cope during droughts or can be uprooted by storms. Although *Armillaria* are white-rot fungi, lignin remains largely unaffected until after the cellulose and hemicellulose have been destroyed along with the other carbohydrates. Suberised cell walls are also broken down, even though bark is generally resistant. Although it is doubtful if any woody plant is truly immune, some woody plants are more tolerant to disease development than others. It is not known which genes determine these physiological or biochemical characteristics. The source of the substrate on which the inoculum grew is also considered significant. Inoculum on hardwood is often regarded as more virulent than that on coniferous wood. Hardwood stumps provide a longer lived food base as they often regrow, whereas cut conifers rapidly rot and disappear. The resistance mechanisms of a host and the success of *Armillaria* are influenced by stress from a variety of sources. Until comparatively

Armillaria *Root Rot: Biology and Control of Honey Fungus*

recently, our concept of the taxonomy of *Armillaria* considered a miscellany of different pathogenic species for one rather 'variable species'. This mistaken concept of the 'single pathogen' *Armillaria mellea sensu lato* means that many conclusions were actually drawn from miscellaneous composites of quite dissimilar species. These different species, with a spectrum of widely differing virulences, had also interreacted with hosts of differing susceptibilities. The less virulent species appear to need wounds, which more virulent species do not. Roots are the major source of wood inoculum, though debris, as well as the stumps that result from felling, are frequently infected once the host resistance in a living tree is lost. Afterwards, its entire root system may become converted into inoculum. Many successful experimental infections have been obtained using small pieces of wood as the inoculum. Experiments on the survival of small units of inoculum have demonstrated the influence of soil moisture, temperature, and also antagonistic fungi such as *Trichoderma viride*. Longevity may be less important if a regular supply of new hosts is available for colonization. Repeated inspections of a rotted stump are necessary to estimate the age, extent and persistence of a source of active *Armillaria* inoculum. Regrowth from the living stumps of hardwood trees can greatly prolong the survival of the fungus. The colonization of stumps by airborne basidiospores does occur but relatively rarely gives rise to new infection centres. Where hosts are abundant, spread is frequently by root-to-root contact. Generally, spread between hosts often depends on the rhizomorphs that grow through the soil from infested trees or stumps. Rhizomorphs provide a supply route allowing the inoculum to extend, focus infection and ensure its persistence. Rhizomorph growth rate is correlated to inoculum size. Inoculum potential also depends on its proximity to the host and the prevailing environment. Inocula varying from fragments to entire stumps become linked by a rhizomorph network which survive as long as a sequence of sources of food bases continue to be accessible to different parts of the network. The direction of nutrient flow adapts to feed and maintain the entire web-like system. Fresh tips sprout from both cut ends of a rhizomorph. The rhizomorphs of *A. mellea* are produced in successive waves but are not as long-lived or persistent as those of *A. lutea* and *A. cepistipes*. *Armillaria* infection involves both mechanical force and enzymes including suberinase, but the rhizomorphs often entwine around the roots for some time until a tip establishes a route through the bark. Mycelium produced from the rhizomorphs encroach underneath into the wood, often radially along the plates of the parenchyma ray cells. The white rotted wood is often watersoaked and weakened by breakdown of lignin but when it dries it crumbles if compressed. The mycelium and network of rhizomorphs may extend a short way upwards to establish the rot in fresh host tissue further up the trunk. If the tree is blown over in a storm, the trunk can also rot as it lies on the soil. Large trees of some species often survive for many years, but can be killed if its vigour declines through waterlogging, shade, drought, damage due to defoliation, advanced age, other agents, or competition due to a high population density of plants. Wounding of the roots can enhance infection, allowing entry to some species of *Armillaria* that are unable to penetrate intact bark. Following colonization, phenols which suppress the growth of *Armillaria* accumulate in the inner bark. Other phenols stimulate, as do glucose and ethanol. Tissues brown during infection due to discoloration by phenol oxidases. Alkaloids produced by several plant families inhibit *Armillaria* while some other plant constituents, such as the fatty acid fraction of lipids and resin acids, stimulate. The genetic background of an isolate of *Armillaria* governs its virulence and its infection processes but the reaction between it and a host *also* depends on the species of host that is attacked, both interactions are modified by the prevailing environmental conditions. *Armillaria* mainly metabolizes glucose. The carbohydrate and nitrogen levels in the root tissues colonized by *Armillaria* are affected by drought and defoliation. The starch content in the root wood can be considerably diminished by defoliation, which also reduces sucrose levels. A badly drained site can create anaerobic conditions which, in turn, promote defoliation. This reduces transpiration, thus maintaining the wet soil conditions which encourages the production of ethanol by the anaerobes.

The pathogens

The *Armillaria* species commonly known as honey fungus in different parts of the world are notorious tree pathogens. In some forests, as much as 35% of the annual tree mortality is caused by species of *Armillaria* (James *et al.*, 1984). They are able to survive within forest soils for many decades, or even hundreds of years, ceaselessly causing butt rot and slowly killing generation after generation of trees (Smith *et al.*, 1992). These *Armillaria* species have gained this reputation as a diseased tree hardly ever recovers, yet the individual circumstances of its death may vary. Mature trees of some species can survive infection for many years, even decades, albeit dependent on roots that are being steadily attacked and gradually rotted by *Armillaria,* whereas its effects often appear to be devastatingly sudden on many others (Greig and Strouts, 1983; Greig *et al.*, 1991).

Phytotoxic substances seem to be involved as the branches above diseased roots are killed when the infected roots are left intact, but not if severed. Metabolites produced

Figure 6.1. Author examining roots of windthrown *Cedrus deodara* rotted by *Armillaria mellea,* on day after storm at Jealott's Hill, Berkshire.

Figure 6.2. Infected privet bush showing mycelial fans beneath the loose bark, and adventitious roots above the infection.

by *Armillaria* in liquid culture cause plant xylem vessels to block (Thornberry and Ray, 1953). Wilting was induced in tomato seedlings and peach twigs by a dark brown protein-like pigment which penetrated 15–20 mm into vascular tissues. No systemic effects were detected when electrical resistance was measured around actively expanding lesions, but trees killed by *Armillaria* show similar patterns of electrical resistance to those mechanically girdled. The physiological and biochemical consequences of stem girdling or root death require further study as similar foliage and crown symptoms can be induced by other fungal root diseases, winter injury to the roots or root collar, and root suffocation due to flooding.

After the mycelium has penetrated the bark of the roots, crown or lower trunk, it fans out, spreading around underneath, destroying the active phloem and xylem, separating off links to the underlying wood comparable to 'ring barking' carried out by orchardists. The effects are similar. The foliage wilts after the connections of the vascular system are disrupted, followed by an abundant but final crop of fruits or cones.

Once the roots have started to decay, symptoms are usually discernible above ground (Morrison and Pellow, 1998). Even if these are not noticeable, the tree may ultimately succumb to either droughts or storms after the diseased roots cease to function adequately. Many such trees are uprooted, as high winds can easily snap their weakened roots, particularly during storms (Gibbs and Greig, 1990; Macmillan-Browse *et al.*, 1990) and with it the threat of serious damage to people and property, particularly in garden and amenity areas (Fox, 1990).

Slayers or just decayers

The genus *Armillaria* contains many species that are non-pathogenic saprophytic fungi that grow on wood as well as the more conspicuous root pathogens. Like those species that survive predominately or entirely as saprophytes, the pathogens also serve as important decayers and recyclers of dead wood (Cha *et al.*, 1996). The latter are, nevertheless, among the most familiar destroyers of deciduous and coniferous trees and shrubs, as well as many herbaceous plants (Robinson-Bax, 1999; Hughes *et al.*, 1996; Tirro, 1989).

Many species with various modes of life are found worldwide in natural forests, plantations, orchards and amenity plantings. The pathogenic species vary in virulence, some are primary pathogens, others mainly stress-induced secondary invaders. Some species are obligate saprophytes or even mycorrhizal.

Armillaria are classed as white-rot fungi, but until the cellulose and hemicellulose have been destroyed along with the other carbohydrates, lignin is hardly affected for the first 6–12 months of the colonization of wood (Scurti, 1956), then it too is destroyed (Campbell, 1932). It has been reported that suberised cell walls are also broken down (Swift, 1965), even though it is a common observation that bark invariably resists colonization more persistently than the woody xylem that it covers.

HOST RANGE

Various pathogenic species of the genus *Armillaria* can be found in suitable ecological habitats throughout the temperate and tropical parts of the world causing root, collar and butt rots on over 650 hosts (Raabe, 1962). The different *Armillaria* species demonstrate some host specialization (Siepmann and Leibiger, 1989). Common hosts include broad-leaved and coniferous trees and shrubs grown in forests, orchards, parks and gardens (Anderson and Ullrich, 1979; Kile and Watling, 1983; Kim and Ko, 1993; Thanassoulopoulos and Artopeadis, 1991; Blodgett and Worrall, 1992a), as well as vines and some herbaceous hosts (Moore, 1959; Robinson-Bax, 1999).

Robinson-Bax (1999) tested 23 common herbaceous plants to find if they were susceptible hosts, resistant or tolerant 'carriers' of *Armillaria mellea*. Results indicated that only *Polygonum rude* (Polygonaceae) is virtually resistant with the other species showing evidence of widespread infection and colonization. Sea beet *Beta vulgaris* ssp. *maritima* (Chenopodiaceae) is highly susceptible. The other species he infected include the first reports of *Armillaria* root rot on *Alchemilla mollis, Arundinaria pumila, Epimedium, Saxifraga* × *urbium, Lamium, Physalis alkekengi, Cimifuga, Thymus, Seseli osseum, Geranium albanum, Oenothera, Hosta, Vinetoxicum nigrum, Strobilanthes, Valerianella rimosa, Succisella petteri, Pilosella laticeps, Phlox paniculata, Sedum* 'Autumn Joy' and *Pelargonium* 'Multi'. These represent the following plant families – Rosaceae, Poaceae, Berberidaceae, Saxifragaceae, Lamiaceae, Solanaceae, Ranuculaceae, Apiaceae, Onagraceae, Agavaceae, Asclepiadaceae, Acanthaceae, Valerianaceae, Dipsacaceae, Asteraceae, Polemoniaceae, Crassulaceae and Geraniaceae.

Although there are trees and shrubs on which an attack of honey fungus has not yet been recorded in the literature, and some others that are only very rarely attacked, it is unlikely that any woody plant is truly immune. Neither is it known which characteristics determine resistance or tolerance to disease development. Among the plant families considered most susceptible to *Armillaria (*Greig and Strouts, 1983) are the Betulaceae, Rosaceae, Salicaeae, Saxifragaceae, Vitaceae and Tiliaceae. Greig and Strouts (1983) mention a number of woody plants in several other families, not listed above, which consist of some genera that are highly susceptible, together with others that are more tolerant.

The botanical origin of the substrate on which the inoculum grew, as well as that of the host being infected, is also considered important. Inoculum of *Armillaria* growing on hardwood is often thought to be more virulent than that grown on coniferous wood (Redfern and Filip, 1991). In order to test if infection by *Armillaria* is influenced by the type of substrate providing the inoculum, experimental inocula have been prepared from the stems and roots of several woody species as a food base for *Armillaria*. Although coniferous wood (Patton and Riker, 1959; Ono, 1970; Redfern, 1975, 1978; Singh, 1980) initially appears as effective as many hardwoods (Wilbur *et al.*, 1972; Shaw, 1977; Shaw *et al.*, 1981; Morrison, 1982), *A. ostoyae, A. lutea* and *A. mellea* produce less rhizomorphs on pine than plane. Yet the opposite is the case for *A. cepistipes*, and Mwangi *et al.* (1990) discovered chemical factors in *Pinus strobus* inhibitory to *Armillaria ostoyae.* However, Rykowski (1984) also found that hardwood yields more rhizomorphs and encourages more infection by the virulent *A. ostoyae* and *A. mellea* than coniferous wood. Unlike many hardwood stumps, those of most conifers rarely regrow but rapidly rot and disappear, making them a less readily conserved food base (Morrison, 1972; Rishbeth, 1972).

Both Australasian species, *A. limonea* and *A. novae-zelandiae*, are more devastating to plantations of *Pinus radiata*, an introduction from the Californian coast, than to any of the indigenous New Zealand forest trees (MacKenzie and Shaw, 1977; van der Pas, 1981; Benjamin and Newhook, 1984b; MacKenzie, 1987; MacKenzie and Self, 1988; Chou, 1991). Likewise, *A. luteobubalina* shows no predilection for any particular dry sclerophyll eucalypt forest tree (Kellas *et al.*, 1987; Pearce *et al.*, 1986; Shearer and Tippett, 1988).

Host reaction

The physiological origins of symptom development leading to host mortality are largely the responses of the host to the physical disruption to its vascular system by *Armillaria* and its metabolic toxins. The position of the infection on the root system of the host and its activity largely determine the character and pace of symptom development (Morrison and Pellow, 1998). Usually, the first symptoms observed are a decline in shoot growth which leads to a dieback in the foliage of the crown. This stress can, in turn, provoke the production of an abundance of fruits or cones. Meanwhile, symptoms start to appear at the base of the trunk or just above the root collar, such as exudation, cankering and bark cracking, followed by the death of the host.

Symptoms often develop most rapidly on hosts that are growing vigorously, but they may not be noticeable when disease development is slow. *Armillaria* induces changes in the foliage which are consistent with the gradual physical destruction of the host's vascular tissue. Hence, as the growth of the shoots slows down, leaves are lost and change colour.

HOST-INDUCED LYSIS

The chitin and beta-1,3-glucan contained in the hyphal walls of *Armillaria* are vulnerable to lysis by chitinase and beta-1,3-glucanase (Ballesta and Alexander, 1972; Bouveng *et al.*, 1967; Wargo, 1975). However, the stress resulting from defoliation also lowers the activities of these enzymes which are found in the bark and sap of several oaks and sugar maples (Wargo, 1975, 1976). The host-mediated lysis of *Armillaria* hyphae *in vivo* that has also been described in association with certain orchids, such as *Gastrodia elata* (Hamada, 1940; Kusano, 1911; Wang and Shao, 1990; Xu *et al.*, 1990), has been implicated in the digestion of fungi by them (Burges, 1939).

There is a fine equilibrium between wall synthesis and wall lysis which enables growth of the tip of a hypha, but this can burst when the balance shifts towards the lytic stage (Bartnicki-Garcia and Lippman, 1972). Disruption of the growth of hyphae does not imply their total dissolution. The extrahyphal enzymes in host cells which disintegrate hyphal wall components may distort the stability of wall formation, disrupt hyphal-tip growth, contribute to defence against invading fungal pathogens, and hence inhibit their progress (Schlumbaum *et al.*, 1986). *Armillaria* can defend itself against such lysis, as the fungal phenol oxidase enzymes, principally tyrosinase, are associated with their synthesis of melanin (Mayer and Harel, 1979). *Armillaria* produces melanin-like pigments in rhizomorphs and hyphae (Chet and Hüttermann, 1977; Smith and Griffin, 1971) which may reinforce (Bell and Wheeler, 1986) and protect them from lytic enzyme dissolution (Bloomfield and Alexander, 1967).

Wargo (1971) concluded that host-pathogen interactions ultimately depend on the complex individual relationships between fungal species, host species and the environment in which they interact (Blodgett and Worrall, 1992b), including the disturbances induced by stress. However, much of the existing information on the physiological and chemical interactions of *Armillaria* species and their hosts, and the particular circumstances, are still fragmentary. Stress from a variety of sources can influence both the resistance mechanisms of a host and the success of *Armillaria* in penetrating, colonizing and killing it.

Studies using clonal host material and known species and genotypes of *Armillaria* and stressed and non-stressed systems must be conducted to elucidate the resistant and susceptible reactions. Some aspects of the stimulation and control of the penetration and colonization of a substrate have been characterized. Wargo (1971) suggested that our ignorance of the correct taxonomy of *Armillaria* until comparatively recently, means that our knowledge of the relevant physiological interactions is based on a composite of different pathogenic species on hosts with differing susceptibilities (Garraway *et al.*, 1991).

Since some species were not at first recognized to be primarily pathogens and others saprophytes, the true role of an infection court provided by damaged or weakened roots was not clear. Although physically wounded roots can be utilized during infection, their actual importance as regular infection courts has not been entirely substantiated. Depending on the circumstances, wounds appear to be either required (Buckland, 1953) or preferred (Dimitri, 1969) to healthy undamaged roots. Wounding can increase the efficacy and aggressiveness of *A. tabescens* on peach roots (Weaver, 1974) and *A. ostoyae* on balsam fir (Whitney *et al.*, 1989b). Wounds are vital infection courts for less virulent species such as *A. lutea* (Popoola and Fox, 1996), yet Redfern and Filip (1991) argued that in both natural disease outbreaks and inoculation experiments, wounds are superfluous for infection by more virulent species. This was subsequently confirmed experimentally by Popoola and Fox (1996), who compared the requirements of *A. mellea* with those of *A. lutea*. Possibly, since they need to encounter wounds, the rhizomorphs of the less virulent species are more liable to become attached along the surface of the host root than those of the more virulent species (Gregory, 1985).

Several factors, often in combination, influence the success of infection through their effect on inoculum potential. These include the nature, vigour and virulence of the inoculum, particularly the invading mycelium and rhizomorphs, which provide the means of infection and spread in most *Armillaria* species, as well as variable factors such as the effects of the environment on the fungus, the area of its surface in contact with the host, and the quality of the substrate, usually wood, that the latter provides.

SOURCES OF INOCULUM

Despite the huge numbers of basidiospores that are produced, their individual inoculum potential is considered too minuscule to be capable of infecting live roots directly (Rishbeth, 1970). Nonetheless, the cut surfaces of stumps can be colonized by basidiospores (Rishbeth, 1970, 1978, 1988; Stanosz and Patton, 1990a; Pearce and Malajczuk, 1990b)), though the results may often be meagre (Swift, 1972) or negative (Kile, 1983; Leach, 1939; Podger *et al.*, 1978). Although not proven experimentally, rhizomorphs might be formed if basidiospores could colonize dead organic matter (Hartig, 1874). The identification of genets has produced some indirect evidence for some previous successful infection by basidiospores (Hood and Sandberg, 1987; Horner, 1988; Kile, 1983; Ullrich and Anderson, 1978). Successful infection following inoculation by basidiospores may be relatively uncommon (Horner, 1988), yet it is a vital source of genetic diversity and long-range spread, for example onto trees growing on previously unforested land.

By contrast, the importance of rhizomorphs as inoculum can be demonstrated readily, providing their inoculum potential is high enough (Stanosz and Patton, 1990b). Healthy seedlings can even be infected experimentally by a piece of rhizomorph without a substrate if it is sufficiently large to sprout new growing tips (Redfern, 1973; Rykowski, 1984), but entry through wounds may be necessary in some cases (Holdenreider, 1987).

Tree roots are the major source of wood inoculum, though debris (MacKenzie and Shaw, 1977; Rykowski and Sierota, 1988) as well as the stumps that result from felling, are frequently infected once the host resistance in a living tree is lost. If the roots and stumps of living trees become colonized, often the tree is killed and its entire root system converted into inoculum. A stump may be colonized by expansion from a lesion where it was kept in check while alive (Kile, 1980; Leach, 1939). Stumps can

Figure 6.3. *Armillaria* rhizomorph attached to decomposing leaf whitened by loss of lignin.

also be invaded by *Armillaria* which has survived on the roots or arrived as rhizomorphs by vegetative spread from an external source of inoculum. Fragments of infected roots may also become washed to new sites by water (Hewitt, 1936).

Although naturally infected roots have been used (Leach, 1937), many successful experimental infections have been obtained using small pieces of wood as the inoculum. Although some other substrates are effective (Bliss, 1941; Plakidas, 1941; Guyot, 1927), most inocula are prepared by culturing the fungus on woody stem or root segments (Patton and Riker, 1959; Redfern, 1975; Shaw, 1977; Thomas, 1934) as these are reliable, durable and effective.

LONGEVITY OF INOCULUM

Most estimates suggest *Armillaria* species can survive for several decades in the stumps of either broad-leaved and coniferous trees (Pronos and Patton, 1978; Rishbeth, 1972, 1985a; Swift, 1972; Shaw, 1975; Roth *et al.*, 1980; Kile, 1980, 1981). Variations in the longevity of different species of hosts and size of their stumps, together with the substantial variations in ambient environment, including soil type (Blenis *et al.*, 1989), preclude useful comparisons of the persistence of different *Armillaria* species based solely on limited observations in the field. Preliminary experiments on the survival of small units of inoculum have demonstrated the influence of soil moisture (Pearce and Malajczuk, 1990a), temperature (Bliss, 1946), and also antagonistic fungi such as *Trichoderma viride* (Garrett, 1957). The size of the inoculum may be less important than the species of wood (Singh, 1980; Patton and Riker, 1959).

Genets of *Armillaria ostoyae* have continued to survive for several centuries on a site by colonizing a succession of substrates (Shaw and Roth, 1976). In this case, the pathogen can continue to expand the perimeter of the disease focus by infecting fresh hosts or the trees that regenerate within it. In the latter case, this may happen after the inoculum had become quiescent within the stumps of the original hosts, enabling a resurgence of disease. The extensive rhizomorph systems of less virulent species may also be an important factor, ensuring their persistence as they probably have less chance of acquiring fresh substrates (Jennings, 1990). Pathogens may benefit, particularly among susceptible species, by surviving in the gaps caused by disease until new hosts become available. Longevity may be less important if a regular supply of stumps is available for colonization, for example as a result of regular cutting and thinning in intensively managed forests.

It is not difficult to underestimate the age, extent and persistence of a source of active *Armillaria* inoculum based solely on a single inspection of a rotted stump. Regrowth from the living stumps of hardwood trees can greatly prolong the survival of the fungus.

INFECTION

Little is known for certain about the primordial mode of entry of pathogenic *Armillaria* species into uninfested stands of trees in natural ecosystems. However, it is evident that in clear felled forests, the colonization of stumps by airborne basidiospores does occur. When mycelium of *Armillaria* was found on freshly cut wood exposed to air, Molin and Rennerfelt (1959) considered it to result from the deposition of airborne spores. However, although earlier attempts at artificial inoculation failed (Leach, 1939; Gibson and Goodchild, 1961), Rishbeth (1964) obtained a few successful inoculations of both pine and hardwood stumps. He concluded that, whereas colonization by spores might be relatively rare, nonetheless it was likely to be vital in giving rise to new infection centres, as well as establishing genets with novel genetic potential.

Although where hosts are abundant, spread is frequently by contact between healthy and infested roots, spread between hosts often depends on the rhizomorphs that grow through the soil from infested trees or stumps (Molin and Rennefelt, 1959). Rhizomorphs benefit many species of *Armillaria* since they offer a supply route for extending the spread of inoculum, focus infection, and ensure their persistence. The major determinants of inoculum potential in those *Armillaria* species that produce rhizomorphs are size of the inoculum source, proximity of it to the host, and the influence of the environment (Garrett, 1970). Under environmental circumstances which prevent or restrict rhizomorph formation, or in those few species that do not possess rhizomorphs or where they are not functional, infection is restricted to those areas on a host root in direct contact with the inoculum.

Although experiments with species that lack rhizomorphs were not included, Garrett (1956, 1958) and Rykowski (1984) used model systems incorporating small woody inocula and potato tubers (Gregory, 1984) in soil to find that rhizomorph growth rate is correlated to inoculum size. Whereas the extent of infection in potato tubers increased with inoculum size, it decreased with the distance between inoculum and tuber. Rhizomorph growth rate also started to decline due to both depletion of nutrients in the inoculum source and also competition for nutrients between the main apex and branches of the rhizomorph system. The number, length and weight of rhizomorph systems produced from inocula, and the number of apices on those systems, may be used to calculate the inoculum potential infection threat in different soils (Rykowski, 1984). In practice, however, infection is not simply associated with specific point sources. Inocula are seldom discrete and often vary in size from fragments of a single root to entire stumps linked by a rhizomorph network. As long as a sequence of sources of food bases continue to be accessible to different parts of the network, it should survive almost indefinitely, providing the direction of nutrient flow carries on switching around to feed and maintain the entire system (Redfern, 1973). This web-like behaviour seems difficult to reconcile with those experiments which demonstrate that translocation only occurs towards growing tips (Anderson and Ullrich, 1982; Schütte, 1956) or where nutrients absorbed by growing tips were found not to be translocated towards the food base (Morrison, 1975). However, no experiments have yet attempted to represent either the functioning of an intact network or the potential of individual rhizomorphs in a network to periodically alternate the directions of translocation during fruiting or the depletion of food reserves (Anderson and Ullrich, 1982). Changing the direction of translocation can be demonstrated by either severing a rhizomorph that still forms part of a network, fresh rhizomorph tips will sprout from both cut ends, or removing a segment of a thick rhizomorph as substitute rhizomorph tips will also sprout from it to replace both cut ends (Hintikka, 1974; Redfern, 1973; Rykowski, 1984). There is little support for Nechleba (1915), who proposed that the rhizomorphs instinctively 'find their way' towards living roots to colonize them.

Although they often establish complex inocula incorporating several active stumps, the rhizomorphs of *A. mellea* are not as long-lived or persistent as those of *A. lutea* and *A. cepistipes*, and are produced in successive waves (Guillaumin *et al.*, 1989).

Rhizomorphs entwine around the roots until a suitable crevice or other entrance through the bark is located. After the tip of a rhizomorph has penetrated through the bark on the surface of the root, it can begin to encroach underneath. Mycelium produced from the tips of the rhizomorph beneath the bark invades into the wood radially along the plates of the parenchyma ray cells. If this mycelium is able to extend throughout the root system, the death of the tree is eventually assured. The wood invaded by the fungus mycelium is weakened by a white rot caused by the widespread breakdown of lignin, often the rotten tissues of the roots are watersoaked but dry out to become brittle if exposed and dried in the air. At early stages, the striated fans of cottonwool-like mycelium are most obvious below the bark of the roots or collar, but occasionally the mycelium and network of rhizomorphs extend a short way upwards to establish the rot in fresh host tissue further up the trunk. Colonization of the trunk can also become extensive if the tree trunk lies on the soil, for example after it has been blown over in a storm.

Although large trees of some species can be killed, most become more resistant with age and may even survive for many years. A wide range of physical, chemical, and physiological factors also influence whether or not a particular tree is infected by *Armillaria*. These factors include waterlogging, shading or insufficient light, drought, defoliation, advanced age, declining vigour of the host plant, damage due to other agents (pollution, nutrient deficiency, insects, fungi, etc.), high population density of plants, high amounts of glucose and fructose (Wilbur *et al.*, 1972), tannin (Cheo, 1982). Low organic matter (Singh, 1983), low soil pH (Gramss, 1983), phenolic hydroxystilbenes in pines, chitinase and β-1,3-glucanase found in the stem and root of some *Quercus* species can provide some degree of resistance to plants.

Armillaria infection is both mechanical and enzymatic. Both bark and secondary periderm succumb to penetration by *Armillaria* hyphae (Rykowski, 1975; Thomas, 1934) which grow faster and invade around the periderm (Rykowski, 1975) or penetrate directly through it, probably by enzymatic activity (Arthaud *et al.*, 1980; Rykowski, 1975; Thomas, 1934; Swift, 1965; Zimmermann and Seemüller, 1984) to degrade suberin. When concentrated enzyme prepared from culture supernatants incubated with suberin for 16 hr reduced its dry weight by up to 1%, analysis by gas chromatography indicated that most of the components of the aliphatic monomers had been lost (Kolattukudy *et al.*, 1981).

As penetration of the outer bark appears similar in most hosts, only the subsequent colonization of the inner bark and cambial zone tissues may differentiate those plants that are highly susceptible from those that are more resistant (Thomas, 1934). However, it is not known whether all species of *Armillaria* are capable of successfully penetrating the outer bark as most historical studies used a single isolate of species that was unknown, although in some, names can subsequently be assigned with some confidence. Wounding of the roots can enhance infection, allowing entry to some species of *Armillaria* that are unable to penetrate intact bark (Popoola and Fox, 1996).

Although rhizomorphs of some sort are formed in soil by most *Armillaria* species, and initially they were considered essential (van Vloten, 1936), infection at root contacts has either been observed (Kawada *et al.*, 1962; Prihoda, 1957; Zeller, 1926) or inferred (Marsh, 1952; Molin and Rennerfelt, 1959) for many of these species.

Some of the more recently described species reveal substantial differences in rhizomorph production (Gregory, 1985; Redfern, 1975), which may indicate major ecological consequences. To an avirulent or poorly virulent pathogen, the main advantage of widely distributing its inoculum is that at least part of it will be in an advantageous place when a potential food source eventually becomes accessible. A saprophyte needs an extensive rhizomorph network system capable of enmeshing roots to be ready to colonize stumps or substrates on living trees with declining resistance. If a species is sufficiently virulent, this is not necessary as it is capable of spreading among its susceptible hosts by root contact. Hence, a pathogen may be more affected by the distribution of tree roots than by factors influencing rhizomorphs.

Armillaria infiltrates and kills the living cambial tissues after their resistance reactions have been destroyed. *Armillaria* then supports itself by its ability to decay woody tissues (Cha *et al.*, 1996), building up inoculum potential ready to infect and kill when the tree subsequently becomes weakened by stress.

Although physical barriers cannot prevent infection by *Armillaria*, chemical resistance by preformed constituents in the bark or as mobilized constituents in response to penetration can delay the penetration and infection of root tissue.

PHENOLS

After fungal colonization, phenol accumulation occurs predominately in the inner bark regions (Ostrofsky *et al.*, 1984; Wargo, 1988). Some phenols present in both coniferous and deciduous hosts suppress the growth of *Armillaria in vitro*. When isolates of *A. mellea*, *A. lutea*, *A. ostoyae* and *A. sinapina* were challenged with hydrolyzable tannin (tannic acid, gallotannin) and gallic acid, they were both stimulated and inhibited depending on the phenol, the concentration of glucose, and the presence or absence of ethanol in the growth media (Wargo, 1980). Generally, compared to the growth of the untreated control, gallic acid inhibited, whereas hydrolyzable tannin stimulated, depending whether the phenolics were oxidized or not (this could be determined by assessing the browning of the medium). Growth was stimulated greatly if oxidation occurred readily, but was inhibited if the isolate failed to fully oxidize the phenol. Oxidation could be initiated or accelerated by the addition of glucose and ethanol. Compared to isolates of *A. ostoyae*, those of *A. lutea* oxidized gallic acid and grew better in its presence with or without ethanol.

Although certain phenols are inhibitory, other phenols may stimulate the production of rhizomorphs and mycelium. This has been reviewed by Garraway *et al.* (1991), who also contend that Armillatox, a proprietary phenolic fungicide based on a formulation of cresylic acids, is ineffective as a control agent, even though it inhibits the formation rhizomorphs from blocks of wood (Rahman, 1978). Differences in the growth of isolates of *Armillaria* on culture media amended with various phenolics have been analyzed to distinguish species (Rishbeth, 1982, 1986). Growth patterns depend on whether tannic acid or its hydrolysed form, gallic acid, was used as the phenol amendment (Shaw, 1985), implying differences in the permeability of the cell membranes of the fungi. Attempts to correlate virulence with differences in growth on media amended with gallic acid, with and without ethanol (Wargo, 1980), face the same obstacle.

The browning of tissues during infection, due to discoloration by phenol oxidases, is common during infection and colonization by *Armillaria* spp. (Rykowski, 1975; Thomas, 1934; Wargo, 1977, 1984) involves both fungal and host polyphenol oxidases, and their actions are not usually separated. Decay of wood by *Armillaria* is slightly different to nearly all other white-rot fungi as lignin degradation in laboratory tests was more restricted than cellulose degradation in tests carried out by Campbell (1932). However, when Scurti (1956) grew *Armillaria in vitro* on pure cellulose and pure lignin, lignin was degraded, not cellulose, possibly reflecting the metabolic differences between the different species or isolates of *Armillaria*.

Other constituents of host plants

Although Greathouse and Rigler (1940) found that alkaloids from several plant families inhibited the growth of *Armillaria in vitro*, some other plant constituents stimulate, such as the fatty acid fraction of lipids from the roots of various conifers and peach (Moody and Weinhold, 1972a,b). Pine resin is readily metabolized (Draczynska-Lusiak and Siewinski, 1989) and resin acids from ponderosa pine can be twice as stimulatory to the growth of rhizomorphs as the fatty acids from the same amount of root tissue (Shaw, 1975). Resin acids have been added to media for cultural pairing tests (Shaw and Roth, 1976). Breakdown products of the resin acids appear to be the stimulatory factors, because abietic acid only raised the production of rhizomorphs after it was sterilized by autoclaving, not filtration.

GENETIC CONTROL OF VIRULENCE

The genetics of the particular isolate of *Armillaria*, albeit modified by the environment, must dictate the metabolic control of pathogen virulence, the physiology of the infection processes leading to the development of disease within the host tree or resistance and the host-pathogen interreactions (Daly, 1976). Garraway *et al.* (1991) state that the reaction between a host and *Armillaria* depends on the species and perhaps genotype of *Armillaria* that is attacking, as well as the species of host that is attacked and the environmental conditions under which host and fungus are growing. Unfortunately, the genetic backgrounds of the isolates of *Armillaria* in most previous reports of interactions with hosts needs to be re-evaluated as their authors were unaware that the pathogen was not one species but several (Davidson and Rishbeth, 1988; Rishbeth, 1982, 1985b).

STRESSFUL ENVIRONMENTAL CONDITIONS

In a stressful environment, the genetics of the host and the pathogen determine whether the host develops a susceptible or resistant response to a fungal pathogen. Several sources of stress can upset the stability of the interaction between the host and its root disease pathogens. Some stresses clearly influence the progress of the disease and some may interact (Entry *et al.*, 1991).

Colonization of root tissues in stressed trees succeeds due to the ability of *Armillaria* to oxidize phenols and the inability of the tree to prevent it. Hence, in a weakened, deciduous tree, living tissues near an infection are often discoloured before they are colonized, whereas in an otherwise healthy tree, surrounding tissues are rarely 'browned' as the pathogen is initially confined to wounded and necrotic tissue. This discoloration ahead of actual colonization is probably due to the extracellular secretion of laccase and peroxidase (Wargo, 1983, 1984), with results analogous to the necrosis seen in hypersensitivity reactions (Goodman *et al.*, 1986). The highly reductive state in contiguous healthy tissues appears to inhibit and contain the enzymes induced by *Armillaria* that cause necrosis, unlike the stressed tissues which cannot, so necrosis begins and spreads.

Although drought often increases the incidence and severity of *Armillaria* root disease (Przezborski, 1987), it is not clear how either drought or waterlogging affect the pathogen either in the soil as rhizomorphs or in wood as mycelium. Neither is it clear how turgor pressure in the rhizomorph influences its penetration into the root bark or how extremes of moisture influence this relationship.

The carbohydrate and nitrogen levels in the root tissues colonized by *Armillaria* are affected by drought and defoliation (Gregory and Wargo, 1986; Parker, 1979; Parker and Houston, 1971; Parker and Patton, 1975; Wargo, 1972; Wargo *et al.*, 1972; Twery *et al.*, 1990). The starch content in the root wood can be considerably diminished by defoliation, which also reduces sucrose levels in the bark and cambium of sugar maple roots (Wargo, 1972, 1981b).

In spring, when carbohydrates are mobilized for growth, reducing-sugar concentrations in defoliated trees can be 3–4 times higher than the normal spring high in non-defoliated trees, and can sometimes reach 4–5 times higher (Wargo, 1971). This is crucial to *Armillaria*, which mainly metabolizes glucose (Wargo, 1972; Garraway, 1975; Wargo, 1981a). Growth on glucose or its polymers, such as maltose and starch, can be 1.5–3 times greater than on other carbon sources (Wargo, 1981a). Defoliation also increases amino acids, both individually and in total (Weinhold and Garraway, 1966). Ethanol stimulates rhizomorph production and growth of *Armillaria* (Weinhold,

1963; Weinhold and Garraway, 1966). A badly drained site can create anaerobic conditions, which in turn promote defoliation because it reduces transpiration, thus perpetuating the wet soil conditions encouraging the production of significant amounts of ethanol by the anaerobes (Coutts and Armstrong, 1976).

Stress-induced chemicals in roots may also influence the oxidation of phenols, and hence the behaviour of *Armillaria*. For example, adding more glucose to the medium reduces or reverses the growth of *Armillaria* inhibited by gallic acid (Wargo, 1980). Glucose and nitrogen added to bark extracts also stimulated growth by aiding the oxidation of phenols as carbon sources or maybe growth regulators (Wargo, 1983). The former is more likely since the growth of *Armillaria* significantly declined if oxidation of the phenols is inhibited by adding a reducing agent (Wargo and Harrington, 1991).

Even though *A. mellea* is considered primary pathogen on most of its hosts (Morrison *et al.*, 1989), Wargo (1980) has argued that host plants often require some stress to predispose them to infection. On the other hand, waterlogging inhibits rhizomorph formation (Guillaumin and Leprince, 1979), probably due to the low oxygen and high carbon dioxide levels in the soil (Redfern, 1973; Hintikka, 1974; Morrison, 1976; Rishbeth, 1978; Redfern and Filip, 1991).

CYTOLOGY AND GENETICS

As well as aiding species identification and explaining the caryological cycles, studies on the cytology and genetics of *Armillaria* have demonstrated the value of the biological species concept for the genus (Guillaumin *et al.*, 1991).

Four mating-factor combinations occur at approximately equal frequencies when single-spore isolates from one basidiome are paired with each other. Since there are many different alleles in the population and a particular pairing of nonsiblings from different sites is not likely to be identical, most pairings within a large population are compatible. Like other bifactorially heterothallic basidiomycetes, there are probably several dozen different mating-factor alleles in the species of *Armillaria*. All temperate species of *Armillaria* investigated (Guillaumin, 1986; Guillaumin *et al.*, 1983; Kile, 1983; Kile and Watling, 1988; Korhonen, 1978; Ullrich and Anderson, 1978) possess the same bifactorial sexual incompatibility system apart from *Armillaria ectypa* (Korhonen, unpubl.; Guillaumin, unpubl. in Shaw and Kile, 1991).

The Buller phenomenon is a mating between a monokaryon and a dikaryon which donates compatible haploid nuclei to the monokaryon which is dikaryotized (Raper, 1966). In *Armillaria*, this can be observed when a fluffy haploid mycelium is paired with a crustose dikaryotic isolate of the same species, the morphology of the former gradually changes to crustose, indicating dikaryotization (Anderson and Ullrich, 1982; Korhonen, 1978, 1983).

Interfertility tests have become a common method for routine identification of species and for the differentiation of unknown isolates into groups (Korhonen, 1978; Anderson and Ullrich, 1979). In these mating tests, a haploid tester strain (monospore isolates) is used that represents each of the species to which the isolate might belong. After the unknown isolate has been paired with all the tester strains, the mating reactions are assessed by the appearance of the mycelium. Although the unmated haploid cultures are typically fluffier than the crustose dikaryotic cultures, considerable variation may occur in colony morphology depending on the species, isolate, and culture conditions (Guillaumin *et al.*, 1991).

In Europe, the results of mating tests have shown that the biological species and the morphological species match perfectly, many have also been confirmed by their physiological, morphogenetic and biochemical characteristics. Elsewhere, in many cases this verification has yet to be confirmed.

INTRASPECIES VARIATION

Like most other wild organisms, the diversity within individual species of *Armillaria* has been studied much less than the variation between species, which has proved of such value in establishing the present species concepts. Some intraspecific variation in the characteristics of the morphology of the basidiomes, the patterns of rhizomorph branching, and levels of virulence is readily observed. There are also individual differences in physiology and biochemistry. Polymorphisms in isoenzyme profiles (Lin *et al.*, 1989; Morrison *et al.*, 1985) and restriction fragment patterns in nuclear (Anderson *et al.*, 1987, 1989; Anderson and Smith, 1989) and mitochondrial DNA (Jahnke *et al.*, 1987; Anderson and Smith, 1989) confirm abundant subspecific diversity.

References

ANDERSON, J.B. AND ULLRICH, R.C. (1979). Biological species of *Armillaria* in North America. *Mycologia* **71**, 402–414.

ANDERSON, J.B., ULLRICH, R.C. AND ROTH, L.F. [and others] (1979). Genetic identification of clones of *Armillaria mellea* in coniferous forests in Washington. *Phytopathology* **69**, 1109–1111.

ANDERSON, J.B. AND ULLRICH, R.C. (1982). Translocation in rhizomorphs of *Armillaria mellea*. *Experimental Mycology* **6**, 31–40.

ANDERSON, J.B., PETSCHE, D.M. AND SMITH, M.L. (1987). Restriction fragment polymorphisms in biological species of *Armillaria mellea*. *Mycologia* **79**, 69–76.

ANDERSON, J.B. AND SMITH, M.L. (1989). Variation in ribosomal and mitochondrial DNAs in *Armillaria* species. Sex and evolution in *Armillaria*. In: *Proceedings of the 7th international conference on root and butt rots*, August 9–16, 1988, Vernon and Victoria, B.C.. Ed. D.J. Morrison, pp 60–71. Victoria, B.C.: International Union for Forestry Research Organizations.

ANDERSON, J.B., BAILEY, S.S. AND PUKKILA, P. (1989). Variation in ribosomal DNA among biological species of *Armillaria*, a genus of root-infecting fungi. *Evolution* **43**, 1652–1662.

ARTHAUD, J., DAVID, A. AND FAYE, M. [and others] (1980). Processus d'infection par *Armillariella ostoyae* Romagn. de racines de *Pinus pinaster* Sol. isolees et cultivees sur un milieu synthetique. *Bulletin de la Société Mycologique de France* **96**, 262–269.

BALLESTA, J.P.G. AND ALEXANDER, M. (1972). Susceptibility of several basidiomycetes to microbial lysis. *Transactions of the British Mycological Society* **58**, 481–487.

BARTNICKI-GARCIA, S. AND LIPPMAN, E. (1972). The bursting tendency of hyphal tips of fungi: presumptive evidence for a delicate balance between wall synthesis and wall lysis in apical growth. *Journal of General Microbiology* **73**, 487–500.

BELL, A.A. AND WHEELER, M.H. (1986). Biosynthesis and functions of fungal melanins. *Annual Review of Phytopathology* **24**, 411–451.

BENJAMIN, M. AND NEWHOOK, F.J. (1984b). The relative susceptibility of various *Eucalyptus* spp. and *Pinus radiata* to *Armillaria* grown in different food bases. In: *Proceedings of the 6th international conference on root and butt rots of forest trees*, August 25–31, 1983,

Melbourne, Victoria, and Gympie, Australia. Ed. G.A. Kile, pp 140–147. Melbourne, Australia.

BENNELL, A.S., WATLING, R. AND KILE, G. (1985). Spore ornamentation in *Armillaria* (Agaricales). *Transactions of the British Mycological Society* **83**, 447–455.

BLENIS, P.V., MUGALA, M.S. AND HIRATSUKA, Y. (1989). Soil affects *Armillaria* root rot of lodgepole pine. *Canadian Journal of Forest Research* **19** (12), 1638–1641.

BLISS, D.E. (1941). Artificial inoculation of plants with *Armillaria mellea*. *Phytopathology* **31**, 859. Abstract.

BLISS, D.E. (1946). The relation of soil temperature to the development of Armillaria root rot. *Phytopathology* **36**, 302–318.

BLISS, D.E. (1951). The destruction of *Armillaria mellea* in citrus soils. *Phytopathology* **41**, 665–683.

BLODGETT, J.T. AND WORRALL, J.J. (1992a). Distributions and hosts of *Armillaria* species in New York. *Plant Disease* **76** (2), 166–170.

BLODGETT, J.T. AND WORRALL, J.J. (1992b). Site relationships of *Armillaria* species in New York. *Plant Disease* **76** (2), 170–174.

BLOOMFIELD, B.J. AND ALEXANDER, M. (1967). Melanins and resistance of fungi to lysis. *Journal of Bacteriology* **93**, 1276–1280.

BOUVENG, H.O., FRASER, R. AND LINDBERG, B. (1967). Polysaccharides elaborated by *Armillaria mellea* (Tricholomataceae), II. Water-soluble mycelium polysaccharides. *Carbohydrate Research* **4**, 20–31.

BUCKLAND, D.C. (1953). Observations on *Armillaria mellea* in immature Douglas fir. *Forestry Chronicle* **29**, 344–347.

BURGES, A. (1939). The defensive mechanism in orchid mycorrhiza. *New Phytologist* **38**, 273–283.

CAMPBELL, W.G. (1932). The chemistry of the white rots of wood, III. The effect on wood substance of *Ganoderma applantum* (Pers.) Pat, *Fomes fomentarius* (Linn.) Fr., *Polyporus adustus* (Willd.) Fr., *Pleurotus ostreatus* (Jacq.) Fr., *Armillaria mellea* (Vahl) Fr., *Trametes pini* (Brot.) Fr., and *Polystictus abeitinus* (Dicks.) Fr. *Biochemistry Journal* **26**, 1829–1838.

CHA, J.Y., TAMAI, Y., MIYAMOTO, T. AND IGARASHI, T. (1996). Histological characteristics and decay ability of wood of *Betula platyphylla* Sukatchev var. *japonica* (Miq.) Hara and *Abies sachalinensis* (Fr. Schm.) Mast. of S. Hokkaido by *Armillaria* in Japan. (In Japanese.) *Research Bulletin of the Hokkaido University Forests* **53** (2), 235–244.

CHEO, P.C. (1982). Effects of tannic acid on rhizomorph production by *Armillaria mellea*. *Phytopathology* **72**, 676–679.

CHET, I., AND HÜTTERMANN, A. (1977). Melanin biosynthesis during differentiation of *Physarum polycephalum*. *Biochemica et Biophysica Acta* **499**, 148–155.

CHOU, C.K.S. (1991). Perspectives of disease threat in large-scale *Pinus radiata* monoculture – the New Zealand experience. *European Journal of Forest Pathology* **21** (2), 71–81.

COUTTS, M.P. AND ARMSTRONG, W. (1976). Role of oxygen transport in the tolerance of trees to waterlogging. In: *Tree physiological yield improvement*. Eds. M.G.R. Cannell and F.T. Last, pp 361–385. New York: Academic Press.

DALY, J.M. (1976). The carbon balance of diseased plants: changes in respiration, photosynthesis and translocation. In: *Physiological plant pathology, physiology of host response to infection*, Vol. 4, Chapter 5. Eds. R. Heitefuss and P.H. Williams, pp 450–474. Berlin and New York: Springer-Verlag.

DAVIDSON, A.J. AND RISHBETH, J. (1988). Effect of suppression and felling on infection of oak and Scots pine by *Armillaria mellea*. *European Journal of Forest Pathology* **18**, 161–168.

DIMITRI, L (1969). Untersuchungen uber die unteriridischen Eintrittspforten der wichtigsten Rotfauleerreger bei der Fichte (*Picea abies* Karst.). [The subterranean infection courts for the chief fungi causing red rot on Norway spruce.] *Forstwissenschaftliches Centralblatt* **88**, 281–308.

DRACZYNSKA-LUSIAK, B. AND SIEWINSKI, A. (1989). Enantioselectivity of the metabolism of some monoterpenic components of coniferous tree resin by *Armillariella mellea* (honey fungus). *Journal of Basic Microbiology* **29** (5), 269–275.

ENTRY, J.A., CROMACK, K., JR., HANSEN, E. AND WARING, R. (1991). Response of western coniferous seedlings to infection by *Armillaria ostoyae* under limited light and nitrogen. *Phytopathology* **81** (1), 89–94.

FOX, R.T.V. (1990). Diagnosis and control of *Armillaria* honey fungus root rot of trees. *Professional Horticulture* **4**, 121–127.

GARRAWAY, M.O. (1975). Stimulation of *Armillaria mellea* growth by plant hormones in relation to the concentration and type of carbohydrate. *European Journal of Forest Pathology* **5** (1), 35–43.

GARRAWAY, M.O., HÜTTERMANN, A. AND WARGO, P.M. (1991). Ontogeny and Physiology. In: *Armillaria Root Disease.* Eds. C.G. Shaw, III and G.A. Kile, pp 21–47. *Forest Service Handbook No. 691*. Washington, D.C.: USDA.

GARRETT, S.D. (1956). Rhizomorph behaviour in *Armillaria mellea* (Vahl) Quél. II. Logistics of infection. *Annals of Botany* **20**, 193–209.

GARRETT, S.D. (1957). Effect of a soil microflora selected by carbon disulphide on survival of *Armillaria mellea* in woody tissues. *Canadian Journal of Microbiology* **3**, 135–149.

GARRETT, S.D. (1958). Inoculum potential as a factor limiting lethal action by *Trichoderma viride* (Fr.) on *Armillaria mellea* (Vahl) Quél. *Transactions of the British Mycological Society* **41**, 157–164.

GARRETT, S.D. (1970). *Pathogenic Root-Infecting Fungi.* 294pp. Cambridge: Cambridge University Press.

GIBBS, J.N. AND GREIG, B.J.W. (1990). Survey of parkland trees after the great storm of October 16, 1987. *Arboricultural Journal* **14** (4), 321–347.

GIBSON, I.A.S. AND GOODCHILD, N.A. (1961). *Armillaria mellea* in Kenya tea plantations. In: *Report of the 6th Commonwealth Mycological Conference*, 1960, pp 39–40, pp 54–55. Kew, Surrey, England: Commonwealth Mycological Institute.

GOODMAN, R.N., HIRALY, Z. AND WOOD, H.R. (1986). Secondary metabolites. In: *The biochemistry and physiology of plant disease*, Chapter 6, pp 211–244. Columbia, MO: University of Missouri Press.

GRAMSS, G. (1983). Examination of low-pathogenicity isolates of *Armillaria mellea* from natural stands of *Picea abies* in Middle-Europe. *European Journal of Forest Pathology* **13**, 142–151.

GREATHOUSE, G.A. AND RIGLER, N.E. (1940). The chemistry of resistance of plants to *Phymatotrichum* root rot, V. Influence of alkaloids on growth of fungi. *Phytopathology* **30**, 475–485.

GREGORY, S.C. (1984). The use of potato tubers in pathogenicity studies of *Armillaria* species. In: *Proceedings of the 6th international conference on root and butt rots of forest trees*, August 25-31, 1983, Melbourne, Victoria, and Gympie, Queensland, Australia. Ed. G.A. Kile, pp 148–160. Melbourne: International Union of Forestry Research Organizations.

GREGORY, S.C. (1985). The use of potato tubers in pathogenicity studies of *Armillaria* isolates. *Plant Pathology* **34**, 41–48.

GREGORY, S.C. (1989). *Armillaria* species in northern Britain. *Plant Pathology* **38**, 93–97.

GREGORY, R.E. AND WARGO, P.M. (1986). Timing of defoliation and its effects on bud development, starch reserves, and sap-sugar concentrations in sugar maple. *Canadian Journal of Forest Research* **16**, 10–17.

GREIG, B.J.W. AND STROUTS, R.G. (1983). *Honey fungus*. Arboricultural Leaflet 2. Revised. 16pp. Great Britain: Her Majesty's Stationery Office; Department of the Environment, Forestry Commission.

GUILLAUMIN, J.J. (1986). *Contribution à l'étude des Armillaires phytopathogènes, en particulier du groupe* Mellea: *cycle caryologique, notion d'espèce, role biologique des espèces.* 270pp. These. Univ. Claude Bernard-Lyon I.

GUILLAUMIN, J.J. AND LEPRINCE, S. (1979). Influence de divers types de matière organique sur l'initiation et la croissance des rhizomorphes d'*Armillaria mellea* (Vahl) Karst. dans le sol. [Influence of different organic materials on the initiation and growth of rhizomorphs of *Armillaria mellea* in soil.] *European Journal of Forest Pathology* **9**, 355–366.

GUILLAUMIN, J.J., BERTHELAY, S. AND SAVIN, V. (1983). Etude de la polarité sexuelle des Armillaires du groupe Mellea. *Cryptogamie Mycologie* **4**, 301–319.

GUILLAUMIN, J.J., MOHAMMED, C. AND BERTHELAY, S. (1989). *Armillaria* species in the nothern temperate hemisphere. In: *Proceedings of the 7th International Conference on Root and Butt Rots*, August 9–16, 1988, Vernon and Victoria, B.C. Ed. D.J. Morrison, pp 27–43. Victoria, B.C.: International Union of Forestry Research Organizations.

GUILLAUMIN, J.J., ANDERSON, J.B. AND KARHONEN, K. (1991). In: *Armillaria Root Disease. Forest Service Handbook No. 691*. Eds. C.G. Shaw, III and G.A. Kile, pp 10–20. Washington, D.C.: USDA.

GUYOT, R. (1927). Mycélium lumineux de l'Armillaire. *Compte-Rendus de la Societé de Biologie* **96**, 114–116.

HAMADA, M. (1940). Physiologisch-morphologische Studien uber *Armillaria mellea* (Vahl) Quél. mit besonderer Rucksicht auf die Oxalsaure-bildung. Ein Nachtrag zur Mykorrhiza von *Galeola septentrionalis* Reichb.f. *Japanese Journal of Botany* **10**, 388–463.

HARTIG, R. (1874). Wichtige Krankheiten der Waldbaume. Beitrage zur Mycologie und Phytopathologie fur Botaniker, und Forstmanner. 127pp. Berlin: Springer. [Important Diseases of Forest Trees. Contributions to mycology and phytopathology for botanists and foresters. Phytopathological Classics No. 12, 1975. St. Paul, MN: American Phytopathological Society.]

HEWITT, J.L. (1936). A survey concerning a native pathogen *Armillaria mellea*. Bulletin 25. *California Department of Agriculture*, 226–234.

HINTIKKA, V. (1974). Notes on the ecology of *Armillariella mellea* in Finland. *Harstenia* **14**, 12–31.

HOLDENRIEDER, O. (1987). Simple inoculation techniques for *Armillaria* by rhizomorphs. *European Journal of Forest Pathology* **17**, 317–320.

HOOD, I.A. AND SANDBERG, C.J. (1987). Occurrence of *Armillaria* rhizomorph populations in the soil beneath indigenous forests in the Bay of Plenty, New Zealand. *New Zealand Journal of Forestry Science* **17**, 83–99.

HORNER, I.J. (1988). *Armillaria* root-rot of kiwifruit. In: *5th International Congress of Plant Pathology*, August 20–27, 1988, Kyoto, Japan, 204. Abstract.

HUGHES, C.N.G., WEST, J.S. AND FOX, R.T.V. (1996). Control of broad-leaved docks by *Armillaria mellea*. In: *Proceedings of the IX International Symposium on Biological Control of Weeds*, Stellenbosch, South Africa, 19–26 January, 1996. Eds. V.C. Moran and J.H. Hoffmann, pp 531-534. Rondebosch, South Africa: University of Cape Town.

JAHNKE, H-D., BAHNWEG, G. AND WORRALL, J.J. (1987). Species delimitation in the *Armillaria mellea* complex by analysis of nuclear and mitochondrial DNAs. *Transactions of the British Mycological Society* **88**, 572–575.

JAMES, R.L., STEWART, C.A. AND WILLIAMS, R.E. (1984). Estimating root disease losses in the northern Rocky Mountain national forests. *Canadian Journal of Forest Research* **14**, 652–655.

JENNINGS, D.H. (1990). The ability of Basidiomycete mycelium to move nutrients through the soil ecosystem. *Nutrient cycling in terrestrial ecosystems: field methods, application and interpretation.* Eds. A.F. Harrison, P. Ineson and O.W. Heal, pp 233–245, 15 ref. Barking, U.K.: Elsevier Applied Science Publishers Ltd.

KAWADA, H., TAKAMI, M. AND HAMA, T. (1962). [A study of Armillaria root rot of larch. Effects of soil conditions on its occurrence and some information of field observation.] (In Japanese.) *Meguro: Bulletin of the Government Forest Experiment Station* **143**, 39–98.

KELLAS, J.D., KILE, G.A. AND JARRETT, R.G. [and others] (1987). The occurrence and effects of *Armillaria luteobubalina* following partial cutting in mixed eucalypt stands in the Wombat Forest, Victoria. *Australian Forest Research* **17**, 263–276.

KILE, G.A. (1980). Behaviour of an *Armillaria* in some *Eucalyptus obliqua - Eucalyptus regnans* forests in Tasmania and its role in their decline. *European Journal of Forest Pathology* **10**, 278–296.

KILE, G.A. (1981). *Armillaria luteobubalina*: a primary cause of decline and death of trees in mixed species eucalypt forests in central Victoria. *Australian Forest Research* **11**, 63–77.

KILE, G.A. (1983b). Identification of genotypes and the clonal development of *Armillaria luteobubalina* Watling and Kile in eucalypt forests. *Australian Journal of Botany* **31**, 657–671.

KILE, G.A. AND WATLING, R. (1983). *Armillaria* species from south-eastern Australia. *Transactions of the British Mycological Society* **81**, 129–140.

KILE, G.A. AND WATLING, R. (1988). Identification and occurrence of Australian *Armillaria* species, including *A. pallidula* sp.nov. and comparative studies between them and non-Australian tropical and Indian *Armillaria*. *Transactions of the British Mycological Society* **91** (2), 305–315.

KIM, H.J. AND KO, M.K. (1993). Cultivation of *Armillaria mellea* mushrooms on sawdust media and their host specificity (in Korean with English figures and tables). *Research Reports of the Forestry Research Institute Seoul* **47**, 129–139.

KOLATTUKUDY, P.E., ESPELIE, K.E. AND SOLIDAY, C.L. (1981). Hydrophobic layers attached to cell walls. Cutin, suberin and associated waxes. In: *Encyclopedia of plant physiology*, N.S., vol. 13B. Eds. W. Tanner and E.A. Loewus, pp 225–254. Berlin: Springer Verlag.

KORHONEN, K. (1978). Interfertility and clonal size in the *Armillaria mellea* complex. *Karstenia* **18**, 31–42.

KORHONEN, K. (1983). Observations on nuclear migration and heterokaryotization in *Armillaria*. *Cryptogamie, Mycologie* **4**, 79–85.

KUSANO, S. (1911). Gastrodia elata and its symbiotic association with *Armillaria mellea*. *Imperial University of Tokyo; Journal of the College of Agriculture* **4**, 1–65.

LEACH, R. (1937). Observations on the parasitism and control of *Armillaria mellea*. *Proceedings of the Royal Society of London, Series B* **121**, 561–573.

LEACH, R. (1939). Biological control and ecology of *Armillaria mellea* (Vahl) Fr. *Transactions of the British Mycological Society* **23**, 320–329.

LIN, P., PUMAS, M.T. AND HUBBES, M. (1989). Isozyme and general protein patterns of *Armillaria* spp. collected from the boreal mixedwood forest of Ontario. *Canadian Journal of Botany* **67**, 1143–1147.

MACKENZIE, M. (1987). Infection changes and volume loss in a 19-year-old *Pinus radiata* stand affected by *Armillaria* root rot. *New Zealand Journal of Forestry Science* **17** (1), 100–108.

MACKENZIE, M. AND SHAW, C.G., III (1977). Spatial relationships between *Armillaria* root-rot of *Pinus radiata* seedlings and the stumps of indigenous trees. *New Zealand Journal of Forestry Science* **7**, 374–383.

MACKENZIE, M. AND SELF, N.M. (1988). *Armillaria* in some New Zealand second rotation pine stands. In: *Proceedings of the 36th annual Western International Forest Disease Work Conference*, September 19-23, 1988, Park City, UT. Comp. B.J. van der Kamp, pp 82–87.

MACMILLAN-BROWSE, P., RUSSELL, T.D., BROWSE, P.M. AND MACMILLAN-BROWSE, P. (1990). The wind storm of January 25th, 1990. 1. A Preliminary report on damage to the garden at Wisley. 2. Westonbirt Arboretum. *Arboricultural Journal* **14** (2), 97–106.

MARSH, R.W. (1952). Field observations on the spread of *Armillaria mellea* in apple orchards and in a blackcurrent plantation. *Transactions of the British Mycological Society* **35**, 201–207.

MAYER, A.M. AND HAREL, E. (1979). Polyphenol oxidases in plants. *Phytochemistry* **18**, 193–215.

MOLIN, N. AND RENNERFELT, E. (1959). Honungsskivlinger, *Armillaria mellea* (Vahl) Quél., som parasit på barrträd (in Swedish). *Meddelanden från Statens Skogsforskningsinstitut* **48**, 1–26.

MOODY, A.R. AND WEINHOLD, A.R. (1972a). Fatty acids and naturally occurring plant lipids as stimulants of rhizomorph production in *Armillaria mellea*. *Phytopathology* **62**, 264–267. Abstract.

MOODY, A.R. AND WEINHOLD, A.R. (1972b). Stimulation of rhizomorph production by *Armillaria mellea* with lipid from tree roots. *Phyopathology* **62**, 1347–1350.

MOORE, W.C. (1959). British parasitic fungi: A host-parasite index and guide to British literature on the fungus diseases of cultivated plants. xvi + 430pp. Cambridge: Cambridge University Press.

MORRISON, D.J. (1972). *Studies on the biology of* Armillaria mellea. Ph.D. dissertation. 169pp. Cambridge: University of Cambridge. [Internal Report BC–30. Victoria, B.C.: Canadian Forestry Service, Pacific Forest Research Centre. 169pp.]

MORRISON, D.J. (1975). Ion uptake by rhizomorphs of *Armillaria mellea*. *Canadian Journal of Botany* **53**, 48–51.

MORRISON, D.J. (1976). Vertical distribution of *Armillaria mellea* rhizomorphs in soil. *Transactions of the British Mycological Society* **66**, 393–399.

MORRISON, D.J. (1982). Variation among British isolates of *Armillaria mellea. Transactions of the British Mycological Society* **78**, 459–464.

MORRISON, D.J. (1989). Pathogenicity of *Armillaria* species is related to rhizomorph growth habit. In: *Proceedings of the 7th international conference on root and butt rots*, August 9-16, 1988, Vernon and Victoria, B.C. Ed. D.J. Morrison, pp 584–589. Victoria, B.C.: International Union for Forestry Research Organizations.

MORRISON, D.J., THOMSON, A.J. AND CHU, D. [and others] (1985). Isozyme patterns of *Armillaria* intersterility groups occurring in British Columbia. *Canadian Journal of Microbiology* **31**, 651–653.

MORRISON, D.J., THOMSON, A.J., CHU, D., PEET, F.G. AND SAHOTA, T.S. (1989). Variation in isozyme patterns of esterase and polyphenol oxidase among isolates of *Armillaria ostoyae* from British Columbia. *Canadian Journal of Plant Pathology* **11**, 229–234.

MORRISON, D. AND PELLOW, K. (1998). Relationship between above-ground indicators and below-ground infection for *Armillaria* root disease in the southern interior of British Columbia. In: *Proceedings of the Ninth International Conference on Root and Butt Rots*, Carcans-Maubuisson, France, September 1–7, 1997. (Les Colloques, no.89). Eds. C. Delatour, J.J. Guillaumin, B. Lung-Escarmant and B. Marçais, pp 131–135. Paris, France: INRA.

MWANGI, L.M., LIN, D. AND HUBBES, M. (1990). Chemical factors in *Pinus strobus* inhibitory to *Armillaria ostoyae. European Journal of Forest Pathology* **20** (1), 8–14.

NECHLEBA, A. (1915). Die Hallimasch: studien beobechtungen und hypothesen. [The honey agaric.] *Forstwissenschaftliches Centralblatt.* **59**, 384–392.

ONO, K. (1970). Effect of soil conditions on the occurrence of Armillaria root rot of Japanese larch. (In Japanese.) *Meguro Bulletin of the Government Forest Experiment Station* **229**, 123–219. [*Review of Plant Pathology* **50**, 2001.]

OSTROFSKY, D., IV, SHORTLE, C., IV AND BLANCHARD, R.O. (1984). Bark phenolics of American beech (*Fagus grandifolia*) in relation to the beech bark disease. *European Journal of Forest Pathology* **14**, 52–59.

PARKER, J. (1979). Effects of defoliation and root height above a water table on some red oak root metabolites. *Journal of the American Society of Horticultural Science* **104**, 417–421.

PARKER, J. AND HOUSTON, D.R. (1971). Effects of repeated defoliation on root and root collar extractives of sugar maple trees. *Forest Science* **7**, 91–95.

PARKER, J. AND PATTON, R.L. (1975). Effects of drought and defoliation on some metabolites in roots of black oak seedlings. *Canadian Journal of Forest Research* **5**, 457–463.

PATTON, R.F. AND RIKER, A.J. (1959). Artificial inoculations of pine and spruce trees with *Armillaria mellea. Phytopathology* **49**, 615–622.

PEARCE, M.H., MALAJCSUK, N. AND KILE, G.A. (1986). The occurrence and effects of *Armillaria luteobubalina* in the karri (*Eucalyptus diversicolor* F. Muell.) forests of Western Australia. *Australian Forest Research* **16**, 243–259.

PEARCE, M.H. AND MALAJCSUK, N. (l990a). Factors affecting the growth of *Armillaria luteobubalina* rhizomorphs in soil. *Mycological Research* **94**, 38–48.

PEARCE, M.H. AND MALAJCZUK, N. (1990b). Stump colonization by *Armillaria luteobubalina* and other wood decay fungi in an age series of cut-over stumps in karri (*Eucalyptus diversicolor*) regrowth forests in south-western Australia. *New Phytologist* **115** (1), 129–138.

PLAKIDAS, A.G. (1941). Infection with pure cultures of *Clitocybe tabescens. Phytopathology* **31**, 93–95.

PODGER, F.D., KILE, G.A. AND WATLING, R. [and others] (1978). Spread and effects of *Armillaria luteobubalina* sp. nov. in an Australian *Eucalyptus regnans* plantation. *Transactions of the British Mycological Society* **71**, 77–87.

POPOOLA, T.O.S. AND FOX, R.T.V. (1996). Effects of root damage on honey fungus. *Arboricultural Journal* **20**, 329–337.

PRIHODA, A. (1957). Nakazazivych smrku vaclavkou. *Les* (Bratislava) **13**, 173–176.

PRONOS, J. AND PATTON, R.F. (1978). Penetration and colonization of oak roots by *Armillaria mellea* in Wisconsin. *European Journal of Forestry* **8**, 259–267.

PRZEZBORSKI, A. (1987). Effect of a long drought on the occurrence of root rots caused by *Fomes annosus [Heterobasidion annosum]* and *Armillaria mellea*. [Wplyw dlugotrwalej suszy na wystepowanie chorob korzeniowych (*Fomes annosus* Fr.) Cke, *Armillaria mellea* (Vahl/ Quél.)] (In Polish.) Wplyw suszy lat 1982–1984 na drzewa. *Sylwan* **131** (4), 43–52.

RAABE, R.D. (1962a). Host list of the root rot fungus, *Armillaria mellea*. *Hilgardia* **33**, 25–88.

RAHMAN, A. (1978). The effect of Armillatox in the mycelial growth and rhizomorph production by *Armillaria mellea* in culture. *European Journal of Forest Pathology* **8**, 75–83.

RAPER, J.R. (1966). *Genetics of sexuality in higher fungi*. 283pp. New York: Ronald Press.

REDFERN, D.B. (1973). The growth and behaviour of *Armillaria mellea* rhizomorphs in soil. *Transactions of the British Mycological Society* **61**, 569–581.

REDFERN, D.B. (1975). The influence of food base on the rhizomorph growth and pathogenicity of *Armillaria mellea* isolates. In: *Biology and Control of soil-borne plant pathogens*. Ed. G.W. Bruehl, pp 69–73. U.S.A.: The American Phytopathology Society.

REDFERN, D.B. AND FILIP, G.M. (1991). Inoculum and Infection. In: *Armillaria Root Disease*. Eds. C.G. Shaw, III and G.A. Kile, pp 48–61. *Forest Service Handbook No. 691*. Washington, D.C.: USDA.

RISHBETH, J. (1964). Stump infection by basidiospores of *Armillaria mellea*. *Transactions of the British Mycological Society* **47**, 460.

RISHBETH, J. (1970). The role of basidiospores in stump infection by *Armillaria mellea*. In: *Root diseases and soil-borne pathogens: Proceedings of the symposium*, July, 1968, Imperial College, London. Eds. T.A. Toussoun, R.V. Bega and P.E. Nelson, pp 141–146. Berkeley: University of California Press.

RISHBETH, J. (1972). The production of rhizomorphs by *Armillaria mellea* from stumps. *European Journal of Forest Pathology* **2**, 193–205.

RISHBETH, J. (1978). Infection foci of *Armillaria mellea* in first-rotation hardwoods. *Annals of Botany* **42**, 1131–1139.

RISHBETH, J. (1982). Species of *Armillaria* in southern England. *Plant Pathology* **31**, 9–17.

RISHBETH, J. (1985a). *Armillaria*: resources and hosts. In: *Developmental biology of higher fungi*. Eds. D. Moore, L.A. Casselton and D.A. Wood, pp 87–101. Cambridge: Cambridge University Press.

RISHBETH, J. (1985b). Infection cycle of *Armillaria* and host response. *European Journal of Forest Pathology* **15**, 332–341.

RISHBETH, J. (1986). Some characteristics of English *Armillaria* species in culture. *Transactions of the British Mycological Society* **85**, 213–218.

RISHBETH, J. (1988). Stump infection by *Armillaria* in first-rotation conifers. *European Journal of Forest Pathology* **18**, 401–408.

ROBINSON-BAX, C. (1999). *A survey of the herbaceous hosts of Armillaria mellea and possible integrated control*. Horticulture M.Sc. Dissertation. U.K.: University of Reading.

ROTH, L.F., ROLF, L. AND COOLEY, S. (1980). Identifying infected ponderosa pine stumps to reduce costs of controlling Armillaria root rot. *Journal of Forestry* **78**, 145–151.

RYKOWSKI, K. (1975). Modalité d'infection des pins sylvestres par l'*Armillaria mellea* (Vahl) Karst. dans les cultures forestières. [Infection patterns in Scots pine plantations by *Armillaria mellea* (Vahl) Karst.] *European Journal of Forest Pathology* **3**, 65–82.

RYKOWSKI, K. (1984). Niektore troficzne uwarunkowania patogenicznosci *Armillaria mellea* (Vahl) Quél. w uprawach sosnowych. [Some trophic factors in the pathogenicity of *Armillaria mellea* in Scots pine plantations.] (In Polish.) *Prace Instytutu Badawczego Lesnictwa* **640**, 1–140.

RYKOWSKI, K. AND SIEROTA, Z. (1988). Logging waste as a source of infection risk. [Odpady zrebowe jako zrodlo zagrozenia infekcyjnego.] (In Polish.) *Las Polski* **18**, 7–9.

SCHLUMBAUM, A., MAUCH, F. AND VOGELI, U. (1986). Plant chitinases are potent inhibitors of fungal growth. *Nature* **324**, 365–367.

SCHÜTTE, K.H. (1956). Translocation in the fungi. *New Phytologist* **55**, 164–182.

SCURTI, J.C. (1956). Sulla demolizione della cellulosa e della lignina per opera du funghi lignicoli. (In Italian.) *Nuovo Giornale Botanico Italiano, N.S.* **63**, 411–412.

SHAW, C.G., III. (1975). *Epidemiological insights into* Armillaria mellea *root rot in a managed ponderosa pine forest*. 201pp. Ph.D. dissertation. Corvallis, OR: Oregon State University.

SHAW, C.G., III. (1977). *Armillaria* isolates from pine and hardwoods differ in pathogenicity to pine seedlings. *Plant Disease Reporter* **61**, 416–418.

SHAW, C.G., III. (1980). Characteristics of *Armillaria mellea* on pine root systems in expanding centers of root rot. *Northwest Science* **54**, 137–145.

SHAW, C.G., III. (1985). *In vitro* responses of different *Armillaria* taxa to gallic acid, tannic acid, and ethanol. *Plant Pathology* **34**, 594–602.

SHAW, C.G., III AND ROTH, L.F. (1976). Persistence and distribution of a clone of *Armillaria mellea* in a ponderosa pine forest. *Phytopathology* **66**, 1210–1213.

SHAW, C.G., III, MACKENZIE, M. AND TOES, E.H.A. [and others] (1981). Cultural characteristics and pathogenicity to *Pinus radiata* of *Armillaria novae-zelanidae* and *A. limonea*. *New Zealand Journal of Forestry Science* **11**, 65–70.

SHAW, C.G., III AND KILE G.A. (eds.) (1991). *Armillaria Root Disease*. 233pp. *Forest Service Handbook No. 691*. Washington, D.C.: USDA.

SHEARER, B.L. AND TIPPETT, J.T. (1988). Distribution and impact of *Armillaria luteobubalina* in the *Eucalyptus marginata* forest of South-western Australia. *Australian Journal of Botany* **36**, 433–445.

SIEPMANN, R. AND LEIBIGER, M. (1989). On the host specialization of *Armillaria* species. [Uber die Wirtsspezialisierung von Armillaria-Arten.] (In German.) *European Journal of Forest Pathology* **19** (5–6), 334–342.

SINGH, P. (1980). *Armillaria mellea* root rot: artificial inoculation and development of the disease in greenhouse. *European Journal of Forest Pathology* **10**, 420–431.

SINGH, P. (1983). *Armillaria* root rot: influence of soil nutrients and pH on the susceptibility of conifer species to the disease. *European Journal of Forest Pathology* **13**, 92–101.

SMITH, A.M. AND GRIFFIN, D.M. (1971). Oxygen and the ecology of *Armillariella elegans* Heim. *Australian Journal of Biological Sciences* **24**, 231–262.

SMITH, M.L., BRUHN, J.N. AND ANDERSON, J.B. (1992). The fungus *Armillaria bulbosa* is among the largest and oldest living organisms. *Nature* (London) **256**, 428–431.

STANOSZ, G.R. AND PATTON, R.F. (1990a). Stump colonization by *Armillaria* in Wisconsin aspen stands following clearcutting. *European Journal of Forest Pathology* **20** (6–7), 339–346.

STANOSZ, G.R. AND PATTON, R.F. (1990b). Quantification of *Armillaria* rhizomorphs in Wisconsin aspen sucker stands. *European Journal of Forest Pathology* **21** (1), 5–16.

SWIFT, M.J. (1965). Loss of suberin from bark tissue rotted by *Armillaria mellea*. *Nature* **207**, 436–437.

SWIFT, M.J. (1972). The ecology of *Armillaria mellea* (Vahl ex Fries) in the indigenous and exotic woodlands of Rhodesia. *Forestry* **45**, 67–86.

THANASSOULOPOULOS, C.C. AND ARTOPEADIS, M.C. (1991). Some previously unreported hosts of *Armillaria* root rot. *Plant Disease* **75** (1), 101.

THOMAS, H.E. (1934). Studies on *Armillaria mellea* (V. Quél.), infection, parasitism and host resistance. *Journal of Agricultural Research* **48**, 187–218.

THORNBERRY, H.H. AND RAY, B.R. (1953). Wilt-inducing protein-like pigment from *Armillaria mellea* isolated from peach roots. *Phytopathology* **43**, 486. Abstract.

TIRRO, A. (1989). Characterization of several isolates of *Armillaria mellea* from prickly-pear cactus. [Caracterizzazione di alcuni isolati di *Armillaria mellea* da piante di ficodindia.] (In Italian.) *Micologia Italiana* **18** (3), 111/7–111/11.

TWERY, M.J., MASON, G.N., WARGO, P.M. AND GOTTSCHALK, K.W. (1990). Abundance and distribution of rhizomorphs of *Armillaria* spp. in defoliated mixed oak stands in western Maryland. *Canadian Journal of Forest Research* **20** (6), 674–678.

ULLRICH, R.C. AND ANDERSON, J.B. (1978). Sex and diploidy in *Armillaria mellea*. *Experimental Mycology* **2**, 119–129.

VAN DER PAS, J.B. (1981). A statistical appraisal of *Armillaria mellea* root rot in New Zealand plantations of Pinus radiata. *New Zealand Journal of Forestry Science* **11**, 23–36.

VAN VLOTEN, H. (1936). Onderzoekingen over *Armillaria mellea* (Vahl) Quél. *Fungus Wageningen* **8**, 20–23.

WANG, B.H. AND SHAO, J.B. (1990). New techniques of cultivating *Gastrodia elata*. (In Chinese.) *Zhongguo Shiyongjun* [Edible Fungi of China] **9** (5), 36–37.

WARGO, P.M. (1971). Seasonal changes in carbohydrate levels in roots of sugar maple. Res. Pap. NE-213, 8pp. Upper Darby, PA: U.S. Department of Agriculture, Northeastern Forest Experiment Station.

WARGO, P.M. (1972). Defoliation-induced chemical changes in sugarmaple roots stimulate growth of *Armillaria mellea. Phytopathology* **62**, 1278–1283.

WARGO, P.M. (1975). Lysis of the cell wall of *Armillaria mellea* by enzymes from forest trees. *Physiological Plant Pathology* **5**, 99–105.

WARGO, P.M. (1976). Lysis of fungal pathogens by tree produced enzymes – a possible disease resistance mechanism in trees. In: *Proceedings of the 23rd northeastern forest tree improvement conference*, August 4-7, 1975, New Brunswick, NJ. Ed. P.W. Garrett, pp 19–23.

WARGO, P.M. (1977). *Armillariella mellea* and *Agrilus bilineatus* and mortality of defoliated oak trees. *Forest Science* **23**, 485–492.

WARGO, P.M. (1980). Interaction of ethanol, glucose, phenolics and isolate of *Armillaria mellea. Phytopathology* **70**, 480. Abstract.

WARGO, P.M. (1981a). Defoliation and secondary-action organism attack: with emphasis on *Armillaria mellea. Journal of Arboriculture* **7**, 64–69.

WARGO, P.M. (1981b). Defoliation and tree growth. In: *The gypsy moth: research toward integrated pest management.* Eds. C.C. Doane and M.L. McManus, pp 225–240. Tech. Bull. 1584. Washington, D.C.: U.S. Department of Agriculture, Animal and Plant Health Inspection Service.

WARGO, P.M. (1983). The interaction of *Armillaria mellea* with phenolic compounds in the bark of roots of black oak. *Phytopathology* **73**, 838. Abstract.

WARGO, P.M. (1984). Changes in phenols affected by *Armillaria mellea* in bark tissue of roots of oak, *Quercus* spp. In: *Proceedings of the 6th international conference on root and butt rots of forest trees*, August 25–31, 1983, Melbourne, Victoria, and Gympie, Queensland, Australia. Ed. G.A. Kile, pp 198–206. Melbourne: International Union of Forestry Research Organizations.

WARGO, P.M. (1988). Amino nitrogen and phenolic constituents of bark of American beech, *Fagus grandifolia*, and infestation by beech scale, *Cryptococcus fagisuga. European Journal of Forest Pathology* **18**, 279–290.

WARGO, P.M., PARKER, J. AND HOUSTON, D.R. (1972). Starch content of defoliated sugar maple. *Forest Science* **18**, 203–204.

WARGO, P.M. AND HARRINGTON, T.C. (1991). Host stress and susceptibility. In: *Armillaria Root Disease.* Eds. C.G. Shaw, III and G.A. Kile, pp 88–101. *Forest Service Handbook No. 691*. Washington, D.C.: USDA.

WEAVER, D.J. (1974). Effect of root injury on the invasion of peach roots by isolates of *Clitocybe tabescens. Mycopathologia et Mycologia Applicata* **52**, 313–317.

WEINHOLD, A.R. (1963). Rhizomorph production by *Armillaria mellea* induced by ethanol and related compounds. *Science* **142**, 1065–1066.

WEINHOLD, A.R. AND GARRAWAY, M.O. (1966). Nitrogen and carbon nutrition of *Armillaria mellea* in relation to growth-promoting effects of ethanol. *Phytopathology* **56**, 108–112.

WHITNEY, R.D., IP, D.W. AND IRWIN, R.N. (1989b). Survival of *Armillaria obscura* (Pers.) Herink inoculated into roots of field-grown *Abies balsamea* (L.) Mill. treated with simulated acid rain. In: *Proceedings of the 7th international conference on root and butt rots of forest trees*, August, 1988, Vernon and Victoria, B.C. Ed. D.J. Morrison, pp 492–502. Victoria, B.C.: International Union of Forestry Research Organizations.

WILBUR, W., MUNNECKE, D.E. AND DARLEY, E.F. (1972). Seasonal development of *Armillaria* root rot of peach as influenced by fungal isolates. *Phytopathology* **62**, 567–570.

XU, J.T. AND MU, C. (1990). The relation between growth of *Gastrodia elata* protocorms and fungi. (In Chinese.) *Acta Botanica Sinica* **32** (1), 26–31.

ZELLER, S.M. (1926). Observations on infections of apple and prune roots by *Armillaria mellea* Vahl. *Phytopathology* **16**, 479–484.

ZIMMERMANN, W. AND SEEMÜLLER, E. (1984). Degradation of raspberry suberin by *Fusarium solani* f. sp. *pisi* and *Armillaria mellea. Phytopathologische Zeitschrift* **110**, 192–199.

SECTION 4
Control

7 The Extent of Losses and Aims for Managing *Armillaria*

ROLAND T.V. FOX

The University of Reading, School of Plant Sciences, Department of Horticulture and Landscape (Crop Protection), 2 Earley Gate, Reading, Berkshire RG6 6AU, U.K.

Synopsis

Accurate detection and diagnosis are essential prerequisites to predict whether losses are, or are likely to be, substantial. Based on this forecast, the most effective control method can be chosen, if one is necessary. The cost of treatment can thus be saved if the *Armillaria* present has been correctly identified as avirulent. Losses in horticulture are frequently considered sufficiently important to warrant the cost of removal of tree stumps and infected roots from orchards, vineyards, gardens and woodland in order to reduce subsequent disease and risk of windthrow. Soil in an infested commercial orchard may warrant the cost of fumigation, or even removal and replacement, prior to replanting. Fungicide drenches and biological control measures are still being developed independently and as an integrated treatment, largely for amenity plantings. Despite considerable losses in some exotic pine plantations, the control of *Armillaria* root rot is not considered vital in commercial forests if most of the timber from infected trees can still be recovered.

Detection and diagnosis to predict losses

In order to reduce losses, infection must be controlled as promptly as possible (Turner and Fox, 1988; Fox *et al.*, 1991). Yet, although the extent of decay from an established fungal rot can readily be mapped by metal probes with or without electronic devices, their presence first has to be discovered. Some signs of infection, such as unusually sparse foliage, can be recognized by an experienced arboriculturalist, and may even be detected from the air (Hunt *et al.*, 1971; Shaw and Kile, 1991). Alternatively, dogs can be trained to bark when they smell rotten wood (Swedjemark, 1989).

Rotten trees become more easily detected at the late stages of colonization, but by then they are difficult, if not impossible, to treat (Swift, 1970; Bray, 1970; Pawsey and Rahman, 1974, 1976a,b; Filip and Roth, 1977).

Armillaria *Root Rot: Biology and Control of Honey Fungus*

Early diagnosis is vital to ensure the successful control of any plant disease. In common with most other soil diseases, it is very difficult to recognize the presence of a pathogen before the appearance of its above-ground symptoms, by which time it is then extremely difficult, if not impossible, to control the disease.

Traditionally, in order to confirm *Armillaria* root disease, the root collar and lower bole of the tree is examined for signs specific to *Armillaria* species like mycelial fans, rhizomorphs, and, eventually, the characteristic tawny toadstool basidiomes. Rhizomorphs often form when the edges of the mycelial fans covering the food base are prevented from advancing any further because of contacting either loose bark or the soil.

Although fruiting bodies are useful for identifying the location of stumps and roots within centres of disease, they are much more seasonal than decay and, occasionally, the wood being actively decayed by some, but not all, *Armillaria* species is bioluminescent. Positive proof of the presence of honey fungus at earlier stages generally still depends on being able to trace rhizomorphs that develop around dead hosts. Later, these may form a smothering network around a prospective host and then grow under the bark (Hood and Sandberg, 1987). Detection of the latter may indicate that the host is at risk from infection and closer inspection may reveal effective multiple penetration into the root cambium, as for example during successful colonization by *A. mellea*. However, rhizomorphs may not be in evidence following infection by some species of *Armillaria*, such as *A. ostoyae*, which nonetheless can cause serious rotting, bringing about significant losses of several economically important conifers through windthrow or direct kill. Also, even *Armillaria mellea* often infects roots by root-to-root contact.

Until comparatively recently, *Armillaria* root rot was considered to be caused by a single species, *A. mellea*, albeit with an extremely varied morphology and levels of virulence. Now the genus has been divided into over thirty distinct species. Some of them are virtually harmless saprophytes, while others such as *A. mellea*, *A. ostoyae*, *A. novae-zelandiae* and *A. luteobubalina* are highly virulent pathogens, and still others, such as *A. lutea,* attack trees under stress, as well as infecting some susceptible herbaceous plants. So great are the differences in the behaviour of the different species that an accurate identification generally determines the level of likely losses, and hence whether a control measure would be required or not.

However, many of their initial and even later symptoms are non-specific, including reduction of shoot growth, changes in foliage characteristics, crown dieback, stress-induced reproduction, basal stem indicators, and death so non-specific that it could be induced by a number of biological and non-biological causes.

Forest soils may often contain several different species of *Armillaria* causing similar symptoms (Morrison, 1989), resulting in further confusion (Guillaumin, 1988). When several similar species of *Armillaria* co-habit the same environment, as they frequently do, the only feasible quick ways to separate virulent pathogens from harmless saprophytes until recently were based on the morphology of any rhizomorphs or basidiocarps present. Without the presence of characteristic fruiting bodies (Pegler and Gibson, 1972; Watling *et al.*, 1982; Roll-Hansen, 1985; Greig *et al.*, 1991) that may be present in some years and then only for a few days in autumn, it is often very difficult to detect the presence of honey fungus in an old dead stump when it no longer has living foliage to exhibit symptoms, or the pathogen is deep seated (Shigo and Tippett, 1981; Fox, 1990c).

Although fruiting bodies can be induced *in vivo* and *in vitro* (Raabe, 1984; Fox and Popoola, 1990), they form too slowly (Intini, 1993) to be used routinely and the timing of their appearance can be unpredictable. Like nearly all root diseases, *Armillaria* root rots are hard to identify or quantify, particularly by the amateur, when roots may be deeply buried in soil which frequently obscures any visible symptoms, or even its sporadic luminosity (Guyot, 1927). If a rotten root lacks rhizomorphs that can be used in diagnosis (Morrison, 1989), *Armillaria* will go undetected. In this situation, conventional identification techniques often rely on culturing some of the mycelium present in the rotten wood. Even if successful, this process of isolation onto agar may take days or even weeks to complete and is highly prone to contamination problems as *Armillaria* is relatively slow growing.

Conventional identification techniques involve isolation of the fungus mycelium present and culturing on agar, but this takes a long time and the cultures are very prone to contamination. Although widely used by taxonomists, successful *in vitro* mating tests, where pure cultures of an unknown isolate are plated next to cultures of a known species (Korhonen, 1978; Rishbeth, 1982; Wargo and Shaw, 1985) require laboratory conditions and are slow and frequently difficult to interpret (Gregory, 1989).

Cultures of *Armillaria* spp. show considerable intraspecific variation, so interfertility tests with haploid or diploid testers are generally necessary, even though these take experience to interpret. As yet, no biochemical methods, such as those investigated by Hütterman *et al.* (1984) or Wargo and Shaw (1985), have proved of much benefit.

Without reliable diagnosis of *Armillaria mellea*, the expense and labour of control measures will be wasted if the rot is caused by a harmless saprophyte (Fox, 1990b). It is possible to differentiate pathogens from saprophytes readily by baiting with strawberry (Fox and Popoola, 1990) or logs (Mallett and Hiratsuka, 1985); even potato tubers are susceptible (Gregory, 1984). Baiting can be effective in mapping the extent of ground that is infected, but the method cannot give immediate results.

Speed is always crucial when planting bare rooted trees and is frequently critical when a control agent has to be selected. Few diagnostic tests are as quick as inspecting a specimen visually for symptoms. However, in practice any time saved in performing the diagnosis will be wasted if sampling is slow or the result is not immediately available. Even though many foliar diseases are often fairly instantly recognized by farmers and growers, providing the symptoms are sufficiently conspicuous, the identification of a pathogen infecting a root is rarely as simple, particularly when soil still obscures the symptoms. Lesions on the roots of a tree therefore take far longer to find than those on its foliage.

The most important criterion for any diagnostic test is reliability (Fox, 1993). Even an apparently straightforward method, such as the visual inspection of plants for symptoms, is critically dependent on the general condition of the specimen of root that has been sampled, an absence of other pathogens or saprophytes, as well as a minimal level of varietal and phenotypic variation between samples (Fox, 1992). In many routine situations, this potentially simple method has proven inherently unreliable.

To a lesser extent, the success of associated microscopical techniques also depends on the quality of the specimens, especially where the pathogen is embedded in relatively tough wood. Rotted root material for examination which is contaminated with other micro-organisms can greatly frustrate efforts to isolate pure cultures (Nobles, 1948; Rishbeth, 1986; Whitney *et al.*, 1978), as well as delaying any subsequent testing with other cultures for anastomosis and interfertility (Korhonen, 1978). Since the outcome of investigations based on such tests is easily affected by often subtle changes in circumstance, their efficacy is less dependable than those methods based on fundamental biochemical differences, such as the immunological properties of proteins and the hybridization of nucleic acids. Another disadvantage of these traditional methods is their greater need for information and expertise, rather than equipment and reagents, which tend to reduce cost-effectiveness.

While the time taken to complete a diagnosis is easily measured, other advantages of one method over another are less clear cut, but a comparison reveals some critical differences, many of which are common to other soil-borne pathogens (Fox, 1990a). Several highly sensitive modern methods for the rapid diagnosis of soil-borne pathogens that have been adapted from other branches of biology are reliable and accurate (Fox 1990a, 1993; Dusunceli and Fox, 1992; Duncan and Torrance, 1992), yet require no expertise. Some, such as gel electrophoresis (Morrison *et al.*, 1984; Poon, 1988), are not only slower and more laborious than the routine inspection for rhizomorphs by an expert, but are also not very specific as there is little variation between species of *Armillaria*, even though these can readily be distinguished from other fungi that rot wood. Neither contamination nor the presence of soil or wood affects methods based on immunology or nucleic acid hybridization (Fox, 1993). However, apart from immunological techniques, most other laboratory diagnostic procedures have so far proved ill-suited for detecting soil-borne fungi in the field as they are neither sufficiently flexible nor portable (Fox 1990a).

To be effective, a diagnostic test must be simple, accurate, rapid and safe to perform, yet be sensitive enough to avoid 'false positives'. Since DNA not proteins form the basis of its action, DNA hybridization is so highly sensitive that it should detect a single nucleus or mitochondrion. Nonetheless, although DNA hybridization has been useful for a number of genetic and ecological studies, such as that for identifying the extent of different clones of *Armillaria lutea* (Smith *et al.*, 1992), it has not been adapted for use in the field. Although not yet available for *Armillaria* spp., simple monoclonal antibody ELISA kits have been produced commercially and used widely for other soil-borne pathogens (Miller *et al.*, 1988). Some of the latter can detect or diagnose diseases safely and clearly using antigen coated magnetic beads and chromogens conjugated to specific antibodies. The choice of the most appropriate experimental procedures for detecting root pathogens such as *Armillaria* should widen as innovation continues to solve current problems. Even so, nucleic acid hybridization seems destined to remain for somewhat longer largely in the hands of academics and the advisory or consultancy services.

With the advent of the modern diagnostic techniques based on enzyme-linked immunosorbent assay (ELISA) and nucleic acid hybridization, however, rapid identification of the disease has become possible. Several immunological methods (Lung-Escarmant and Dunez, 1979, 1980; Lung-Escarmant *et al.*, 1985) based on polyclonal antiserum have been developed but are not well suited for routine serological detection of *Armillaria* since they are not as uniform as monoclonal antibodies (Koehler and Milstein, 1975). Monoclonal antibodies may be obtained with much higher specificity for particular *Armillaria* species and hence achieve higher reproducibility. It is also possible to maintain hybridomas and thus produce monoclonal antibodies cheaply, uniformly and indefinitely (Halk and de Boers, 1985;

Barnes, 1986; Fox, 1990a). An experimental protocol using selected monoclonal antibody ELISA has therefore been tested for individualizing isolates of *Armillaria* (Fox and Hahne, 1989) which can distinguish between the major European *Armillaria* pathogens in a simple ELISA assay. However, since the ELISA test depends on quantitative reactions, a qualitatitive test has now been developed using molecular biology of nucleic acids based on the PCR test (Manley, 1992). There would be major research implications if this sensitive test could be more portable by binding the specific antibodies to either inert granules or sticks, making it feasible to use in the field. This could also lead to the development of a simple cheap kit such as those already widely available in medicine. Infection in the roots of a tree by pathogenic *Armillaria* species could then be detected unambiguously and quickly, allowing treatment or destruction long before it develops into a dangerous rot or spreads to other trees.

Immunological assays are also entering a period of great change as relatively cheap, easy-to-use kits are being developed which allow low levels of disease to be monitored on the spot under field conditions. This might enable the initial stages of infection by basidiospores to be observed (Shaw and Kile, 1991). Since these kits are so sensitive, not only could the life cycle (Wargo and Shaw, 1985) and patterns of spread (Rishbeth, 1986) be validated, but it would be possible to treat lower inoculum levels of pathogens than previously. Consequently, if *Armillaria* can be detected earlier, this should permit more effective control and the use of fungicides may be avoided where no pathogens are detected.

Formerly difficult to diagnose pathogens, such as *Armillaria* and other soil-borne organisms which are traditionally rarely easily quantified even when recognized, could become a more commercially viable market for fungicides and other control measures.

Economics of losses vs. control

Under forest conditions, normally only cultural control is attempted to lessen damage. In primary or ancient woodland, *Armillaria* may even be considered a useful thinning agent and recycler of nutrients (Hagle and Shaw, 1991). Even in managed timber forest situations, the disease, although common, may not cause severe economic damage if only the butt is decayed. However, great importance is placed on the aesthetic damage to gardens and amenity areas, where the value of each established tree is high. Additionally, a host with weakened roots may be windthrown, causing substantial damage to surrounding plantings, buildings and people. Economic loss may also occur in orchards, plantations, and vineyards, where the crop is of high value.

The value of the losses that would be sustained in the absence of control must be the criterion used to decide whether control is worthwhile and what type should be selected. Although an assessment of the consequences to timber yields and amenity value may prove an expensive exercise, it is essential before any control is considered. As well as the financial costs, the environmental costs of alternative methods of control must be justified. The ability of a method to control disease in a particular circumstance is another critical consideration. Hagle and Shaw (1991) remind that, although stump removal to reduce inoculum may greatly improve productivity where this is the principal problem or where a fast growing crop is being grown, it may have much less effect with a crop endangered by secondary disease spread. It is therefore rarely easy to estimate the amount and whereabouts of inoculum that must be removed to obstruct the expansion of the pathogen, and the cost of its removal that is warranted economically as its entire eradication by mechanical, biological or chemical programmes is both unlikely (Williams *et al.*, 1989), and probably of questionable longterm benefit, once spore infection is taken into account (Rishbeth 1978). Any disturbance of the soil is also likely to stimulate rhizomorph production, thus endangering newly planted trees (Morrison, 1976; Redfern, 1970). It has also been found that extracting woody debris may also damage mycorrhizae as well (Harvey *et al.*, 1981), but this must be weighed against the benefits. Any control venture requires extended evaluation over decades in the case of fruit orchards, and a century or more for forests. Hagle and Shaw (1991) recommend that, where the disease is spreading from secondary inoculum, monitoring may be necessary for at least 20–30 years to evaluate efficacy, and may need to be carried over into subsequent rotations (Shaw *et al.*, 1991).

Hagle and Shaw (1991) also suggest that it might be possible to escape the losses caused by epidemics of *Armillaria* root rot by maintaining the natural balance and diversity of those forests where the fungus is a minor pest and natural thinning agent but where significant losses would result if these forests were to be cleared for exotic plantations (Kile *et al.*, 1991; Hood *et al.*, 1991). Hence, the ecological losses as well as the economic costs of converting indigenous forests to plantations of exotics should be balanced against any profits derived. By contrast, there are a number of trees that can tolerate *Armillaria* sufficiently to survive either to a full, or almost full, lifespan, while still managing to maintain satisfactory productivity, even though it may thin out young stands (Filip *et al.*, 1989; Morrison, 1981; Rishbeth, 1972) and recycle nutrients in old stands (Durrieu *et al.*, 1985; Kile *et al.*, 1991).

Sometimes, losses are severe during the early years of the establishment of a plantation but subside later. Often this is because the secondary inoculum of the infected living host is less effective than the stumps or other buried woody material that comprise the primary inoculum. This pattern occurs in *Pinus radiata* plantations in New Zealand (Roth *et al.*, 1979), young, naturally regenerated *Pinus ponderosa* stands and plantations in western North America (Hadfield *et al.*, 1986; Hagle and Goheen, 1988; Morrison, 1981), and first-rotation conifer plantations on cleared hardwood sites in Europe (Hartig, 1873b; Nechleba, 1915; Pawsey, 1973).

Unless there is secondary spread of disease, even a high mortality following establishment is often insufficient to justify expenditure on control (Morrison, 1981; Hadfield *et al.*, 1986; Hood *et al.*, 1991), even in some sites in New Zealand where 50% of *Pinus radiata* crops may be lost during the first 5 years after planting (van der Pas, 1981b). In these cases, it may only be economically feasible to reduce primary inoculum (Hagle and Shaw, 1991). Alternative strategies, including increasing planting densities to compensate for these losses, may be more economical and environmentally acceptable than attempting either stump removal or chemical treatment to lessen the levels of initial inoculum. In this case, the surviving patches of trees in the plantation may need to be thinned later. By contrast, the cost of losing a few, or perhaps even a single tree or vine in orchards, amenity plantings, and vineyards, may justify more expensive control measures, such as the removal of inoculum. However, any decision to implement control requires accurate cost-benefit information (Hood *et al.*, 1991).

Choosing appropriate plants

Although some trees like holly and yew are much more tolerant than apple or lilac, they are not truly resistant like the grasses, some herbaceous plants, and some climbers like honeysuckle and probably ivy (but possibly not tree ivy). So, for complete confidence, turf an infected area with grass or plant other resistant non-woody or starchy rooted plants as no truly woody plant has yet been demonstrated to be immune to infection by *Armillaria*. Nonetheless, relative differences in resistance of many species with woody and starchy roots can be observed in forests, orchards, and parks, as well as in controlled inoculation experiments. Resistant rootstocks have been developed.

Mere fungal presence is not sufficient cause to treat. Some well adapted native forest trees may be sufficiently tolerant for *Armillaria* to act as a thinning agent in young stands and as a nutrient recycler in old stands. Host resistance to *Armillaria* root rot involves the genetics of both the host and the pathogen, as well as environmental influences. It also can involve managing mixtures of genotypes with varying levels of resistance. Some host species with superior resistance or tolerance to infection in one location may be more susceptible in other locations. Even within a locality, or even in a hedge or orchard, discrepancies in resistance can be seen, and intraspecific, or even individual, variation in adaptation to sites may be as great as interspecific variation within the natural ranges of any two or more species.

Advantages and disadvantages of eradication

Physical removal of inoculum by removing diseased trees and uprooting even neighbouring uninfected stumps has been recommended. This involves destroying stumps and root remnants by deeply digging through the soil. In France, trenches over a metre deep are dug to isolate the infected plants from healthy parts of a vineyard, or a fruit orchard is used to control spread of *Armillaria* root disease. This method has been adapted for cocoa and coffee plantations in Africa. Laying a plastic barrier in a trench and then back-filling it with the removed soil has been used for controlling the disease in kiwifruit orchards in New Zealand.

All these methods of physical control are laborious and mostly impracticable, especially in established orchards. The quantity and location of inoculum that must be removed to prevent disease build-up and spread, and the cost of removal may not be justified by the future value of the crop. Complete eradication of the fungus is also improbable and of doubtful value, and re-invasion by the fungus is possible. A major impediment in the chemical and biological control of *Armillaria* is the inability of the control agents to reach the site of inoculum inside wood in natural infections in sufficiently active state. The pathogen has evolved highly sophisticated mechanisms of protection against outside deleterious effects. These include the production of antibiotics and the formation of pseudosclerotia. After becoming established in the roots, the vegetative mycelium of *Armillaria* develops a protective layer of thick-walled fungus cells, the pseudosclerotial envelope, about the body of the infected wood and the white mycelial fans in the cambium region. The rhizomorphs growing from the apices of the pseudosclerotium are also covered by this protective layer. In order for any chemical or biological control agent to eradicate the fungus, they would have to enter these resistant structures before it can act.

References

BARNES, L.W. (1986). The future of phytopathological diagnostics. *Plant Disease* **70**, 180.

BRAY, V. (1970). Using creosote for treating Honey fungus. *Journal of the Royal Horticultural Society* **95**, 27–28.

DUNCAN, J.M. AND TORRANCE, L. (1992). *Techniques for the rapid detection of plant pathogens.* 235pp. Oxford: Blackwell.

DURRIEU, G., BENETEAU, A. AND NIOCEL, S. (1985). *Armillaria obscura* dans l'écosystème forestier de Cerdagne. *European Journal of Forest Pathology* **15**, 350–353.

DUSUNCELI, F. AND FOX, R.T.V. (1992). The accuracy of methods for estimating the size of *Thanatephorus cucumeris* populations in soil. *Soil Use & Management* **8**, 21–26.

FILIP, G.M. AND ROTH, L.F. (1977). Stump infections with soil fumigants to eradicate *Armillaria mellea* from young-growth ponderosa pine killed by root rot. *Canadian Journal of Forest Research* **7**, 226–231.

FILIP, G.M., GOHEEN, D.J., JOHNSON, D.W. AND THOMPSON, J.H. (1989). Precommercial thinning in a ponderosa pine stand affected by *Armillaria* root disease: 20 years of growth and mortality in central Oregon. *Western Journal of Applied Forestry* **4** (2), 58–59.

FOX, R.T.V. (1990a). Rapid Methods for Diagnosis of Soil Borne Plant Pathogens. In: *Soil-borne Diseases*. Ed. D. Hornby. Special Issue. *Soil Use & Management* **6**, 179–184.

FOX, R.T.V. (1990b). Diagnosis and control of *Armillaria* honey fungus root rot of trees. *Professional Horticulture* **4**, 121–127.

FOX, R.T.V. (1990c). Fungal foes in your garden, 10. Honey fungus root rot. *Mycologist* **4**, 192.

FOX, R.T.V. (1992). Honey Fungus detected. *Arboricultural Association Journal* **16**, 317– 326.

FOX, R.T.V. (1993). *Principles of Diagnostic Techniques in Plant Pathology*. Wallingford: CAB International.

FOX, R.T.V. AND HAHNE, K. (1989). Prospects for the rapid diagnosis of *Armillaria* by monoclonal antibody ELISA. In: *Proceedings of the 7th international conference on root and butt rots,* August 9–16, 1988, Vernon and Victoria, B.C. Ed. D.J. Morrison, pp 458–468. Victoria, B.C.: International Union of Forestry Research Organizations.

FOX, R.T.V. and POPOOLA, T.O.S. (1990). Induction of fertile basidiocarps in *Armillaria bulbosa. Mycologist* **4**, 70–72.

FOX, R.T.V., MCQUE, A.M. AND OBANYA OBORE, J. (1991). Prospects for the integrated control of *Armillaria* root rot of trees. In: *Biotic Interactions and Soil-borne Diseases.* Ed. A.B.R. Beemster, pp 154–159. Amsterdam: Elsevier Science Publishers.

GREGORY, S.C. (1984). The use of potato tubers in pathogenicity studies of *Armillaria* species. In: *Proceedings of the 6th international conference on root and butt rots of forest trees*, August 25–31, 1983, Melbourne, Victoria, and Gympie, Queensland, Australia. Ed. G.A. Kile, pp 148–160. International Union of Forestry Research Organizations.

GREGORY, S.C. (1989). *Armillaria* species in northern Britain. *Plant Pathology* **38**, 93–97.

GREIG, B.J.W., GREGORY, S.C. AND STROUTS, R.G. (1991). Honey fungus. *Forestry Commission Bulletin 100*, 11pp. London: Her Majesty's Stationery Office, Department of the Environment.

GUILLAUMIN, J.J. (1988). The *Armillaria mellea* complex. *Armillaria mellea* (Vahl.) Kummer *sensu stricto*. In: *European Handbook of Plant Diseases*. Eds. I.M. Smith, J. Dunez, D.H. Phillips, R.A. Lelliot and S.A. Archer, pp 520–523. Oxford: Blackwell Scientific Publications.

GUYOT, R. (1927). Mycélium lumineux de l'Armillaire. *Compte Rendus de la Societé de Biologie* **96**, 114–116.

HADFIELD, J.S., GOHEEN, D.J. AND FILIP, G.M. (1986). *Root diseases in Oregon and Washington conifers*. R6-FPM25086, 27pp. Portland, OR: U.S. Department of Agriculture, Forest Service, Pacific Northwest Region, Forest Pest Management.

HAGLE, S.K. AND GOHEEN, D.J. (1988). Root disease response to stand culture. In: *Proceedings of the future forests of the intermountain west: a stand culture symposium*. Gen. Tech. Rep. INT-243. Ogden, U.T.: USDA, Forest Service, Intermountain Forest and Range Experiment Station: pp 303–309.

HAGLE, S.K. AND SHAW, C.G. III (1991). Avoiding and Reducing Losses from Armillaria Root Disease In: *Armillaria Root Disease*. Eds. C.G. Shaw, III and G.A. Kile. *Forest Service Agriculture Handbook No. 691*. Washington, D.C.: USDA.

HALK, E.L. AND DE BOER, S.H. (1985). Monoclonal antibodies in plant disease research. *Annual Review of Phytopathology* **23**, 321–350.

HARTIG, R. (1873b). Vorlaufige Mittheilung uber den Parasitismus von *Agaricus melleus* und dessen Rhizomorphen. [Preliminary report on the parasitism of *Agaricus melleus* and its rhizomorphs.] *Botanische Zeitung* **31**, 295–297.

HARVEY, A.E., JURGENSEN, M.F. AND LARSEN, M.J. (1981). Organic reserves: Importance to ectomycorrhizae in forest soils of western Montana. *Forest Science* **27**, 442–445.

HOOD, I.A. AND SANDBERG, C.J. (1987). Occurrence of *Armillaria* rhizomorph populations in the soil beneath indigenous forests in the Bay of Plenty, New Zealand. *New Zealand Journal of Forestry Science* **17**, 83–99.

HOOD, I.A., REDFERN, D.B. AND KILE, G.A. (1991). *Armillaria* in planted hosts. In: *Armillaria Root Disease*. Eds. C.G. Shaw, III and G.A. Kile. *Forest Service Agriculture Handbook No. 691*. Washington, D.C.: USDA.

HUNT, R.S., PARMETER, J.R., JR. AND COBB, FW., JR. (1971). A stump treatment technique for biological control for forest root pathogens. *Plant Disease Reporter* **55**, 659–662.

HÜTTERMANN, A., FEIG, R. AND TROJANOWSKI, J. (1984). Biochemical capabilities and regulation as a basis for species differentiation and ecological behavior of *Armillaria* species. In: *Proceedings of the 6th international conference of root and butt rots of forest trees*, August 25–31, 1983, Melbourne, Victoria, and Gympie, Queensland, Australia. Ed. G. Kile, pp 57–72. International Union of Forestry Research Organizations.

INTINI, M.G. (1993). Development of *Armillaria* carpophores. *Mycologist* **7**, 16–24.

KILE, G.A., MCDONALD, G.I. AND BYLER, J.W. (1991). In: *Armillaria Root Disease*. Eds. C.G. Shaw, III and G.A. Kile. *Forest Service Agriculture Handbook No. 691*. Washington, D.C.: USDA.

KOEHLER, G. AND MILSTEIN, C. (1975). Continuous culture of fused cells secreting antibody of predetermined specificity. *Nature* **256**, 495–497.

KORHONEN, K. (1978). Interfertility and clonal size in the *Armillariella mellea* complex. *Karstenia* **18**, 31–42.

LUNG-ESCARMANT, B., DUNEZ, J. AND MONSION, M. (1978). La differenciation sérologique des formes typique et ostoyae d'Armillaire (*Armillaria mellea*), une preuve supplementaire de la valeur du critere immunologique dans la taxonomie des Champignons. *Comptes Rendus Hebdomadaires des Seances de l'Academie des Sciences, Paris, Serie P* **287**, 475–478.

LUNG-ESCARMANT, B. AND DUNEZ, J. (1979). Differentiation of *Armillariella* and *Clitocybe* species by the use of immunoenzymatic ELISA procedure. *Annales de Phytopathologie* **11**, 515–518.

LUNG-ESCARMANT, B. AND DUNEZ, J. (1980). Les propriétés immunologiques, une critère possible de classification de l'Armillaire. *Annales de Phytopathologie* **12**, 57–70.

LUNG-ESCARMANT, B., MOHAMMED, C. AND DUNEZ, J. (1985). Nouvelles méthodes de détermination des Armillaires européens: immunologie et electrophorse en gel de polyacrylamide. *European Journal of Forest Pathology* **15**, 278–288.

MALLETT, K.I. AND HIRATSUKA, Y. (1985). The 'trap-log' method to survey the distribution of *Armillaria mellea* in forest soils. *Canadian Journal of Forest Research* **15**, 1191–1193.

MANLEY, H. (1992). *Detection of Armillaria root rot*. Validation Report. University of Reading.

MILLER, S.A., RITTENBURG, J.H. PETERSEN, F.P. AND GROTHAUS, G.D. (1988). Application of rapid, field-usable immunoassays for the diagnosis and monitoring of fungal pathogens in plants. *Proceedings 1988 Brighton Crop Protection Conference - Pests and Diseases* 2, pp 795–804.

MORRISON, D.J. (1976). Vertical distribution of *Armillaria mellea* rhizomorphs in soil. *Transactions of the British Mycological Society* **66**, 393–399.

MORRISON, D.J. (1981). Armillaria root disease. A guide to disease diagnosis, development and management in British Columbia. Information Report BC-X-203. *Environment Canada, Canadian Forestry Service* 1–16.

MORRISON, D.J. (1989). Pathogenicity of *Armillaria* species is related to rhizomorph growth habit. In: *Proceedings of the 7th international conference on root and butt rots*, August 9–16, 1988, Vernon and Victoria, B.C. Ed. D.J. Morrison, pp 584–589. Victoria, B.C.: International Union for Forestry Research Organizations.

MORRISON, D.J., THOMSON, A.J. AND CHIN, D. (1984). Characterisation of *Armillaria* intersterility groups by isozyme patterns. In: *Proceedings of the 6th international conference on root and butt rot of forest trees*, August 25–31, 1983, Melbourne, Victoria, and Gympie, Australia. Ed. G.A. Kile, pp 2–11. International Union of Forestry Research Organizations.

NECHLEBA, A. (1915). Per Hallimasch: studien beobechtungen und hypothesen. [The honey agaric.] *Forstwissenschaftliches Centralblatt* **59**, 384–392.

NOBLES, M.K. (1948). Identification of cultures of wood-rotting fungi. *Canadian Journal of Research, Sect. C* **26**, 281–431.

PAWSEY, R.G. (1973). Honey fungus: recognition, biology and control. *The Arboricultural Association Journal* **2**, 116–126.

PAWSEY, R.G. AND RAHMAN, M.A. (1974). Armillatox field trials. *Gardeners Chronicle* **175**, 29–31.

PAWSEY, R.G. AND RAHMAN, M.A. (1976a). Chemical control of infection by honey fungus, *Armillaria mellea*: a review. *Arboricultural Journal* **2**, 468–479.

PAWSEY, R.G. AND RAHMAN, M.A. (1976b). Field trials with Armillatox against *Armillaria mellea*. *Pest Articles and News Summaries* **22**, 49–56.

PEGLER, D.N. AND GIBSON, I.A.S. (1972). *Armillariella mellea*. Commonwealth Mycological Institute Descriptions of Pathogenic Fungi and Bacteria No. 321. 2pp.

POON, O-S. (1988). *Detecting specific proteins for separating and identifying Armillaria species*. M.Sc. Thesis. U.K.: University of Reading.

RAABE, R.D. (1984). Production of sporophores of *Armillaria mellea* in isolated and pure culture. *Phytopathology* **74**, 855.

REDFERN, D.B. (1970). The ecology of *Armillaria mellea*: rhizomorph growth through soil. In: *Root diseases and soil-borne pathogens: Proceedings of the symposium*, July, 1968. Imperial College: London. Eds. T.A. Toussoun, R.V. Bega and P.E. Nelson, pp 147–149. Berkeley: University of California Press.

RISHBETH, J. (1970). The role of basidiospores in stump infection by *Armillaria mellea*. In: *Root diseases and soil-borne pathogens: Proceedings of the symposium*, July, 1968. Imperial College: London. Eds. T.A. Toussoun, R.V. Bega and P.E. Nelson, pp 141–146. Berkeley: University of California Press.

RISHBETH, J. (1972). Resistance to fungal pathogens of tree roots. *Proceedings of the Royal Society of London, Series B* **181**, 333–351.

RISHBETH, J. (1978) Infection foci of *Armillaria mellea* in first-rotation hardwoods. *Annals of Botany* **42**, 1131–1139.

RISHBETH, J. (1982). Species of *Armillaria* in southern England. *Plant Pathology* **31**, 9–17.

RISHBETH, J. (1986). Some characteristics of *Armillaria* species in culture. *Transactions of the British Mycological Society* **85**, 213–218.

ROLL-HANSEN, F. (1985). The *Armillaria* species in Europe. *European Journal of Forest Pathology* **15**, 22–31.

ROTH, L.F., SHAW, C.G., III AND MACKENZIE, M. (1979). Early patterns of Armillaria root rot in New Zealand pine plantations converted from indigenous forest - an alternative interpretation. *New Zealand Journal of Forestry Science* **9**, 316–323.

SHAW, C.G., III AND KILE, G.A. (1991). *Armillaria Root Disease. Forest Service Agriculture Handbook No. 691*. Washington, D.C.: USDA.

SHAW, C.G., III, STAGE, A.R. AND MCNAMEE, P. (1991). Modeling the dynamics, behavior and impact of Armillaria root disease. In: *Armillaria Root Disease*. Eds. C.G. Shaw, III and G.A. Kile, pp 150–156. *Forst Service Agriculture Handbook No. 691*. Washington, D.C.: USDA.

SHIGO, A.L. AND TIPPETT, J.T. (1981). Compartmentalization of decayed wood associated with *Armillaria mellea* in several tree species. *Res. Pap. NE488. USDA, Forest Service*. 20pp.

SMITH, M.L., BRUHN, J.N. AND ANDERSON, J.B. (1992). The fungus *Armillaria bulbosa* is among the largest and oldest living organisms. *Nature (London)* **256**, 428–431.

SWEDJEMARK, G. (1989). The use of sniffing dogs in root rot detection. In: *Proceedings of the 7th international conference on root and butt rots*, August 9–16, 1988, Vernon and Victoria, B.C. Ed. D.J. Morrison, pp 180–182. Victoria, B.C.: International Union of Forestry Research Organizations.

SWIFT, M.J. (1970). *Armillaria mellea* (Vahl ex Fries) Kummer in central Africa: studies on substrate colonisation relating to the mechanism of biological control by ring-barking. In: *Root diseases and soilborne pathogens: Proceedings of the symposium*, July, 1968. Imperial College: London. Eds. T.A. Toussoun, R.V. Bega and P.E. Nelson, pp 150–152. Berkeley: University of California Press.

TURNER, J.A. AND FOX, R.T.V. (1988). Prospects for the successful chemical control of *Armillaria* species. *Proceedings 1988 Brighton Crop Protection Conference - Pests and Diseases* 1, 235–240.

VAN DER PAS, J.B. (1981). Impact and control of Armillaria root rot in New Zealand pine plantations. In: *Root and butt rots in Scotch pine stands*, 1981, Poznan, Poland, pp 69–77. Polish Academy of Sciences, Poland: International Union of Forestry Research Organizations.

WARGO, P.M. AND SHAW, C.G., III. (1985). *Armillaria* root rot: the puzzle is being solved. *Plant Disease* **69**, 826–832.

WATLING, R., KILE, G.A. AND GREGORY, N.M. (1982). The genus *Armilliaria* - nomenclature, typification, and the identity of *Armillaria mellea* and species differentiation. *Transactions of the British Mycological Society* **78**, 271–285.

WHITNEY, R.D., MYREN, D.T. AND BRITNELL, W.E. (1978). Comparison of malt agar with malt agar plus orthophenylphenol for isolating *Armillaria mellea* and other fungi from conifer roots. *Canadian Journal of Forest Research* **8**, 348–351.

WILLIAMS, R.E., SHAW, C.G., III AND WARGO, P.M. (1989). *Armillaria mellea* Armillaria root disease. Forest Insect and Disease Leaflet 78 (rev.). *USDA, Forest Service*. 8pp.

8
Cultural Methods to Manage *Armillaria*

ROLAND T.V. FOX

The University of Reading, School of Plant Sciences, Department of Horticulture and Landscape (Crop Protection), 2 Earley Gate, Reading, Berkshire RG6 6AU, U.K.

Synopsis

Unlike forestry, in amenity and commercial horticulture attempts are often made to remove tree stumps and infected roots to reduce subsequent disease and windthrow. Soil may also be removed around the infection. Replanting may be delayed for several years. Soil fumigants have been used, but present materials require specialist handling and seem destined to be banned as they have been implicated with global warming. Soil drenches of formalin might have been a useful alternative, but this is now considered too hazardous. Other eradicant and protectant fungicides have given disappointing results in the field. Although many field soils already contain fungi which colonize *Armillaria* and reduce its viability, the application of cultures can provide protection to young trees. The action of some antagonistic micro-organisms may be compatible with certain fungicides. However, intervention should be delayed until a pathogen has been correctly identified, as similar harmless species of *Armillaria* and other macrofungi frequently co-habit the same environment as pathogens. Hence, accurate diagnosis is an essential prerequisite for deciding on the most effective control method to choose or, indeed, whether any form of control should be practised.

Overview of the management of *Armillaria*

In 1985, Schütt remarked that, whereas biological knowledge about *Armillaria* had increased markedly since the time of Hartig, the efficiency of control measures, with some exceptions, had hardly improved. Nonetheless, as a result of improved information about the biology of *Armillaria*, Shaw and Kile (1991) appreciate that the degree of destruction and the ways that root disease are expressed (Morrison *et al.*, 1991) are largely determined by the species and genotypes of *Armillaria* (Shaw *et al.*, 1981; Guillamin and Lung, 1985; Rishbeth, 1982; Kile and Watling, 1988; Roll-Hansen, 1985; Intini, 1989a) and their inoculum characteristics (Redfern and Filip, 1991). The inherent host resistance or tolerance (Thomas and Raphael, 1935), host adaptation to site (Intini, 1989a; Singh and Richardson, 1973), stand structure and species

Armillaria *Root Rot: Biology and Control of Honey Fungus*

composition, management history and site factors (Redfern, 1978; Blenis *et al.*, 1989; Gregory *et al.*, 1991) also directly affect the pathogen.

Convenient, cost-effective methods of control without significant economic, social or ecological drawbacks are needed to reduce the heavy losses in some orchards and amenity plantings (Hood *et al.*, 1991). Hagle and Shaw (1991) suggest a number of ways to avoid or reduce severe reductions in yield from *Armillaria* root disease epiphytotics in forests by good husbandry, where only properly applied cultural control methods are efficient and effective. They also consider that, despite substantial losses in some orchards and amenity plantings (Hood *et al.*, 1991), there actually may be no convenient or cost-effective methods of control free from significant economic, social or ecological costs. Nonetheless, progress in the accurate verification of the species involved (Sierra *et al.*, 1999) and its level of virulence if a pathogen (Fox, 1999), and models of its epidemiology (Kile *et al.*, 1991) ensure a clearer impression of the current and potential condition of the disease and possible options for its control or avoidance (Fox, 1999).

Economics of losses and control

Usually, only cultural control is attempted under forest conditions to lessen damage. In managed forests, damaging levels of disease can be avoided by management practices that include planting only on sites considered disease-free, cultural manipulation, chemical application, biological methods and integrated biological methods (Hagle and Shaw, 1991; Kile *et al.*, 1991). In ancient and other indigenous forests, *Armillaria* can cause three types of disease depending on the species of pathogen and host. Many tree and shrub species are attacked and killed by virulent pathogens (Filip, 1977; Gibson, 1960; Kile, 1981; MacKenzie and Shaw, 1977). Alternatively, *Armillaria* survives primarily in chronic, but normally non-aggressive, mild infections on roots. *Armillaria* can also cause butt rots which may or may not be related to other types of disease (Morrison *et al.*, 1991; Gregory *et al.*, 1991).

In primary or ancient woodland, *Armillaria* may even be considered a useful thinning agent and recycler of nutrients (Hagle and Shaw, 1991). Even in managed timber forest situations, the disease, although common, may not cause severe economic damage if only the butt is decayed. However, great importance is placed on the aesthetic damage to gardens and amenity areas, where the value of each established tree is high. Additionally, a host with weakened roots may be windthrown, causing substantial damage to surrounding plantings, buildings and people.

In horticultural plantations, orchards, vineyards and amenity plantings, the virulence of pathogenic *Armillaria* species also varies from aggressive to benign, but the options for control may be different to those possible in forests as their generally higher value can often compensate for higher costs (Hood *et al.*, 1991). Ideally, all tree stumps and infected roots should be removed from areas of woodland or orchard to reduce subsequent disease and windthrow (Fox, 1990), as well as removal of soil around the infection. Alternatively, the existing soil can be replaced after it has been thoroughly sifted to remove even the smallest infected root pieces where the crop is of high value (Horner, 1988).

Deciding whether to intervene or not

The species of *Armillaria* found on a decaying stump, diseased tree or shrub is the chief factor influencing the decision to attempt control (Rishbeth, 1983). In an illustration from Hagle and Shaw (1991), while control is not necessary when *A. lutea* is discovered to have spread widely from a stump in a garden without attacking any other trees or shrubs, it would be advisable if, instead, *A. mellea* had spread from a stump and killed neighbouring trees. However, until recently, the only feasible ways to separate virulent pathogens from harmless saprophytes were based on the morphology of any rhizomorphs or basidiocarps present, or interfertility tests carried out with isolated cultures. Although widely used by taxonomists, successful *in vitro* mating requires laboratory conditions and is slow and frequently difficult to interpret, since pure cultures have to be plated next to isolates of a known species. Several molecular methods based on immunolology or nucleic acid techniques have been developed, but need to be made easier if cheap kits are to be made available to arboriculturalists. Now the identification of *Armillaria* species is becoming more routine (see chapters 4 and 5), diagnostics are likely to be regularly employed before a control programme is advised. Infections in the roots of a tree by pathogenic *Armillaria* species should then be detected unambiguously and the levels of infection assessed, so it can be treated or destroyed before developing into a dangerous rot or spreading to other trees.

In managed forests, damaging levels of disease can be avoided by management practices that include planting only on sites considered disease-free (Horner, 1990; Baker, 1972; Heaton and Dullahide, 1989), cultural manipulation, chemical application, biological methods and integrated biological methods (Hagle and Shaw, 1991; Kile *et al.*, 1991). In ancient and other indigenous forests, *Armillaria* can cause three types of disease depending on the species of pathogen and host. Many tree and shrub species are attacked and killed by virulent pathogens (Filip, 1977; Gibson, 1960; Kile, 1981; MacKenzie and Shaw, 1977). Alternatively, *Armillaria* survives primarily in chronic, but normally non-aggressive, mild infections on roots. *Armillaria* can also cause butt rots which may or may not be related to other types of disease (Morrison *et al.*, 1991; Gregory *et al.*, 1991).

In unmanaged woodland, while the first disease type may require drastic measures to suppress the epiphytotic, the other two types may cause little damage, especially if the resistance or tolerance of infected trees is maintained and not stressed by drought, insect attack, other diseases, or human activity (Wargo and Harrington, 1991).

Drawbacks of methods involving cultural procedures

Direct control of *Armillaria* is difficult to achieve due to the extensive network of rhizomorphs and mycelium in the soil, sometimes at great depth or protected inside dead wood. Destroying toadstools is unlikely to affect local disease levels very much.

Choosing a suitable site

Another inexpensive way to reduce losses is by matching the species of tree to a site with soil that does not predispose the host in any way (Sokolov, 1964; Ono, 1970; Gramss, 1983) or is still heavily contaminated by pathogenic *Armillaria* species from infections that occurred in the previous stand. However, some more productive habitat types can have greater root disease severity than those which are less productive (Byler *et al.*, 1990; McDonald *et al.*, 1987). The vegetation initially growing on a site prior to clearance to establish an orchard or plantation may display variations in root disease hazard (Leach, 1939; Gibson, 1960; Shaw and Calderon, 1977; Hendrickson, 1925; Cooley, 1943; Proffer *et al.*, 1987).

Methods that alter predisposing environmental conditions

Efforts to reduce disease should concentrate on removing the predisposing condition where there is one (Wargo and Harrington, 1991), as the fungus is often a more successful pathogen when the host is stressed. Thus, the overall health of the tree should be considered when control methods are under consideration.

Many other diseases can be controlled by altering the environmental factors which determine the infective activity of the pathogen. Although those affecting *Armillaria* are still rather poorly understood, they are thought to include waterlogging, shading, drought, defoliation, advanced age and declining vigour of the host plant and infection by other diseases (Hood, Redfern and Kile, 1991), damage due to other agents (pollution, insects, fungi, etc.) and a high population density of trees. These factors are complicated in their action and appear to interact. They not only facilitate infection by the virulent species *A. mellea* and *A. ostoyae*, but also allow weakly virulent pathogens such as *A. lutea* to attack a tree which would have resisted the infection if healthy. Popoola and Fox (1991) tried to manage these potentially devastating diseases more effectively by evaluating the role of water stress on the host in *Armillaria* infections.

Until we know more, it is probably advisable to ensure that healthy young trees are planted adequately spaced to allow sufficient light in an appropriate well drained soil, taking care to avoid damage due to abiotic causes and treat other diseases and pests. While the maintenance of healthy, vigorous trees may reduce the risks of *Armillaria* infection, over-zealous pruning should be particularly avoided. It is also prudent to replant previously infected areas with trees which are considered less susceptible to *Armillaria* infection (Greig and Strouts, 1983; Greig *et al.*, 1991), grass or resistant herbaceous plants (Robinson-Bax, 1999).

It is possible to partly modify the pH, organic content and nutrient status of soil in orchard plantations. Young Scots pine plantations benefit much less from fertilizer application than chronically infected trees (Rykowski, 1981). Fertilizers can both boost the fungistatic effects of extracts from periderm and phloem tissues, as well as encourage saprophytic growth due to the improved nutritional value of the wood (Rykowski, 1983). Additional potassium can significantly reduce damage from *Armillaria* in banana plantations in Malawi (Spurling and Spurling, 1975). Adding fertilizers or other soil amendments may not protect forest trees if their nutritional requirements are not known.

There is no point in treating plants infected by *Armillaria* if they suffer injury from exposure to deposits of atmospheric pollution or photochemical oxidants that are, in many cases, not only difficult to confirm but also troublesome to prevent. However, where *Armillaria* is virulent, infection often leads to rapid death, even if the plants were initially vigorous.

Methods involving eradication

PRUNING

Although excessive pruning of the foliage and branches should be avoided as it may result in root death (Popoola, 1991; Wargo and Harrington, 1991), pruning and girdling diseased roots (Kendall, 1931) or drying and aerating the root collar (Munnecke *et al.*, 1976; Kendall, 1931) may be worthwhile (Shaw and Roth, 1978, 1980). This may be made easier if a water jet is used (Levitt, 1947). Exposing root collars in this way can eradicate *Armillaria* (Sokolov, 1964). However, orchard workers could be injured if large holes are left uncovered (Rackham *et al.*, 1966).

STUMP REMOVAL

Physical removal of inoculum by removing diseased trees, uprooting stumps, even neighbouring uninfected stumps, by digging through the soil deeply is often recommended, though all these methods of physical control are laborious and mostly impracticable, especially in established orchards. The quantity and location of inoculum that must be removed to prevent disease build-up and spread, and the cost of removal, may not be justified by the future value of the crop. Complete eradication of the fungus is also improbable and of doubtful value, and re-invasion by the fungus is possible. Soil disturbance may also stimulate fresh rhizomorph production, increasing the risk of disease to newly planted trees. Excessive removal of woody debris from a soil may be detrimental to antagonistic mycorrhizae.

On amenity sites, an attempt to control *Armillaria* by cultural practices has often been recommended by removing stumps and roots whenever it is feasible (Pawsey, 1973). Ideally, all tree stumps and infected roots should be removed from areas of woodland or orchard to reduce subsequent disease and windthrow. The grubbing out and burning or removal of tree stumps reduces inoculum and uncolonized potential substrate. This can reduce the number of infection foci and, more importantly, reduces the energy available to rhizomorphs as the remaining food bases from which they grow are the relatively small remnants of roots left in the soil after grubbing (Gregory, 1989). The smaller pieces of inoculum will have a much shorter life-span than large stumps (Shaw, Stage and McNamee, 1991) but, even then, can pose a threat. The chipping of windthrown trees for the production of mulching material should therefore only be carried out after an inspection of the root system for signs of the disease.

Stump removal can be costly and time consuming in forests and plantations, but may often be totally impracticable to carry out in gardens. Stump removal may involve winches and excavators, or the stump may be chipped using a machine specifically designed for this purpose. All large pieces of wood must be removed and burned and the site should not be replanted for at least 12 months to allow the fungus to starve. In situations where the stump cannot be removed, or is on a neighbour's land, adjacent plants may be protected from rhizomorph infection by physical barriers such as buried polythene sheeting, deep trenches, and even regular digging of soil (which would sever the rhizomorphs from their food bases). Stump and root removal is most effective when the disease is caused only by the inoculum that was already originally present, but ceases once the secondary inoculum only remains. Where secondary inoculum is prominent, the disease tends to occur as well-defined patches rather than scattered diffusely throughout the stands. These clusters permit the decayed stumps and roots to be removed more completely.

In parts of Australia, where *A. luteobubalina* can kill half of the newly planted fruit trees, orchard site preparation begins by ripping to 35 cm to remove the larger roots, followed by ripping again to 25 cm to remove the thinner roots, then any remaining roots are removed by hand (Heaton and Dullahide, 1989). Root debris should decay rapidly in the tropics, so in Kenya, when pits are dug to extract infected coffee roots, they are either left open for several months to provide a short fallow, or they are treated with a soil sterilant before being filled (Baker, 1972).

GIRDLING

Since the nutritional quality of stumps and roots will influence the longevity of the pathogen, ring-barking has been used to restrict the spread of *Armillaria* in African tea plantations (Leach, 1937), possibly because it lowers the concentrations of carbohydrates stored within the trees' roots so much that they were insufficient to support *Armillaria* but adequate for other saprophytic fungi. However, in Britain, Redfern (1968) found that the roots of mature oaks that had been ring-barked one year prior to felling were more rapidly decayed by *Armillaria* than those felled without prior treatment. The greater production of rhizomorphs in European soil than in African soils (Wiehe, 1952) may explain this discrepancy, but Lanier (1971) showed that girdling old Scots pine and beech one year before felling did reduce the subsequent attacks on young pines. Thus, ring-barking may prevent the invasion of stumps from external inoculum sources, but does not inhibit, and probably may enhance, spread from pre-existing lesions (Swift, 1970).

TRENCHES AND BARRIERS

Removal of soil around the infection can also reduce inoculum levels but the construction of physical barriers has met with mixed results, probably because infected roots and rhizomorphs can be present in the soil at considerable depths. Physical barriers should go down to a depth of 45 cm, or more in well drained soils (Gregory, 1989). These barriers are laborious to make and need to be regularly inspected, and protection by digging must be repeated annually. In France, trenches over a metre deep are dug to isolate the infected plants from healthy parts of a vineyard or a fruit orchard, and thus control spread of *Armillaria* root disease. This method has been adapted for cocoa and coffee plantations in Africa. Laying a plastic barrier in a trench and then backfilling it with the removed soil has been used for controlling the disease in kiwifruit orchards in New Zealand.

Methods based on tolerance and resistant rootstocks

Where an infected area cannot be avoided, the use of more resistant plants or rootstocks may also reduce the impact of *Armillaria*. The lists of more and less susceptible woody plants, such as those published by Greig and Strouts (1983) and Greig *et al.* (1991), are helpful, even though they are inevitably a rough guide largely based on anecdotal evidence. No truly woody plant has yet been demonstrated to be immune to infection by *Armillaria*.

According to Greig and Strouts (1983), those plant families with genera consistently among the least affected by *Armillaria* include representatives of the Actinidiaceae, Anacardiaceae, Araliaceae, Aristolochiacae, Berberidaceae, Bignoniaceae, Buxaceae, Calycanthaceae, Caprifoliaceae, Celastraceae Cercidiphyllaceae, Cistaceae, Clethraceae, Compositae, Cornaceae, Elaeagnaceae, Fagaceae, Polygonaceae, Gramineae, Guttiferae, Labiatae, Lardizabalaceae, Lauraceae, Liliaceae, Lythraceae, Magnoliaceae, Malvaceae, Moraceae, Myricaceae, Myrtaceae, Nyssacea, Papavaraceae, Passifloraceae, Pittosporaceae, Plumbaginaceae, Protaceae, Punicaceae, Rhamnaceae, Rutaceae, Sapindaceae, Scrophulariacaeae, Simarubaceae, Stachyuraceae, Staphyleaceae, Sterculiaceae, Styracaceae, Symplocaceae, Tamaricaceae, Taxaceae, Theaceae, Ulmaceae and Verbenaceae.

It may prove significant that several of the notably tolerant families consist of genera of plants that are thin stemmed climbers (Actinidiaceae, Araliaceae, Aristolochiacae, Caprifoliaceae, Passifloraceae), grasses or bamboos (Gramineae). Some contain potent herbal medicines or other bioactive substances (Cistaceae, Compositae, Labiatae, Lauraceae, Myrtaceae, Papavaraceae, Rhamnaceae, Rutaceae, Styracaceae, Taxaceae, Theaceae, Ulmaceae and Verbenaceae). Others have exceptionally dense wood (Buxaceae, Fagaceae). On the other hand, many types of plant have frequently been recorded as being hosts of honey fungus, and some woody plants are considered very susceptible to infection.

Grass and some herbaceous ornamental plants and vegetables can be planted on infected areas but it is not always realized that many herbaceous plants, such as those with starchy roots and tubers like *Iris*, strawberry (Fox and Popoola, 1990) and potato, are prone to attack. Some herbaceous plants like strawberry are so susceptible they may be killed within a few weeks, enabling them to be used as cheap 'live bait' to reveal where *Armillaria* is present or absent in the soil, as clonal material is far quicker and easier to propagate than trees (Fox and Popoola, 1990). Large trees may show a gradual decline, or even an increased final fruit yield, but with strawberries there is often little or no evidence of infection until the plants finally wilt and die in a short space of time. Nonetheless, relative differences in resistance of many species with woody and starchy roots can be observed in forests, orchards, and parks, as well as in controlled inoculation experiments.

Host resistance to *Armillaria* root rot involves the genetics of both the host and the pathogen, as well as environmental influences. It can also involve managing mixtures of genotypes with varying levels of resistance. Some host species with superior resistance or tolerance to infection in one location may be more susceptible in other locations. Even within a locality, or even a hedge or orchard, discrepancies in resistance can be seen, and intraspecific or even individual variation in adaptation to sites may be as great as interspecific variation within the natural ranges of any two or more species.

Although some trees like holly and yew are much more tolerant than apple or lilac, they are not truly resistant, like many grasses, some herbaceous plants and some climbers. So for complete confidence, replant an infected area with turf grass or other resistant non-woody or starchy rooted plants. Of course, host resistance cannot cure a diseased tree or act as a protectant, but may be valuable as part of an integrated programme of different complementary cultural practices, offering some management of the disease, though rarely total control.

Even though stands of some trees may be thinned out when young, some can tolerate *Armillaria* sufficiently to survive, either to a full, or almost full, lifespan, maintaining satisfactory productivity (Filip *et al.,* 1989; Morrison, 1981; Rishbeth, 1972a) and helping recycle nutrients (Durrieu *et al.*, 1985; Mason *et al.*, 1989; Kile *et al.*, 1991). Hence, the mere presence of *Armillaria* in a forest or orchard where it has not resulted in significant harm may not justify its eradication.

Resistant rootstocks selected for plum, pear and grape have been recommended for *Armillaria* resistance (Heaton and Dullahide, 1989). While most *Prunus* species are severely damaged, plum, prune and apricot are generally more tolerant (Thomas *et al.*, 1948). Resistant *Prunus* rootstocks have been selected (Raabe, 1966). Since plum is generally more resistant than peach, almond, apricot, and cherry, the only practicable control is given by such resistant rootstocks resulting from interspecific crosses between diploid plum and peach (Guillaumin *et al.*, 1989). These resistant rootstocks appear to satisfy both *Armillaria* resistance and other cultural demands. There is also considerable variation in the susceptibility of rootstocks of citrus to *Armillaria* (Rhoads, 1948). Loblolly pine, which tolerates *Armillaria,* can be used as a rootstock with a heterospecific graft of slash pine (Armitage and Barnes, 1968).

Probably, there is scope to improve the resistance of populations or clones by disease genetic modification (Hubbes, 1987). While producing species that are more resistant to *Armillaria* may be economically practicable, in some cases they may be best used integrated with other control methods, such as planting mixtures of species with differing resistance to reduce the secondary spread (Morrison *et al.*, 1988).

Instead of replanting the whole of previously infected areas with trees which are considered less susceptible to *Armillaria* infection (Greig and Strouts, 1983; Greig *et al.*, 1991), grass or resistant herbaceous plants (Robinson-Bax, 1999), Hagle and Shaw (1991) recommend growing natural mixtures and densities of local tree species by selecting locally adapted seed sources (Hadfield *et al.*, 1986; Morrison, 1981; Williams *et al.*, 1989). Even the genetic variation within a plantation of non-indigenous trees can be exploited (Lung-Escarmant and Taris, 1989).

The inadequate adaptation of an exotic species to a specific site may influence its susceptibility to *Armillaria* root disease (McDonald, 1990). Using cultural methods to match site conditions (Corns and Annas, 1986) and habitat type (Daubenmire, 1952) can avoid substantial losses due, for example, to drought (Kessler and Moser, 1974). Although *Pinus radiata* planted in New Zealand suffers extensive destruction by *Armillaria* root disease (Shaw and Calderon, 1977), it is not particularly affected in its natural range in western North America (Raabe, 1979), probably due to different *Armillaria* species, inoculum loads, edaphic, climatic and physical site characteristics.

Growing trees from seeds

Planting bareroot stock has advantages (Page, 1970; Schubert *et al.*, 1970), but one of the disadvantages is significantly greater injury from *Armillaria* than trees originating from seeds (Singh and Richardson, 1973; Kessler and Moser, 1974).

Fallows

Armillaria rarely prevents forests being established on former natural grassland or agricultural land. However, if trees were once present, secondary incculum can be a problem as Rishbeth (1972b) found. Hardwood stumps from trees cut 40 years before could still produce rhizomorphs, albeit at much lower quantity than stumps cut more recently. So, stump removal followed by a short fallow period can often help to reduce the inoculum further. Another way of achieving a similar objective in an orchard is to re-establish the new rows of trees midway between the previous rows, thus avoiding the exact place from which the infected trees had recently been removed (Heaton and Dullahide, 1989).

Increasing the length of rotations can significantly decrease losses (Greig and Strouts, 1983; Marquis and Johnson, 1989; Whitney, 1988b) by harvesting stands or individual trees before they are anticipated to succumb to *Armillaria* root disease (Marquis and Johnson, 1989). This is comparable to training fruit trees in Australia to ensure a prompt return on investment by early cropping to avoid the heavy losses caused by the secondary spread and subsequent loss of trees to *Armillaria* (Heaton and Dullahide, 1989).

Short rotations, like intensively managed fruit orchards, allow the pathogen an insufficient duration for secondary spread from inoculum that has not been removed, but as a form of 'rotation', replanting is often delayed for several years.

Although extended fallow periods might also prove as effective as stump removal in temperate regions, it is generally considered too expensive because of the wasted production time (Roth *et al.*, 1977). However, this overlooks the possibility of using the fallow land for grazing or some arable cropping in the interim. Annual cropping with cereals or alfalfa 4–5 years on the sites of old orchards yields income while most of the nutritional sources for *Armillaria* are depleted (Guillamin, 1977; Mallet *et al.*, 1985).

Thinning

Thinning is unlikely to reduce losses and may increase damage by providing additional food bases where the state of the host is unimportant, but reduced inter-tree competition may effectively reduce disease losses where *Armillaria* is a secondary pathogen (Davidson and Rishbeth, 1988; Wargo and Shaw, 1985; Hagle and Goheen, 1988; Filip *et al.*, 1989; Singh, 1981).

Limitations of chemical control

A major impediment in the chemical and biological control of *Armillaria* is the inability of the control agents to reach the site of inoculum inside wood in natural infections in sufficiently active state. The pathogen has evolved highly sophisticated mechanisms of protection against outside deleterious effects. These include the production of antibiotics and the formation of pseudosclerotia. After becoming established in the roots, the vegetative mycelium of *Armillaria* develops a protective layer of thick-walled fungus cells, the pseudosclerotial envelope, about the mass of infected wood and the white mycelial fans in the cambium region. The rhizomorphs growing from the apices of the pseudosclerotium are also covered by this protective layer. In order for any chemical or biological control agent to eradicate the fungus, they would have to enter these resistant structures before it can act. At present, only the most volatile fumigant chemical fungicides appear to be able to succeed adequately.

Shaw and Roth (1978) stated that, apart from the use of soil fumigation with carbon disulphide, methyl bromide, or chloropicrin after removing woody debris, there is very little experimental evidence to support the effectiveness of the most commonly advocated chemical treatments against *Armillaria*. Hagle and Shaw reiterated this view in 1991. Pawsey and Rahman (1976a) claimed that the justification for employing many of these treatments relied on superficial and subjective criteria. Against this background of scepticism voiced by Pawsey and Rahman (1976a), Shaw and Roth (1978, 1980), Thies and Russell (1984), and West (1999) seek to review the history and the potential of chemical control.

The major problem must be the enormous size of some genets (Smith *et al.*, 1992). Apart from the usual problems of soil treatment, Shaw and Roth (1978, 1980) emphasize the need to appreciate whether a chemical application is needed to protect uninfected plants, eradicate the fungus in infected stumps and roots, or treat or cure infected but still living plants.

Research on new chemicals

New research should investigate the behaviour of chemicals in soil and wood, combined with their fungicidal effects. New treatments for soil may be needed now that the chemical soil fumigants used to destroy *Armillaria* in root fragments throughout the soil in orchards, vineyards and floriculture operations (Kissler *et al.*, 1973; Richardson and Johnson, 1935; Godfrey, 1936; Bliss, 1951; Heaton and Dullahide, 1989) seem likely to be banned for their part in global warming.

Once a potential chemical is identified, its efficacy in the field should be examined with consideration to a number of factors such as: phytotoxicity (which may be less after an application in any dormant season), soil type, effect of temperature and wetness, and the potential for fungicide resistance or premature breakdown. The ideal chemical control should be selectively toxic, so as not to cause any phytotoxicity to the host or produce undesirable residues. Partly for these reasons, downwardly mobile systemic chemicals would be valuable.

West (1999) argues that, in order to test both the protectant and eradicant properties of each potential chemical against infection by rhizomorphs and root-to-root contact, artificial inoculum should be used wherever possible. The ability of the chemicals to eradicate rhizomorphs growing in soil, and to eradicate mycelium from within wooden inocula should also be investigated. Activity against a number of different isolates of *A. mellea* or other major pathogenic species should also be considered.

Selectivity is also of practical importance. At lethal doses, the fumigation of stumps can exclude or eradicate *Armillaria* directly (Bliss, 1951; Rackham *et al.*, 1966; Filip and Roth, 1977), but sub-lethal doses may allow it to be replaced by competing fungi which are less affected by the fumigant (Munnecke *et al.*, 1973; Ohr *et al.*, 1973).

Limitations of biological control

The objective of biological control of plant pathogens is to lessen disease by reducing the inoculum of the pathogen through decreased survival between crops, decreased production or release of viable propagules, or decreased spread by mycelial growth, reducing infection of the host by the pathogen, reducing the severity of attack by the pathogen, or a combination of any of these. Antagonism includes three types of activity, antibiosis and lysis, competition or parasitism and predation (Baker and Cook, 1974).

Although an antagonist should produce inoculum in excess, resist, escape, or tolerate other antagonists, germinate and grow rapidly, and invade and occupy organic substrates, extremes of any of these features may cause it to fail as a biological control agent.

A rhizosphere or wood-inhabiting organism can control *Armillaria* by inhibiting or preventing rhizomorph and mycelial development, by limiting the pathogen to substrate already occupied, by actively pre-empting the substrate, or by eliminating *Armillaria* (perhaps through replacement) from substrate already occupied (Hagle and Shaw, 1991). Shaw and Roth (1978, 1980) remind that the pursuit of these potential advantages should consider the feasibility of the techniques available. Hence, these constraints are outlined more fully by Raziq (1999).

Attempts to reduce inoculum in infected stumps has mainly been by the introduction of microbial antagonists which are present in tree stumps (Rishbeth, 1976; Pearce and Malajczuk, 1990). Although a biocontrol agent may eradicate *Armillaria* or prevent saprophytic colonization of new substrate, there is inevitably a delay following surface inoculation of stumps before the biocontrol agent becomes established in the roots. An *Armillaria* colony can also be extensive, and may be protected inside pseudosclerotia in infected wood (Garrett, 1970). Direct inoculation of the roots is difficult. Although Dowson, Rayner and Boddy (1988a, 1988b), Pearce and Malajczuk (1990) successfully established a number of saprotrophic cord-forming fungi in woodland which replace *Armillaria* spp. from soil and wood and also compete with it for substrate, this technique is only suitable to protect stumps, not living trees. In a similar way, painting the stumps with ammonium sulphamate (AMS) to encourage saprophytic antagonists of *Armillaria* (Rishbeth, 1976) is a successful form of indirect biological control, directed at preventing infection by *Armillaria* basidiospores and increasing the colonization and decay of tree stumps by some cord-forming fungi above ground (Rayner, 1977).

In tropical regions where dead roots decay rapidly, pathogens may be readily replaced by saprophytes if silvicides used to kill host tissues are used to kill trees rapidly before cutting (Mallet *et al.*, 1985). However, rapid killing by herbicides like 2,4,5-T may favour colonization by *Armillaria* (Rishbeth, 1976) and, when assessed ten years later, herbicide killed oak stumps produced more rhizomorphs than oaks which had been girdled, probably for the same reason (Pronos and Patton, 1979).

Preventing soil-borne colonization by *Armillaria* appears more likely to employ fungi which naturally occur in field soil and which act as antagonists to *Armillaria*, thus reducing its viability (Raziq, 1999). These may be especially effective after *Armillaria* has already been weakened by a fungicidal soil treatment that alone is non-lethal. Bliss (1951) and others, such as Ohr and Munnecke (1974), have shown that high levels of *Trichoderma viride* in the soil of Californian citrus and peach orchards after fumigation by carbon disulphide and methyl bromide are associated with reduced viability and colonization by *Armillaria*.

Integration of the control of *Armillaria* root rot of trees and shrubs by combining the most effective of the new generation chemical fungicide drenches together with the inoculation of the pathogen with antagonistic micro-organisms, has been suggested (Fox *et al.*, 1990). Even when unaffected by fungicides, *Armillaria* species grow very slowly in comparison with most saprophytic fungi. Since *Armillaria* species require access to easily available carbohydrate before invading host roots (Popoola, 1991), they could be controlled indirectly by the presence of a saprophyte, especially if such a micro-organism is also an antagonist, often producing chemical substances that are inimical to *Armillaria* or is a parasite. Cultures of antagonistic micro-organisms obtained from a variety of sources, including diseased cultivated mushroom beds, have been evaluated as biological control agents (Raziq, 1999). *Armillaria mellea* and *A. ostoyae* inoculated with *Trichoderma harzianum* and *Dactylium dendroide*s were killed, whereas their growth was only inhibited by *Diehliomyces microsporus*, and *Pseudomonas tolaasi* showed no significant effect (Fox *et al.*, 1990). The micro-organisms tested demonstrated three aspects of antagonism; competition, exploitation, and antibiosis. While all the moulds to some extent were competitive enough to prevent the *Armillaria* isolates from utilizing the nutrients on which they grew, the most effective control was shown by *Dactylium dendroides* and *Trichoderma harzianum*, antagonistic fungi which appear to employ antibiosis against the *Armillaria* isolates, which they also exploit as a food source. *Diehliomyces microsporus* failed to out-compete the *Armillaria* isolates which produced an inimical chemical substance, this also affected *Dactylium dendroides*, but to a limited extent. So competition, the direct rivalry between two species for nutrients in short supply, appears inadequate by itself, even though the effect of such stress on *Armillaria* causes reduced growth, reduced production of antibiotic substances, and increased exudation of substances from the fungus, all of which combine to make a favourable environment for the antagonists.

Ectomycorrhiza are thought to increase plant vigour and increase the resistance of plants to root diseases because the pathogen has to penetrate an external barrier of tightly interwoven hyphae, and thicker cortical cells, which may have another layer of hyphae (Jalali and Jalali, 1991). It has been postulated that the formation of mycorrhiza is incomplete on Sitka spruce, which is exotic in France, and thus cannot provide satisfactory protection against *Armillaria* (Gaudray, 1973). Although some mycorrhizal fungi inhibit *Armillaria in vitro* (Eghbaltalab *et al.*, 1975), it is doubtful if their presence on the fine roots would directly protect the main sites on the collar and the larger roots where *Armillaria* usually infects.

One of several intrinsic advantages biological control of any plant pathogen has over its chemical control (Hunt *et al.*, 1971) is that it is more acceptable to the public. Biological control of *Armillaria* also does not entail either the expense, inconvenience or initial unsightliness of stump and root removal.

Limitations of integrated control

Integration of chemical and biological control may be possible by inoculating sources of *Armillaria* with a variety of micro-organisms, including mycophagous nematodes, such as *Aphelenchus avenae,* while protecting trees at risk with drenches of modern fungicides. Several examples of integrated control of *Armillaria* have been reported. *Trichoderma viride* can replace *Armillaria* in roots after fumigation between tree crops. Unfortunately, such biocidal fumigants are unsuitable for use on established trees as they are too phytotoxic.

When root pieces infected with *A. mellea* were fumigated with carbon disulphide and incubated either without soil or in soil that was previously sterilized, *Armillaria* survived, but when the infected root pieces were buried after fumigation in unsterilized soil or in soil amended with *Trichoderma, A. mellea* was killed and replaced by the antagonist. As pure soil cultures of *Trichoderma* were also able to kill *Armillaria* in unfumigated root inocula, *Armillaria* was not killed by the direct fungicidal action of carbon disulphide but by the increase in the populations of *Trichoderma* spp., which overcome the protective pseudosclerotium then reach every part of the fungus body and kill the mycelium by antibiotic action.

Sub-lethal methyl bromide fumigation prevents the production of antibiotics by *Armillaria.* A similar effect may be caused by heating or drying. These stress factors may be very critical and may concurrently stimulate antagonistic organisms, resulting in further damage to the already weakened *Armillaria* whose metabolism is affected, hindering the repair of any ruptures in the pseudosclerotial walls, breaching the defence mechanisms and ending antibiotic production. So, the simultaneous increase in the growth of *Trichoderma* results in the death of *Armillaria.* Stressed mycelium of *Armillaria* leaks substances that could be attractive or stimulatory to *Trichoderma* spp. and other antagonists of *A. mellea. Trichoderma* spp. are several times more resistant to methyl bromide than *A. mellea.* Thus, *Trichoderma* spp. are to able to survive much higher concentrations of methyl bromide than *A. mellea*, and populations are presumably unaffected, or even increased, by field fumigations with methyl bromide that are lethal to *A. mellea. Trichoderma* spp. are much more tolerant of environmental conditions than *A. mellea.* The responses of the two fungi to temperatures of 30 to 39°C are particularly noteworthy. While growth of *Trichoderma* increased as temperatures increased, *A. mellea* was severely affected and its growth ceased.

The incorporation of mycorrhiza into integrated systems is problematic as *Trichoderma viride* and *T. polysporum* are also antagonistic to mycorrhizal colonization and established mycorrhiza on black spruce (*Picea mariana*) seedlings (Jalali and Jalali, 1991). Also, some of the chemicals which have been used in an attempt to control *Armillaria* are also toxic to mycorrhiza.

Physical methods

FIRE, HEATING AND DRYING

Neither the mycelium nor rhizomorphs of *Armillaria* can withstand exposure drying for long. Unfortunately, the methods of heating and drying devised (Birmingham and Stoke, 1921; Rackham *et al.*, 1966) are not considered sufficiently cost-effective or simple, except to high-value orchard and ornamental crops (Hagle and Shaw, 1991). Exposure to fire after the clear felling of indigenous forest in New Zealand can significantly reduce the amount of viable rhizomorphs (Hood and Sandberg, 1989), but this does not seem to prevent the severe losses from *Armillaria* in pine plantations established on them (Shaw and Calderon, 1977), even though Reaves *et al.* (1990) recovered similar numbers of *Trichoderma* isolates from both soils. However, they found that the species isolated most frequently after burning were also those most antagonistic to *Armillaria*.

ELECTRICITY

Although it is possible to demonstrate the lethal effects of an electrical field on *A. mellea* (Fox and Sanson, 1996), at present there still appear to be a number of obstacles to be mastered before it can be used as a practical control measure against any soil-borne disease pathogen. If this limitation could be overcome, it might be possible to limit or fend off the spread of *Armillaria* using solar powered electrical probes stuck into the soil as a barrier between plants requiring protection and the source of infection. Other related techniques to eradicate *Armillaria* using exposure to radiation, for example from microwaves and radiowaves, also face the challenge of penetrating wood and soil.

Conclusions

STRATEGY FOR CONTROL

Controlling *Armillaria* infections needs to be tackled sensibly, employing well thought out procedures, avoiding a panic reaction to the problem. The prudent stewardship of trees demands careful evaluation of the necessity for control and, if found necessary, the best option to implement. *Armillaria* may be present and cause little damage. Thus, the mere fungal presence of rhizomorphs is not sufficient cause to treat. Some well adapted native forest trees may be sufficiently tolerant for *Armillaria* to act as a thinning agent in young stands and as a nutrient recycler in old stands. However, control may be economically justified by the consequences following establishment and secondary spread of the disease.

As always, the value of losses in the absence of control should determine whether an investment in control of *Armillaria* is justified. Depending on the host, the decision will be based on the yield of potential harvest lost or other features, such as loss of amenity value. It may be necessary to continue monitoring a treatment to evaluate fully any gains from any treatment (Jančařik, 1955). Hagle and Shaw (1991) point out that several decades may be necessary to observe fruit orchards affected by the secondary spread of *Armillaria*, but as they take longer to reach maturity, forests may require a century or more, particularly if carried over into subsequent rotations.

Hagle and Shaw (1991) provide this valuable checklist which should be consulted before embarking on any programme to control *Armillaria* root disease:

1. Critically evaluate disease impact to ensure that the level of loss justifies control. The use of disease models may aid this effort.
2. Control through cultural modifications should be given first priority, particularly in forests. As our current forest management rarely emulates nature's processes,

pathologists must work in direct co-operation with foresters to understand and modify disease-stimulating practices.

3. Utilize resistant or tolerant species, genotypes, or rootstocks, if known, that are compatible with other necessary values. Ensure that the host genotype selected for resistance is suitable for planting on potential sites, and will provide for the planned end use of the fruit or fibre. Pursue opportunities to genetically engineer *Armillaria*-resistant or tolerant species.
4. When establishing new plantations or orchards, exercise care in site selection. Small-scale trials to evaluate disease potential should be established prior to large-scale land clearing and plantation or orchard development. If the site is found to have a high disease hazard, then one must be prepared for costly pre-establishment actions such as inoculum removals by more thorough site preparation, postponement of plantation or orchard establishment for some unknown period, or elimination of the site from further consideration.
5. Maintain the general health of the forest, orchard, or amenity planting by preventing damage from other agents, avoiding adverse sites, and discouraging detrimental human activities.
6. Direct reductions of inoculum levels by physical removal of stumps and roots requires careful economic and ecological analysis.
7. When considering chemical treatments, clearly differentiate among protectants, eradicants, and curatives. Except for high-value fruit or amenity trees, curatives are likely to be uneconomical. Even in orchards and amenity plantings, chemical applications need to be realistically evaluated for their relative cost/benefit. For chemical treatment of stumps, consider compounds that can be translocated, particularly basipetally. Protectants should be inexpensive, easy to handle and apply, nonphytotoxic, fungitoxic or fungistatic, and relatively persistent. Possible environmental and human health hazards require consideration.
8. Fumigation, girdling, and silvicide treatment before felling may be useful methods to employ in preparing land for orchards, ornamentals, and some forestry applications, such as seed orchards and test plantations. Fallowing after such treatment may improve effectiveness, especially where disease spread from secondary inoculum is anticipated.
9. Biological control is desirable but requires further development for practical application in most situations. Research on antagonists, particularly cord-formers, needs to continue, as does work on the various actions (i.e. fire, chemicals) that might be used to alter conditions in a way that favours developing and maintaining populations of desirable, antagonistic organisms.

References

ARMITAGE, F.B. AND BARNES, R.D. (1968). Improvement of exotic pine seed source in Rhodesia. In: *Proceedings of the 6th World Forestry Congress*, June, 1966, Madrid, Spain. *World Forestry Congress*, pp 1799–1803.

BAKER, C.J. (1972). Root rots of coffee trees due to fungal infection. *Kenya Coffee* **37**, 255–261.

BAKER, K.F. AND COOK, R.J. (1974). *Biological control of plant pathogens.* San Francisco: W.H. Freeman and Company.

BIRMINGHAM, W.A. AND STOKES, W.S. (1921). Experiments for the control of *Armillaria mellea*. *Agricultural Gazette of New South Wales* **32**, 649–650.

Blenis, P.V., Mugala, M.S. and Hiratsuka, Y. (1989). Soil affects Armillaria root rot of lodgepole pine. *Canadian Journal of Forest Research* **19**, 1638–1641.

Bliss, D.E. (1951). The destruction of *Armillaria mellea* in citrus soils. *Phytopathology* **41**, 665–683.

Byler, J.W., Marsden, M.A. and Hagle, S.K. (1990). The probability of root disease on the Lolo National Forest, Montana. *Canadian Journal of Forest Research* **20**, 987–994.

Cooley, J.S. (1943). Armillaria root rot of fruit trees in the eastern United States. *Phytopathology* **33**, 812–817.

Corns, I.G.W. and Annas, R.M. (1986). *Field guide to forest ecosystems of west-central Alberta.* 251pp. Edmonton, Alberta: Canadian Forestry Service, Northern Forestry Centre.

Daubenmire, R. (1952). Forest vegetation in Idaho and adjacent Washington, and its bearing on concepts of vegetation classification. *Ecological Monographs* **22**, 301–330.

Davidson, A.J. and Rishbeth, J. (1988). Effect of suppression and felling on infection of oak and Scots pine by *Armillaria mellea. European Journal of Forest Pathology* **18**, 161–168.

Dowson, C.G., Rayner, A.D.M. and Boddy, L. (1988a). Inoculation of mycelial cord-forming basidiomycetes into woodland soil and litter, I. Initial establishment. *New Phytologist* **109**, 335–341.

Dowson, C.G., Rayner, A.D.M. and Boddy, L. (1988b). Inoculation of mycelial cord-forming basidiomycetes into woodland soil and litter, II. Resource capture and persistence. *New Phytologist* **109**, 343–349.

Durrieu, G., Beneteau, A. and Niocel, S. (1985). *Armillaria obscura* dans l'ecosystème forestier de Cerdagne. *European Journal of Forest Pathology* **15**, 350–355.

Eghbaltalab, M., Gay, G. and Bruchet, G. (1975). Antagonisme entre 15 especes de basidiomycètes et 3 champignons pathogènes de racines. *Bulletin de la Société Linnéenne de Lyon* **44**, 203–229.

Filip, G.M. (1977). An *Armillaria* epiphytotic on the Winema National Forest, Oregon. *Plant Disease Reporter* **61**, 708–711.

Filip, G.M. and Roth, L.F. (1977). Stump infections with soil fumigants to eradicate *Armillaria mellea* from young-growth ponderosa pine killed by root rot. *Canadian Journal of Forest Research* **7**, 226–231.

Filip, G.M., Goheen, D.J. and Johnson, D.W. (1989). Precommercial thinning in a ponderosa pine stand affected by *Armillaria* root disease: 20 years of growth and mortality in central Oregon. *Western Journal of Applied Forestry* **4**, 58–59.

Fox, R.T.V. (1990). Diagnosis and control of *Armillaria* honey fungus root rot of trees. *Professional Horticulture* **4**, 121–127.

Fox, R.T.V. (1999). Pathogenicity. In: *Armillaria Root Rot: Biology and Control of Honey Fungus*. Ed. R.T.V. Fox. Andover: Intercept.

Fox, R.T.V. and Popoola, T.O.S. (1990). Induction of fertile basidiocarps in *Armillaria bulbosa. The Mycologist* **4**, 70–72.

Fox, R.T.V., McQue, A.M. and Obanya Obore, J. (1990). Prospects for the integrated control of *Armillaria* root rot of trees. *Abstracts First Conference of the European Foundation for Plant Pathology, Biotic Interactions and Soil-borne Diseases* 3.7.

Fox, R.T.V. and Sanson, S. (1996). Lethal effects of an electrical field on *Armillaria mellea. Mycological Research* **100** (3), 318–320.

Garrett, S.D. (1970). *Pathogenic Root-Infecting Fungi.* Cambridge: Cambridge University Press.

Gaudray, D. (1973). Quelques observations relatives aux mycorhizes de l'Epicéa de Sitka et au rôle qu'elles peuvent avoir lors d'une attaque d'Armillaire. [Observations on the mycorrhizae of Sitka spruce and the role they can play during an attack by *Armillaria mellea.*] *Comptes-Rendus de la Société de Biologie* **167**, 1023–1026.

Gibson, I.A.S. (1960). *Armillaria* root rot in Kenya pine plantations. *Empire Forestry Review* **39**, 94–99.

Godfrey, G.H. (1936). Control of soil fungi by soil fumigation with chloropicrin. *Phytopathology* **26**, 246–256.

Gramss, G. (1983). Examination of low-pathogenicity isolates of *Armillaria mellea* from natural stands of *Picea abies* in Middle-Europe. *European Journal of Forest Pathology* **13**, 142–151.

GREGORY, S.C. (1985). The use of potato tubers in pathogenicity studies of *Armillaria* isolates. *Plant Pathology* **34**, 41–48.

GREGORY, S.C. (1989). *Armillaria* species in northern Britain. *Plant Pathology* **38**, 93–97.

GREIG, B.J.W. AND STROUTS, R.G. (1983). *Honey fungus*. Arboricultural Leaflet 2. Revised. 16pp. Great Britain: Her Majesty's Stationery Office, Department of the Environment, Forestry Commission.

GREIG, B.J.W., GREGORY, S.C. AND STROUTS, R.G. (1991). Honey fungus. *Forestry Commission Bulletin* No. 100 (Ed. 8), vii + 11pp.

GUILLAUMIN, J.J. (1977). Apricot root rot, *Armillaria mellea* (Vahl) Karst. *EPPO Bulletin* **7**, 125–135.

GUILLAUMIN, J.J. AND LUNG, B. (1985). Etude de la specialisation d'*Armillaria mellea* (Vahl) Kumm. et *Armillaria obscura* (Secr.) Herink en phase saprophytique et en phase parasitaire. [Study of the specialization of *Armillaria mellea* (Vahl) Kumm. and *Armillaria obscura* (Secr.) Herink in the saprophytic phase and in the parasitic phase.] *European Journal of Forest Pathology* **15**, 342–349.

GUILLAUMIN, J.J., PIERSON, J. AND GRASSELY, C. (1989). The susceptibility of different *Prunus* species used as stone fruit rootstocks to *Armillaria mellea* (*sensu stricto*). In: *Proceedings of the 7th international conference on root and butt rots*, August 9–16, 1988, Vernon and Victoria, B.C. Ed. D.J. Morrison, pp 197–207. Victoria, B.C.: International Union of Forestry Research Organizations.

HADFIELD, J.S., GOHEEN, D.J. AND FILIP, G.M. (1986). *Root diseases in Oregon and Washington conifers*. R6-FPM25086. 27pp. Portland, OR: USDA, Forest Service, Pacific Northwest Region, Forest Pest Management.

HAGLE, S.K. AND GOHEEN, D.J. (1988). Root disease response to stand culture. In: *Proceedings of the future forests of the intermountain west: a stand culture symposium*. Gen. Tech. Rep. INT-243. Ogden, UT: USDA, Forest Service, Intermountain Forest and Range Experiment Station, pp 303–309.

HAGLE, S.K. AND SHAW, C.G., III. (1991). Avoiding and reducing losses from Armillaria root disease. In: *Armillaria Root Disease*. Ed. C.G. Shaw, III and G.A. Kile, pp 157–173. *Forest Service Handbook No. 691*. Washington, D.C.: USDA.

HEATON, J.B. AND DULLAHIDE, S.R. (1989). Overcoming Armillaria root rot in Granite Belt orchards. *Queensland Agricultural Journal* January–February 1989, 25–27.

HENDRICKSON, A.H. (1925). Oak fungus in orchard trees. Circular 289, pp 1–13. Berkeley: University of California College of Agriculture, Agricultural Experiment Station.

HOOD, I.A. AND SANDBERG, C.J. (1989). Changes in soil populations of *Armillaria* species following felling and burning of indigenous forest in the Bay of Plenty, New Zealand. In: *Proceedings of the 7th international conference on root and butt rots*, August 9–16, 1988, Vernon and Victoria, B.C. Ed. D.J. Morrison, pp 288–296. Victoria, B.C: International Union of Forestry Research Organizations.

HOOD, I.A., REDFERN, D.B. AND KILE, G.A. (1991). *Armillaria* in planted hosts. In: *Armillaria Root Disease*. Eds. C.G. Shaw, III and G.A. Kile. *Forest Service Agriculture Handbook No. 691*. Washington, D.C.: USDA.

HORNER, I.J. (1988). *Armillaria* root-rot of kiwifruit. In: *5th International Congress of Plant Pathology*, August 20–27, 1988, Kyoto, Japan, 204. Abstract.

HORNER, I.J. (1990). No easy miracle cures for *Armillaria* yet. *New Zealand Kiwifruit* **8**, 14–15.

HUBBES, M. (1987). Influence of biotechnology on forest disease research and disease control. *Canadian Journal of Plant Pathology* **9**, 343–348.

HUNT, R.S., PARMETER, J.R., JR. AND COBB, F.W., JR. (1971). A stump treatment technique for biological control for forest root pathogens. *Plant Disease Reporter* **55**, 659–662.

INTINI, M.G. (1989a). Observations on the occurrence of *Armillaria ostoyae* on *Abies alba* (silver fir) in Italy. In: *Proceedings of the 7th International conference on root and butt rots*, August 9–16, 1988, Vernon and Victoria, B.C. Ed. D.J. Morrison, pp 252–256. Victoria, B.C.: International Union of Forestry Research Organizations.

INTINI, M.G. (1989b). Species of *Armillaria* in Italy. In: *Proceedings of the 7th International conference on root and butt rots*, August 9–16, 1988, Vernon and Victoria, B.C. Ed. D.J.

Morrison, pp 355–363. Victoria, B.C.: International Union of Forestry Research Organizations.

JALALI, B.L. AND JALALI, I. (1991). Mycorrhiza in Plant Disease Control. In: *Handbook of Applied Mycology - Vol 1 soil and plants*. Eds. Arora, Rai, Mukerji, Knudsen, pp 131–154. New York: Marcel Dekker, Inc.

JANČAŘIK, V. (1955). Soucasne Zpusoby Boje Proti Vaclave. [Contemporary method of controlling *Armillariella mellea*.] *Lesnicka Prace* **34**, 351–358. [Transl. Environm. Can. NO. OOENV44, 1971].

KENDALL, T.A. (1931). Soil aeration used in treatment for oak root fungus. *Journal of Agriculture, Monthly Bulletin* **20** (2), 165–166.

KESSLER, W. AND MOSER, S. (1974). Moglichkeiten der Vorbeugung gegen Schaden durch Hallimasch in Kiefernkulturen. Possibilities of preventing damage by *Armillaria mellea* in *Pinus sylvestris* plantations. *Beitrage fur die Forstwirtschaft* **8**, 86–89.

KILE, G.A. (1981). *Armillaria luteobubalina*: a primary cause of decline and death of trees in mixed species eucalypt forests in central Victoria. *Australian Forest Research* **11**, 63–77.

KILE, G.A. AND WATLING, R. (1988). Identification and occurrence of Australian *Armillaria* species, including *A. pallidula* sp.nov. and comparative studies between them and non-Australian tropical and Indian *Armillaria*. *Transactions of the British Mycological Society* **91**, 305–315.

KILE, G.A., MCDONALD, G.I. AND BYLER, J.W. (1991). In: *Armillaria Root Disease*. Eds. C.G. Shaw, III and G.A. Kile. *Forest Service Agriculture Handbook No. 691*. Washington, D.C.: USDA.

KISSLER, J.J., LIDER, J.V. AND RAABE, R.D. (1973). Soil fumigation for control of nematodes and oak root fungus in vineyard replants. *Plant Disease Reporter* **57**, 115–119.

LANIER, L. (1971). Application au pin sylvestre d'un essai de traitement per annelation circulaire contre l'Armillaire. [An experiment in ring girdling for protection of *Pinus sylvestris* against *Armillaria*.] *Annales de Phytopathologie* **3** (4), 351. Abstract.

LEACH, R. (1937). Observations on the parasitism and control of *Armillaria mellea*. *Proceedings of the Royal Society of London, Series B* **121**, 561–573

LEACH, R. (1939). Biological control and ecology of *Armillaria mellea* (Vahl) Fr. *Transactions of the British Mycological Society* **23**, 320–329.

LEVITT, H.P.A. (1947). Armillaria root rot control. *New South Wales Agricultural Gazette* **67**, 71.

LUNG-ESCARMANT, B. AND TARIS, B. (1989). Methodological approach to assess host response (resinous and hardwood species) to *Armillaria obscura* infection in the southwest French pine forest. In: *Proceedings of the 7th international conference on root and butt rots*, August 9–16, 1988, Vernon and Victoria, B.C. Ed. D.J. Morrison, pp 226–236. Victoria, B.C.: International Union of Forestry Research Organizations.

MACKENZIE, M. AND SHAW, C.G., III. (1977). Spatial relationships between *Armillaria* root-rot of *Pinus radiata* seedlings and the stumps of indigenous trees. *New Zealand Journal of Forestry Science* **7**, 374–383.

MALLET, B., GEIGER, J.P. AND NANDRIS, D. (1985). Les champignons agents de pourridiés en Afrique de l'Ouest. *European Journal of Forest Pathology* **15**, 263–268.

MCDONALD, G.I. (1990). Relationships among site quality, stand structure, and Armillaria root rot in Douglas-fir forests. In: *Interior Douglas-fir: the species and its management: Proceedings of the symposium*. Pullman: Washington State University, Cooperative Extension.

MCDONALD, G.I., MARTIN, N.E. AND HARVEY, A.E. (1987). *Armillaria* in the Northern Rockies; pathogenicity and host susceptibility on pristine and disturbed sites. Res. Note INT-371. 5pp. Ogden, UT: USDA, Forest Service, Intermountain Research Station.

MARQUIS, D.A. AND JOHNSON, R.L. (1989). Silviculture of eastern hardwoods. In: *The scientific basis for silvicultural and management decisions in the National Forest system*. Ed. R.M. Burns, Gen. Tech. Rep. WO-55, pp 9–17. Washington, D.C.: USDA Forest Service.

MASON, G.W., GOTTSCHALK, K.W. AND HADFIELD, J.S. (1989). Effects of timber management practice on insects and diseases. In: *The scientific basis for silvicultural and management decisions in the National Forest system*. Ed. R.M. Burns. Gen. Tech. Rep. WO-55. Washington, D.C.: USA.

MORRISON, D.J. (1981). Armillaria root disease. A guide to disease diagnosis, development and management in British Columbia. Information Report BC-X-203. *Environment Canada, Canadian Forestry Service*, 1–16.

MORRISON, D.J., WALLIS, G.W. AND WEIR, L.C. (1988). Control of Armillaria and Phellinus root diseases: 20-year results from the Skimikin stump removal experiment. Information Report BC-X-302. 16pp. *Canadian Forestry Service, Pacific Forestry Centre*.

MORRISON, D.J., WILLIAMS, R.E. AND WHITNEY, R.D. (1991). Infection, disease development, diagnosis, and detection. In: *Armillaria Root Disease*. Eds. C.G. Shaw, III and G.A. Kile, pp 62–75. *Forest Service Handbook No. 691*. Washington, D.C.: USDA.

MUNNECKE, D.E., KOLBEZEN, M.J. AND WILBER, W.D. (1973). Effects of methyl bromide or carbon disulfide on *Armillaria* and *Trichoderma* growing on agar medium and relation to survival of *Armillaria* in soil following fumigation. *Phytopathology* **63**, 1352–1357.

MUNNECKE, D.E., WILBUR, W. AND DARLEY, E.F. (1976). Effect of heating or drying on *Armillaria mellea* and *Trichoderma viride* and the relation to survival of *A. mellea* in soil. *Phytopathology* **66**, 1363–1368.

OHR, H.D., MUNNECKE, D.E. AND BRICKER, I.L. (1973). The interaction of *Armillaria mellea* and *Trichoderma* spp. as modified by methyl bromide. *Phytopathology* **63**, 965–975.

OHR, H.D. AND MUNNECKE, D.E. (1974). Effects of methyl bromide on antibiotic production by *Armillaria mellea*. *Transactions of the British Mycological Society* **62**, 65–72.

ONO, K. (1970). Effect of soil conditions on the occurrence of *Armillaria* root rot of Japanese larch. *Meguro Bulletin of the Government Forest Experiment Station* **229**, 123–219. [In: *Japanese Review of Plant Pathology* **50**, 2001.]

PAGE, A.I. (1970). The reestablishment of *radiata* pine at Kaingaroa Forest, 1. Basic studies to find the limitation of artificial and natural seeding. *New Zealand Journal of Forestry* **15**, 69–78.

PAWSEY, R.G. (1973). Honey fungus: recognition, biology and control. *The Arboricultural Association Journal* **2**, 116–126.

PAWSEY, R.G. AND RAHMAN, M.A. (1976a). Chemical control of infection by honey fungus, *Armillaria mellea*: a review. *Arboricultural Journal* **2**, 468–479.

PEARCE, M.N. AND MALAJCZUK, N. (1990). Factors affecting the growth of *Armillaria luteobubalina* rhizomorphs in soil. *Mycological Research* **94**, 38–48.

POPOOLA, T.O. (1991). *Role of stress in Armillaria infections*. Ph.D. Thesis. U.K.: University of Reading.

POPOOLA, T.O.S. AND FOX, R.T.V. (1996). Effects of root damage on honey fungus. *Arboricultural Journal* **20**, 329–337.

PROFFER, T.J., JONES, A.L. AND EHRET, G.R. (1987). Biological species of *Armillaria* isolated from sour cherry Orchards in Michigan. *Phytopathology* **77**, 941–943.

PRONOS, J. AND PATTON, R.F. (1979). The effect of chlorophenoxy acid herbicides on growth and rhizomorph production of *Armillaria mellea*. *Phytopathology* **69**, 136–141.

RAABE, R.D. (1966). Testing plants for resistance to oak root fungus. *California Agriculture* **20**, 12.

RAABE, R.D. (1979). Resistance or susceptibility of certain plants to *Armillaria* root rot. *Cooperative Extension Service Leaflet No. 2591*. 11pp. Berkeley: University of California.

RACKHAM, R.L., WILBUR, C., IV AND MILLER, M.P. (1966). Control of oak root fungus in citrus. *California Plant Diseases* 26. 2pp. Berkeley: University of California, Agricultural Extension Service.

RAYNER, A.D.M. (1977). Fungal colonisation of hardwood stumps from natural sources. *Transactions of the British Mycological Society* **69** (2), 303–312.

RAZIQ, F. (1999). Biological control of *Armillaria*. In: *Armillaria Root Rot: Biology and Control of Honey Fungus*. Ed. R.T.V. Fox. Andover: Intercept.

REAVES, J.L., SHAW, C.G., III AND MAYFIELD, J.E. (1990). The effects of *Trichoderma* spp. isolated from burned and non-burned forest soils on the growth and development of *Armillaria ostoyae* in culture. *Northwest Science* **64**, 39–44.

REDFERN, D.B. (1968). The ecology of *Armillaria mellea* in Britain: biological control. *Annals of Botany* **32**, 293–300.

REDFERN, D.B. (1978). Infection by *Armillaria mellea* and some factors affecting host resistance. *Forestry* **51**, 120–135.

REDFERN, D.B. AND FILIP G.M. (1991). Inoculum and Infection. In: *Armillaria Root Disease.* Eds. C.G. Shaw, III and G.A. Kile, pp 48–61. *Forest Service Handbook No. 691.* Washington, D.C.: USDA.

RHOADS, A.S. (1948). *Clitocybe* root rot of citrus trees in Florida. *Phytopathology* **38**, 44–61.

RICHARDSON, H.H. AND JOHNSON, A.C. (1935). Studies of methyl bromide in greenhouse and vault fumigation. *Tech. Bull. 853.* 20pp. Washington, D.C.: USDA, Forest Service.

RISHBETH, J. (1972a). Resistance to fungal pathogens of tree roots. *Proceedings of the Royal Society of London, Series B* **181**, 333–351.

RISHBETH, J. (1972b). The production of rhizomorphs by *Armillaria mellea* from stumps. *European Journal of Forest Pathology* **2**, 193–205.

RISHBETH, J. (1976). Chemical treatment and inoculation of hardwood stumps for control of *Armillaria mellea. Annals of Applied Biology* **82** (1), 57–70.

RISHBETH, J. (1982). Species of *Armillaria* in southern England. *Plant Pathology* **31**, 9–17.

RISHBETH, J. (1983). The importance of the honey fungus (*Armillaria*) in urban forestry. *Aboricultural Journal* **7**, 217–225.

ROBINSON-BAX, C. (1999). *A survey of the herbaceous hosts of Armillaria mellea and possible integrated control.* Horticulture M.Sc. Dissertation. U.K.: University of Reading.

ROLL-HANSEN, F. (1985). The *Armillaria* species in Europe. *European Journal of Forest Pathology* **15**, 22–31.

ROTH, L.F., SHAW, C.G., III AND ROLPH, L. (1977). Marking ponderosa pine to combine commercial thinning and control of *Armillaria* root rot. *Journal of Forestry* **75**, 644–647.

RYKOWSKI, K. (1981). The influence of fertilizers on the occurrence of *Armillaria mellea* in Scots pine plantations, I. Evaluation of the health of fertilized and non-fertilized plantations and the variability, *A. mellea* in the areas investigated. *European Journal of Forest Pathology* **11**, 108–119.

RYKOWSKI, K. (1983). The influence of fertilizers on the occurrence of *Armillaria mellea* in Scots pine plantations, III. The spread of A. *mellea* mycelium inside fertilized and non-fertilized pine roots. *European Journal of Forest Pathology* **13**, 77–85.

SCHUBERT, G., HEIDMANN, L.J. AND LARSON, M. (1970). Artificial reforestation practices for the southwest. *Agriculture Handbook No. 370.* 25pp. Washington, D.C.: USDA, Forest Service.

SCHÜTT, P. (1985). Control of root and butt rots: limits and prospects. *European Journal of Forest Pathology* **15**, 357–363.

SHAW, C.G., III AND CALDERON, S. (1977). Impact of Armillaria root rot in plantations of *Pinus radiata* established on sites converted from indigenous forest. *New Zealand Journal of Forestry Science* **7**, 359–373.

SHAW, C.G., III AND ROTH, L.F. (1978). Control of Armillaria root rot in managed coniferous forests. *European Journal of Forest Pathology* **8**, 163–174.

SHAW, C.G., III AND ROTH, L.F. (1980). Control of Armillaria root rot in managed coniferous forests. In: *Proceedings of the 5th International conference on problems of root and butt rot in conifers*, August, 1978, Kassel, Federal Republic of Germany. Ed. L. Dimitri, pp 245–258.

SHAW, C.G., III, MACKENZIE, M. AND TOES, E.H.A. (1981). Cultural characteristics and pathogenicity to *Pinus radiata* of *Armillaria novae-zelanidae* and *A. limonea. New Zealand Journal of Forestry Science* **11**, 65–70.

SHAW, C.G., III AND KILE, G.A. (1991). *Armillaria Root Disease. Forest Service Handbook No. 691.* Washington, D.C.: USDA.

SHAW, C.G. III, STAGE, A.R. AND MCNAMEE, P. (1991). Modeling the dynamics, behavior and impact of Armillaria root disease. In: *Armillaria Root Disease.* Eds. C.G. Shaw, III and G.A. Kile, pp 150–156. *Forest Service Handbook No. 691.* Washington, D.C.: USDA.

PÉREZ-SIERRA, A., WHITEHEAD, D. AND WHITEHEAD, M. (1999). Investigation of a PCR-based system for the routine identification of British *Armillaria* species. *Mycological Research* **103**, 1631–1636.

SINGH, P. (1981). *Armillaria mellea*: distribution and hosts in Newfoundland and Labrador. *Canadian Plant Disease Survey* **61**, 31–36.

SINGH, P. AND RICHARDSON, J. (1973). *Armillaria* root rot seeded and planted areas in Newfoundland. *Forestry Chronicle* **49**, 180–182.

SMITH, M.K., BRUHN, J.N. AND ANDERSON, J.B. (1992). *Armillaria bulbosa* is among the largest and oldest living organisms. *Nature* **356**, 428–431.

SOKOLOV, D.V. (1964). *Kornevaya gnil' ot Openki bor'ba s nel.* [Root rot caused by *Armillaria mellea* and its control.] (In Russian.) Moscow, Izadatel'stvo Lesnaya Promyshlennost'. [Canada Department of Forestry. 235pp.]

SPURLING, D. AND SPURLING, A.T. (1975). Field trials on Dwarf Cavendish bananas in southern Malawi, II. Fertilizers and mulching. *Acta Horticulturae* **49**, 263–267. [*Horticultural Abstracts* **46**, 5091.]

SWIFT, M.J. (1970). *Armillaria mellea* (Vahl ex Fries) Kummer in central Africa: studies on substrate colonisation relating to the mechanism of biological control by ring-barking. In: *Root diseases and soil-borne pathogens: Proceedings of the symposium*, July 1968, London, Imperial College. Eds. T.A. Toussoun, R.V. Bega and P.E. Nelson. Berkeley: University of California Press.

THIES, W.G. AND RUSSELL, K.W. (1984). Controlling root rots in coniferous forests of northwestern North America. In: *Proceedings of the 6th international conference on root and butt rot of forest trees*, August 25–31, 1983, Melbourne, Victoria, and Gympie, Australia. Ed. G.A. Kile, pp 379–386. Melbourne, Australia: International Union of Forestry Research Organizations.

THOMAS, H.E., THOMAS, H. EARL AND ROBERTS, C. (1948). Rootstock susceptibility to *Armillaria mellea. Phytopathology* **38**, 152–154.

THOMAS, P.J. AND RAPHAEL, T.D. (1935). *Armillaria* control in the orchard. *The Tasmanian Journal of Agriculture* **6**, 1–6.

WARGO, P.M. AND SHAW, C.G., III (1985). *Armillaria* root rot: the puzzle is being solved. *Plant Disease* **69**, 826–832.

WARGO, P.M. AND HARRINGTON, T.C. (1991). Host stress and susceptibility. In: *Armillaria Root Disease. Forest Service Handbook No. 691*. Ed. C.G. Shaw, III and G.A. Kile, pp 88–101. Washington, D.C.: USDA.

WHITNEY, R.D. (1988b). Armillaria root damage in softwood plantations in Ontario. *Forestry Chronicle* **64**, 345–351.

WIEHE, P.O. (1952). The spread of *Armillaria mellea* (Fr.) Quél. in tung orchards. *The East African Agricultural Journal* **18**, 67–72.

WILLIAMS, R.E., SHAW, C.G., III AND WARGO, P.M. (1989). *Armillaria Root Disease. Forest Insect and Disease Leaflet 78* (rev.). 8pp. Washington, D.C.: USDA Forest Service.

WEST, J.S. (1999). Chemical Control of *Armillaria*. In: *Armillaria Root Rot: Biology and Control of Honey Fungus*. Ed. R.T.V. Fox. Andover: Intercept.

9
Chemical Control of *Armillaria*

JONATHAN S. WEST

Institute of Arable Crops Research, Rothamsted, Harpenden, Herts. AL5 2JQ, U.K.

Synopsis

The problem of *Armillaria* root and butt rot of trees is outlined in the perspective of both natural and man-made situations. Factors affecting the management and control of this disease are discussed, with emphasis on the promotion of good tree health, cultural practices, biological control, and chemical control. The chapter includes a review of the chemical control of *Armillaria* with both fumigants and other chemicals, problems faced in the development of control, and techniques used. Finally, suggestions for new research are made.

The problem

Under natural conditions, such as primary or ancient woodland, *Armillaria* may be considered a useful thinning agent and recycler of nutrients (Hagle and Shaw, 1991). Even in managed timber forest situations, the disease, although common, may not cause severe economic damage. However, great importance is placed on the aesthetic damage to gardens and amenity areas, where the value of each established tree is high. Additionally, a host with weakened roots may be windthrown, causing substantial damage to surrounding plantings, buildings and people. Economic loss may also occur in orchards, plantations, and vineyards, where the crop is of high value. There are many different species of *Armillaria*, some of which are only weak pathogens or saprophytes, rather than aggressive pathogens, so, ideally, the identity of the *Armillaria* should be confirmed before any expensive control measures are put in place.

Disease management and control

The fungus is often a more successful pathogen when the host is under such stresses as: drought, waterlogging, mechanical damage, or pruning (Popoola, 1991; Wargo and Harrington, 1991) and infection by other diseases (Hood, Redfern and Kile, 1991). Thus, the overall health of the tree should be considered when control methods are under consideration.

Direct control of *Armillaria* is difficult to achieve due to the extensive network of

Armillaria *Root Rot: Biology and Control of Honey Fungus*

rhizomorphs and mycelium in the soil, sometimes at great depth or protected inside dead wood. The fungus is able to survive for years, or even decades, within a large stump.

COMPLEMENTARY CONTROL PRACTICES

Control by cultural practices has often been attempted. The grubbing out and burning or removal of tree stumps reduces inoculum and uncolonized potential substrate. This can reduce the number of infection foci and, more importantly, reduces the energy available to rhizomorphs as the remaining food bases from which they grow are the relatively small remnants of roots left in the soil after grubbing (Gregory, 1987). The smaller pieces of inoculum will have a much shorter life-span than large stumps (Shaw, Stage and McNamee, 1991) but even then, can pose a threat. The chipping of windthrown trees for the production of mulching material should, therefore, only be carried out after an inspection of the root system for signs of the disease. Stump removal can be costly and time consuming in forests and plantations, but may even be totally impracticable to carry out in gardens. In situations where the stump cannot be removed, or is on a neighbour's land, adjacent plants may be protected from rhizomorph infection by physical barriers, such as buried polythene sheeting, deep trenches, and even regular digging of soil (which would sever the rhizomorphs from their food bases). However, infected roots and rhizomorphs can be present in the soil at considerable depths, so physical barriers should go down to a depth of 45 cm, or more in well drained soils (Gregory, 1987). These barriers are laborious to make and need to be regularly inspected. Protection by digging must be repeated annually.

Where an infected area cannot be avoided, the use of more resistant plants or rootstocks may also reduce the impact of *Armillaria*. Together, these cultural practices may offer management of the disease, but rarely total control.

Biological control methods are currently under investigation. A biocontrol agent may eradicate *Armillaria*, or prevent saprophytic colonization of new substrate. However, following surface inoculation of stumps, there is inevitably a delay before the biocontrol agent becomes established in the roots, and direct inoculation of the roots is difficult (Boddy, 1991). Additionally, an *Armillaria* colony can be extensive, and may be protected inside pseudosclerotia in infected wood (Garrett, 1970). Dowson *et al.* (1988a,b) successfully established a number of saprotrophic cord-forming fungi in woodland soil by direct seeding with 8 cm^3 colonized wood blocks placed in the soil/litter interface. *Phanerochaete laevis* (Fr.) Erikss. & Ryv. in particular was found to actively replace other fungi from inoculum. Boddy (1991) noted that this fungus replaces *Armillaria* spp. from soil and wood, and also competes with other root pathogens for substrate. Pearce and Malajczuk (1990) inoculated tree stumps both with saprophytic cord-forming fungi and *Armillaria luteobubalina*, and found that all the saprotrophs reduced colonization by *Armillaria*. The painting of stumps with ammonium sulphamate (AMS) to encourage saprophytic antagonists of *Armillaria* (Rishbeth, 1976) is a successful form of indirect biological control, directed at preventing infection by *Armillaria* basidiospores. AMS (40% solution) increased colonization and decay by some cord-forming fungi of birch and beech stump roots below ground, and oak stumps above ground (Rayner, 1977). So, for birch, beech, and possibly oak after a longer period, AMS may also help to prevent soil-borne colonization by *Armillaria*.

Ectomycorrhiza increase plant vigour and increase the resistance of plants to root diseases because the pathogen has to penetrate an external barrier of tightly interwoven hyphae, and thicker cortical cells, which may have another layer of hyphae (Jalali and Jalali, 1991).

One problem of biological control is that it may be counterproductive in other respects, so extensive research is required. For example, *Trichoderma viride* and *T. polysporum* were antagonistic to mycorrhizal colonization, and established mycorrhiza on black spruce (*Picea mariana*) seedlings (Jalali and Jalali, 1991). Yet *Trichoderma* spp. have been used as potential biological controls of *Armillaria*. These antagonists often occur in old mushroom compost which can be added to the soil in an *Armillaria* infected area. However, there may also be other pathogens such as *Phytophthora*, or *Pythium* in unsterilized compost. A sterile compost into which the antagonist is added may be a safer solution.

CHEMICAL CONTROL

Many chemicals, some of which are also toxic to the mycorrhiza, or to plant roots directly, have been used in an attempt to control *Armillaria*. Much of the early work did not identify the species of *Armillaria* used in experiments. Many other chemicals have been used without any significant scientifically accepted demonstration of efficacy. A review of chemicals used against *Armillaria* is outlined below.

Fumigants Bliss (1951) reported that carbon disulphide had been used as a fumigant by Horne since 1914 in California. Carbon disulphide has good penetration of soil due to its high vapour pressure. This is important because viable mycelium has been found in citrus roots at a depth of almost 3 m (Bliss, 1951). Additionally, it was noticed by Bliss (1951) that carbon disulphide need not kill *Armillaria* directly, as it allows antagonists such as *Trichoderma viride* to infect and destroy the weakened *Armillaria*. Ohr and Munnecke (1974) found that the fumigant methyl bromide disrupted the antibiotic production of *Armillaria* (as with carbon disulphide), allowing *Trichoderma viride* to attack it. However, Thomas and Lawyer (1939) stated that carbon disulphide is only effective in drier soils, and Anon. (1961) that it is only effective in warmer soils. Greig and Strouts (1983) stated that its effectiveness in the generally colder and wetter soils of Britain was uncertain. Chloropicrin is also used as a fumigant but, again, penetration of the soil is difficult (Hagel and Shaw, 1991). Generally, fumigants are poisonous and require specialist handling. Additionally, Guillaumin (1988) reported that fumigants must be injected at least 60 cm deep in soil, and this, at best, only decreases the number of foci, rather than totally eradicating the disease.

Other chemicals A number of non-systemic chemicals have also been tested including copper (Rayner, 1959), and iodine (Guyot, 1933). Shaw *et al.* (1980) found the activity of sodium pentachlorophenate (NaPCP) to be questionable, and this treatment, as with boric acid (another potential control), also caused phytotoxicity (Shaw and Roth, 1978).

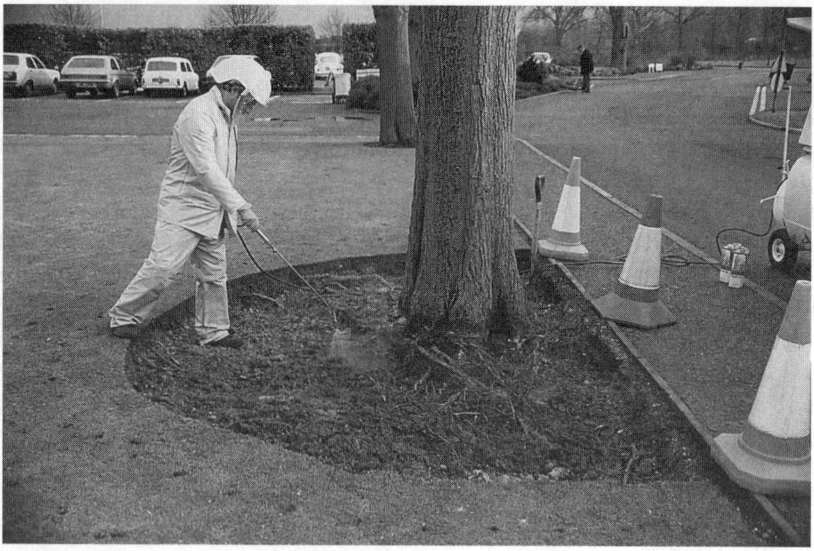

Figure 9.1. Applying cresylic acid formulation to roots of common lime tree *(Tilia x europaea)* attacked by *Armillaria mellea* at Jealott's Hill, Berkshire.

Protection of small (7–15 cm diameter) ponderosa pine to infection by *A. ostoyae* was investigated for a range of chemicals by Filip and Roth (1987). Out of the chemicals tested; benomyl, captan, copper sulphate, iron sulphate, copper wire, methyl isothiocyanate in mixture with 1,2-dichloropropane + 1,3-dichloropropene ('Vorlex'), and chloropicrin; none reduced mortality ten years after the single applications.

Fedorov and Bobko (1988) tested carbendazim ('Derosal'), benomyl ('Fundazol'), thiophanate-methyl ('Topsin' M), potassium (per)manganate VII, cuprosan and copper oxychloride in field trials. They found some chemicals to have a short-term inhibitory effect, but this had washed away after three months. Rhizomorphs were formed from mycelium surviving under the bark, acting as sources of infection.

Heaton and Dullahide (1989) reported that potassium phosphite applied by tree injection gave good results on stone fruit trees in Australia infected with an *Armillaria* species. A similar chemical, aluminium phosphite or fosetyl-Al ('Aliette'), was tested on *Armillaria* by West (1994). It was thought to be worth investigating because of its translocation from leaf to root. This feature is very desirable for a potential chemical control of *Armillaria* as it would be very easy to apply, and, if active against *Armillaria*, should have a curative action. Unfortunately, this chemical only caused a slight delay in the infection of inoculated strawberry plants and willow saplings.

Cheo (1968) tested 20 systemic chemicals on *Armillaria* mycelium growing on potato dextrose agar. Cycloheximide was found to be very effective, being lethal at 20 ppm, while 2,4, dichlorophenoxy acetonitrile was lethal to *Armillaria* at 50 ppm. Acrizane chloride was not lethal, but was fungistatic at 10 ppm. Dimethyl sulphoxide showed only minor inhibition of growth on agar, but is rapidly absorbed and distributed in living materials. These chemicals were promising *in vitro*, but have not been marketed for field use. The performance of chemicals *in vitro* is often very

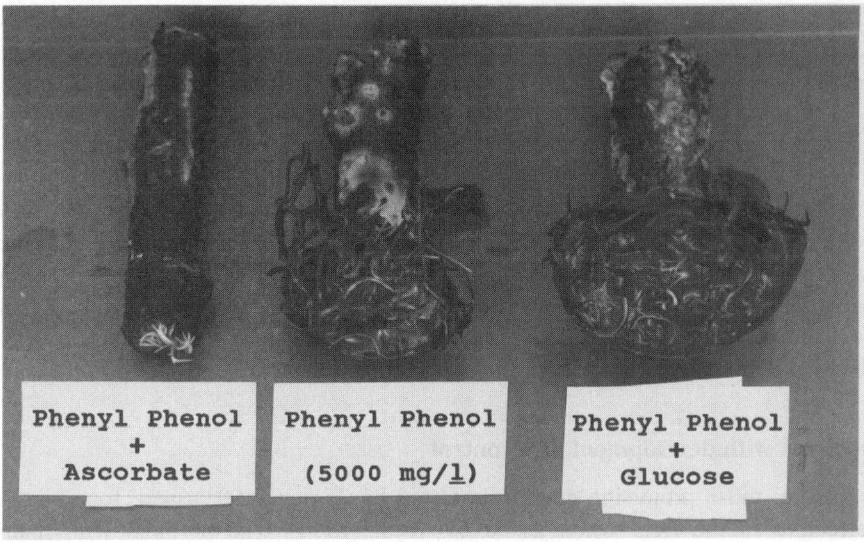

Figure 9.2. Artificial inoculum of *Armillaria mellea* bearing rhizomorphs several weeks after a fungicide application which appeared to kill the surface inoculum.

different from field conditions and this is one reason why chemical control research is so time-consuming.

A refined creosote mixture was investigated by Pawsey and Rahman (1976), and it was marketed as 'Armillatox'. A phenolic emulsion, it was found to have some effect against rhizomorphs, but unable to eradicate the fungus from wood. Rahman (1974) concluded that 'unless the source of infection is eradicated (by removal of stumps, etc.), then the fungus will re-invade once the treatment is allowed to lapse'. A similar mixture of cresylic acids ('Bray's Emulsion') has also been marketed, but, again, its control is not sustained, and there are phytotoxicity problems with both products (Rahman, 1974; Anon., 1990; West, 1994). West (1994) was able to show that, in some instances, these phenolic chemicals stimulate the growth of any *Armillaria* which survived the initially fungicidal chemical, protected inside its woody substrate. It is possible that the fungus could then detoxify the chemical using the same enzymes evolved for the breakdown of phenolic components of lignin. As a result, the fungus could grow more rapidly into treated soil, which would by then also be free from many antagonists.

In order to find an alternative chemical to 'Bray's Emulsion', a new generation of fungicides was investigated by Turner and Fox (1988) *in vitro* and in field trials. The most effective chemicals were hexaconazole and flutriafol (both triazoles), fenpropidin (a piperidine), a guanide, and a phenolic; with 'Bray's Emulsion', (a mixture of cresylic acids) as a standard. The two triazoles and the piperidine are ergosterol-biosynthesis inhibitors (EBIs). The chemicals were tested for eradicant action on mycelium growing on agar, and protectant action *in vitro* on treated sterilized wooden blocks which were exposed to growing mycelium. The degree of activity shown by the chemicals depended on the species of *Armillaria* used (this was not known by Cheo in 1968). *Armillaria mellea* was generally the most resistant species of *Armillaria* to the chemicals under investigation. With the exception of hexaconazole and fenpropidin, *A. mellea* was only eradicated by concentrated doses, over 500 mg/l.

Hexaconazole, followed by fenpropidin, were the most successful protectants; hexaconazole being 100% effective at 100 mg/l.

Additional work by Turner (1991) tested the effects of mixtures of the test fungicides as 1:1 mixes, each of 50 or 500 mg/l. Some mixtures were significantly more effective than a single compound alone. Guazatine and fenpropid in mixture caused 100% inhibition of growth at only 50 ml/l each. It was later realized that guazatine binds to clay particles, thereby making it unavailable to the fungus in many soils. A further search for additional fungicides by Turner (1991) found none that were better than the original six under investigation. Fenpropimorph, penconazole and propiconazole each at 500 mg/ l, gave 98–87% inhibition of growth after ten days incubation of a mycelial disc, which had been drenched for one hour. Other chemicals tested were significantly less effective. Benodanil and copper carbonate had no effect at all.

Problems with development of a control

As before, those promising chemicals tested by Turner (1991) were found to be ineffective in the field when tested by West (1994). The problem with many chemicals is that they fail to reach their target - the mycelium of the pathogen. This may be due to the binding of chemicals to the soil particles or wood surfaces, thereby making them unavailable for fungicidal action. Alternatively, a water soluble chemical is often leached away. This is not a problem in the artificial world of the Petri dish. As new chemicals need to be tested in 'real' conditions, a method is needed which can give realistic results quickly. Strawberry plants are an ideal experimental host as they are cheap and easy to propagate (if clonal material is used, individual variability is reduced), and when inoculated with *Armillaria mellea*, they can become infected and die after only three months at 20°C. However, with both trees and strawberries there is often little or no evidence of infection until the plants finally wilt and die in a short space of time. Larger trees may show a gradual decline, or even an increased final fruit yield.

The use of artificial inoculum billets improves field trials as each plant inoculated has an equal chance of becoming infected apart from any treatment effects. West (1994) found that in experiments with privet plants (*Ligustrum ovalifolium* Hassk.) it took over two years for the first inoculated control plant to die, but as the others were still alive, the trials appeared inconclusive. However, the final root examination showed that a high proportion of the inoculated controls were infected. The inoculation technique had worked very well, and enabled a clear comparison of infection levels between the controls and the treatments. Turner (1991) and West (1994) found that naturally occurring infection was slow and unpredictable, making assessments of chemical effects very difficult, even after several years. In experiments with tree saplings, even inoculated field trials may take two to several years due to the slow disease progression of *Armillaria* and the response of the plants to infection. Infected plants may produce adventitious roots at the soil surface above the zone of infection. These roots could keep the plant supplied with water. Hence, the fungus has to progress to the root collar, or higher, in order to kill the plant by ringing the xylem above the highest roots.

In young privet and willow plants, West (1994) found that death only occurred when the root system was almost completely colonized by *Armillaria*. This may be expedient for *Armillaria* because, once the tree dies, the roots are open to invasion by saprophytes. This could restrict the size of the food base available to *Armillaria* if the *Armillaria* only occupied a small volume at the time of the death of the tree. Size of food base is an extremely important factor in the modelling of *Armillaria* root disease (Shaw, Stage and McNamee, 1991).

A variety of techniques have been used to evaluate the degree of infection caused by *Armillaria*. The incidence of infection (% plants infected) is best studied using many replicates as it is an 'either or' qualitative measurement. Researchers often use seedlings, herbaceous plants or small saplings, for such measurements of the proportion of host plants killed or infected. These may be grown in controlled environments or field conditions. A more quantitative measurement, such as severity of infection (extent of infection in infected plants) requires fewer replicates, but whether area or volume of infection is measured, assessment can be very time consuming. Singh (1980), looking at host susceptibility and symptoms, used 90–96 replicates of 2–4 year-old softwood seedlings, grown in pots in a glasshouse. These were inoculated artificially, disease symptoms were noted, and the percentages of infection and mortality were obtained. A quantitative measurement of disease extent was not made, but incidence of decay in bark and wood of roots, and in the stem, was taken. The number of replicates used enabled a simple qualitative assessment of disease with respect to fungicide treatments. However, the use of smaller plants, such as seedlings, in order to increase the number of replicates may have made the study less relevant to most garden and amenity or forest situations.

ARTIFICIAL INOCULUM

Gregory, Rishbeth and Shaw (1991) noted that some pathogenicity trials had been successfully conducted using comparatively small inocula (1.5–2 cm diameter, by 4–5 cm long) which were placed singly, next to the root collar or major roots. The similar inoculation method used by West (1994) confirms the success of this technique. Rishbeth (1978) stated that the wood of broad-leaved trees is generally a better substrate for rhizomorph production than conifer wood. He considered the stems of hazel *(Corylus avellana* L.) and oak (*Quercus robur* L.) to be the most suitable. Artificial inoculum may be prepared using the following technique.

Hazel billets may be produced from live hazel branches approximately 2.5 cm in diameter, sawn into sections approximately 5–7 cm long (depending on the experiment). These sections (billets) are placed into jars, sealed with aluminium foil and autoclaved at 120°C (1 bar; 15 psi) for 50 minutes. The sterile billets are allowed to cool and then transferred aseptically (in a laminar flow cabinet) to large jars which have a newly established colony of *A. mellea* isolate 1, growing on approximately 1.5 cm depth of 3% malt extract agar in their base. The jars are incubated at 25°C until the mycelium from the established colony had grown up through and around the billets. This takes approximately two months with some isolates of *A. mellea*, and is affected by the dampness of the wood after autoclaving (slightly moist rather than dry is better).

Research on new chemicals

New research should investigate the behaviour of chemicals in soil and wood, combined with their fungicidal effects. The ideal chemical control would be selectively toxic so as not to cause any phytotoxicity to the host. A downwardly mobile systemic chemical would also be desirable.

Once a potential chemical is identified, the field efficacy should be examined with consideration to a number of factors such as: phytotoxicity (which may be less after an application in any dormant season), soil type, effect of temperature and wetness, and the potential for fungicide resistance or premature breakdown. Other factors worthy of investigation include the scale of host resistance, and the continued search for new chemicals. Often, nature provides an idea or basis for novel fungicidal chemicals, such as azostrobin, recently launched by AstraZeneca Agrochemicals (now Syngenta), which was inspired by the strobilurins produced naturally by various forest dwelling fungi and bacteria (Clough *et al.*, 1996).

In order to test potential chemicals indicated by previous work for activity in field conditions, experiments should concentrate on controlling infection by rhizomorphs and root-to-root contact that occurs in an established area of infection. In order to make trials more uniform and reliable, artificial inoculum should be used where possible. Both the protectant and eradicant properties of each chemical should be investigated.

Controlled experiments should also be conducted to understand any limitations to the performance of the chemicals so that improvements may be made.

The chemical activity against different isolates of *Armillaria mellea* should also be considered. Additionally, the ability of the chemicals to eradicate both rhizomorphs growing in soil and mycelium from within wooden inocula should also be investigated.

References

ANON. (1961). *Armillaria root rot.* Advisory leaflet of the Ministry of Agriculture Fisheries and Food. Number 500.

ANON. (1990). *ICI Garden and Professional Products Manual* 1990/91.

BLISS, D.E. (1951). The destruction of *Armillaria (mellea)* in citrus soils. *Phytopathology* **41**, 665–683.

BODDY, L. (1991). Section 18. Importance of wood decay fungi in forest ecosystems. In: *Handbook of Applied Mycology - Vol 1 soil and plants*. Eds. Arora, Rai, Mukerji, Knudsen, pp 507–540. New York: Marcel Dekker, Inc.

CHEO, P.C. (1968). Control of *Armillaria* with systemic chemicals. *Plant Disease Reporter* **52**, 639–641.

CLOUGH, J.M., GODFREY, C.R.A., GODWIN, J.R., JOSEPH, R.S.I. AND SPINKS, C. (1996). Azostrobin: a novel broad-spectrum systemic fungicide. *Pesticide Outlook* **7** (4), 16–20.

DOWSON, C.G., RAYNER, A.D.M. AND BODDY, L. (1988a). Inoculation of mycelial cord-forming basidiomycetes into woodland soil and litter. I Initial establishment. *New Phytologist* **109**, 335–341.

DOWSON, C.G., RAYNER, A.D.M. AND BODDY, L. (1988b). Inoculation of mycelial cord-forming basidiomycetes into woodland soil and litter. II Resource capture. *New Phytologist* **109**, 343–349.

FEDOROV, N.I. AND BOBKO, I.N. (1988). Protection of pine plantations from root rot induced by Honey fungus 1: Chemical control. (In Russian, English Abstract). *Mikologiya i Fitopatologiya* **22** (3), 255–261.

FILIP, G.M. AND ROTH, L.F. (1987). Seven chemicals fail to protect ponderosa pine from Armillaria root disease in central Washington. *U.S. Department of Agriculture, Forest Service, Pacific Northwest Forest and Range Experiment Station.* 8pp.

GARRETT, S.D. (1970). *Pathogenic Root-Infecting Fungi.* Cambridge: Cambridge University Press.

GREGORY, S.C. (1987). Honey Fungus in Gardens. *The Garden. Journal of the Royal Horticultural Society* **112**, 525–529.

GREGORY, S.C., RISHBETH, J. AND SHAW, C.G., III (1991). Pathogenicity and Virulence, Chapter 6. In. *Armillaria Root Disease*. Eds. C.G. Shaw, III and G. Kile, pp 76–87. *Forest Service Agricultural Handbook No. 691.* Washington, D.C.: USDA.

GREIG, B.J.W. AND STROUTS, R.G. (1983). *Honey Fungus. Arboricultural Leaflet 2. Revised.* Great Britain: HMSO, Department of the Environment, Forestry Commission.

GUILLAUMIN, J.J. (1988). The *Armillaria mellea* complex. *Armillaria mellea* (Vahl) Kummer sensu stricto. In: *European Handbook of Plant Diseases*. Eds. I.M. Smith, J. Dunez, D.H. Phillips, R.A. Lelliot and S.A. Archer, pp 520–523. Oxford: Blackwell Scientific Publications.

GUYOT, R. (1933). (De la maladie du rond; de l'influence des foyer d'incendie dans sa propagation. *Rev. gen. Sci.* **44**, 239–47.) In: *Diseases of forest and ornamental trees*. Eds. D.H. Phillips and D.A. Burdekin, 1982. Macmillan Press.

HAGLE, S.K. AND SHAW, C.G., III (1991). Avoiding and Reducing Losses from *Armillaria* Root Disease. In: *Armillaria Root Disease*. Eds. C.G. Shaw, III and G. Kile, Chapter 11, pp 157–173. *Forest Service Agricultural Handbook No. 691*. Washington, D.C.: USDA.

HEATON, J.B. AND DULLAHIDE, S.R. (1989). *Armillaria root rot of stone fruit at Stanthorpe, Queensland.* Paper given at the 7th Australasian Plant Society conference, July 3–7, 1989. Brisbane: University of Queensland. (Programme and Handbook: 68. Abstract 28) [not seen]. In: Hagle and Shaw (1991).

HOOD, I.A., REDFERN, D.B. AND KILE, G.A. (1991). *Armillaria* in Planted Hosts. In: *Armillaria Root Disease*. Eds. C.G. Shaw, III and G. Kile, Chapter 9, pp 122–149. *Forest Service Agricultural Handbook No. 691*. Washington, D.C.: USDA.

JALALI, B.L. AND JALALI, I. (1991). Mycorrhiza in Plant Disease Control. In: *Handbook of Applied Mycology - Vol 1 soil and plants*. Eds. Arora, Rai, Mukerji, Knudsen, pp 131–154. New York: Marcel Dekker, Inc.

OHR, H.D. AND MUNNECKE, D.E. (1974). Effects of methyl bromide on antibiotic production by *Armillaria mellea. Transactions of the British Mycological Society* **62** (1), 65–72.

PAWSEY, R.K. AND RAHMAN, M.A. (1976). Field tests with Armillatox against *A. mellea. Pest Articles and News Summaries* **22** (1), 49–56.

PEARCE, M.H. AND MALAJCZUK, N. (1990). Inoculation of *Eucalyptus diversicolor* thinning stumps with wood decay fungi for control of *Armillaria luteobubalina. Mycological Research* **94,** 32–37.

POPOOLA, T.O.S. (1991). *The Role of Host Plant Stress in Armillaria Root Rot*. Ph.D. Thesis. U.K.: University of Reading.

RAHMAN, M.A. (1974). *Studies on the Effect of Armillatox, a Proprietary Phenolic Emulsion, on the Growth and Infection by Armillaria mellea*. M.Sc. Thesis. U.K.: Oxford University.

RAYNER, A.D.M. (1977). Fungal Colonisation of Hardwood Stumps from natural sources. *Transactions of the British Mycological Society* **69** (2), 303–312.

RAYNER, R.W. (1959). Root rot of coffee and ring barking of shade trees. *Kenya Coffee* **24,** 361–365.

RISHBETH, J. (1976). Chemical Treatment and Inoculation of Hardwood Stumps for Control of *Armillaria mellea. Annals of Applied Biology* **82,** 57–70.

RISHBETH, J. (1978). Effects of Soil Temperature and Atmosphere on Growth of *Armillaria* Rhizomorphs. *Transactions of the British Mycological Society* **70** (2), 213–220.

SHAW, C.G., III AND ROTH, L.F. (1978). Control of Armillaria root rot in managed coniferous forests. *European Journal of Forest Pathology* **8,** 163–174.

SHAW, C.G., III, MACKENZIE, M. AND TOES, E.H.A. (1980). Pentachlorophenol fails to protect seedlings of *Pinus radiata* from Armillaria root rot. *European Journal of Forest Pathology* **10**, 344–349.

SHAW, C.G., III, STAGE, A.R. AND MCNAMEE, P. (1991). Modeling the Dynamics, Behavior, and Impact of *Armillaria* Root Disease. In: *Armillaria Root Disease*. Eds. C.G. Shaw, III and G. Kile, Chapter 10, pp 150–156. *Forest Service Agricultural Handbook No. 691.* Washington, D.C.: USDA.

SINGH, P. (1980). *Armillaria* root rot: Artificial inoculation and development of the disease in the greenhouse. *European Journal of Forest Pathology* **10**, 420–431.

THOMAS, H.E. AND LAWYER, L.O. (1939). The use of carbon bisulphide in the control of *Armillaria* root rot. *Phytopathology* **29**, 827–828.

TURNER, J.A. (1991). *Aspects of the Biology and Control of Armillaria*. Ph.D. Thesis. U.K.: University of Reading.

TURNER, J.A. & FOX, R.T.V. (1988). Prospects for the Chemical Control of *Armillaria* spp. *Brighton Crop Protection Conference - Pests and Diseases 1988*, pp 235–240.

WARGO, P.M. AND HARRINGTON, T.C. (1991). Host Stress and Susceptibility. In: *Armillaria Root Disease*. Eds. C.G. Shaw, III and G. Kile, Chapter 7. *Forest Service Agricultural Handbook No. 691*. Washington, D.C.: USDA.

WEST, J.S. (1994). *Chemical Control of Armillaria root rot*. Ph.D. Thesis. U.K.: University of Reading.

10
Biological and Integrated Control of *Armillaria* Root Rot

FAZLI RAZIQ

Agricultural Research Station, Takhta Band Road, Mingora Swat, NWFP, Pakistan

Summary

For controlling *Armillaria* infections, physical and chemical methods alone are at present inadequate, ineffective, or impractical. There is a need for looking into the possibility of effective biological control, either alone or in integration with another control strategy. Biological control agents (antagonists) of *Armillaria* might function by inhibiting or preventing its rhizomorphic and mycelial development, by limiting it to substrate already occupied, by actively pre-empting the substrate, or by eliminating the pathogen from substrate it has already occupied. The most thoroughly investigated antagonists of *Armillaria* are *Trichoderma* species. Depending on the particular isolate of a *Trichoderma* species, control may be achieved by competition, production of antibiotics, or by mycoparasitism. The level of control is also influenced by the growth and carrier substrate of the antagonist, time of application in relation to the occurrence of the disease, and several environmental conditions. Among a range of the other antagonists tested, several cord-forming fungi and an isolate of *Dactylium dendroides* are important discoveries. Integration of biological and chemical methods can control the disease more effectively. When trying to integrate an introduced antagonist with a fungicide, consideration as to whether the antagonist or the fungicide should be applied first, and the time interval in between, is essential.

The concept of biological control

Biological control represents both the oldest and youngest technology for the control of plant diseases (Cook, 1991). In its widest sense, it includes all the agricultural practices that activate the non-pathogenic microbial population of the rhizosphere or phyllosphere which, in turn, reduces the incidence or severity of the disease. Thus, biological control has been used by man almost since the beginning of organized arable agriculture, through crop rotation, fertilizer application, use of chemicals, and plant breeding (Campbell, 1989). Baker and Cook (1974) defined biological control as 'the reduction of inoculum density or disease-producing activities of a pathogen or

Armillaria *Root Rot: Biology and Control of Honey Fungus*

parasite, in its active or dormant state, by one or more organisms, accomplished naturally or through manipulation of the environment, host, or antagonist, or by mass introduction of one or more antagonists.'

There is now a renewed interest in the study and implementation of biological and integrated control systems in plant pathology. After several decades of little activity, the science of biological control of plant diseases has entered a period of 'information explosion' (Cook, 1991). Commercial interest is also growing. Powell *et al.* (1990) estimated that $10 million a year is spent on all aspects of research on biological control. The increasing concern of the public about the harmful effects of pesticidal chemicals on the environment and, therefore, human and animal health, coupled with advances in the understanding of microbiology and plant pathology has added to the impetus.

According to Baker and Cook (1974), the objective of biological control of plant pathogens is the reduction of disease by:

1. Reduction of inoculum of the pathogen through decreased survival between crops, decreased production or release of viable propagules, or decreased spread by mycelial growth;
2. Reduction of infection of the host by the pathogen;
3. Reduction of severity of attack by the pathogen.

Antagonism of plant pathogens is achieved by three types of activity:

1. Antibiosis and lysis: Antibiosis is the inhibition of one organism by a metabolic product of another. Lysis is a general term for the destruction, disintegration, dissolution, or decomposition of biological materials (Lamana and Malette, 1965).
2. Competition: Competition is the endeavour of two or more organisms to gain the measure each wants from the supply of a substrate, in the specific form and under the specific conditions in which that substrate is presented when that supply is not sufficient for both (Clark in Baker and Snyder, 1965).
3. Parasitism and predation: This is the direct effect of the antagonist on a pathogen. The antagonist may operate by simply using the pathogen as a food source for which it may use enzymes such as chitinases and cellulases to break down the walls of its host (Campbell, 1989).

The need for biological control of *Armillaria* infections

Once infection of a host plant by a pathogenic species of *Armillaria* is diagnosed, or the host plant is considered to be at risk of infection from neighbouring foci of inoculum, one of several control options may be considered. Though the best way of avoiding infections would be planting resistant plants in areas with known or suspected sources of inoculum, unfortunately this option is not reliable. Host resistance to *Armillaria* involves the genetics of both the host and the pathogen, as well as environmental influences. It can also involve managing mixtures of genotypes with varying levels of resistance. Host species with superior resistance to *Armillaria* infections in one geographical location may be susceptible to the pathogen in other locations. Discrepancies in resistance may also be found within very limited areas, and intraspecific variation in adaptation to sites may be as great as interspecific variation within the natural ranges of any two or more species (Hagle and Shaw, 1991).

Physical control methods for established infections are laborious and mostly impracticable, especially in established orchards. Complete eradication of the pathogen is almost difficult to achieve and re-invasion by the fungus is, therefore, possible. Moreover, the costs involved may not be justified by the crop's future value. Soil disturbance may also stimulate fresh rhizomorph production, increasing the risk of disease to newly planted trees (Redfern, 1970; Morrison, 1976). Excessive removal of woody debris from the soil may also adversely affect the mycorrhizae (Maser *et al.*, 1984).

Chemical methods for control of *Armillaria* species have also been found largely ineffective. There is very little experimental evidence to support the use of commonly recommended treatments (Shaw and Roth, 1978). One notable exception is the use of fumigants such as carbon disulphide, methyl bromide and chloropicrin. These fumigants are, however, highly toxic to other soil flora and fauna and may, therefore, be environmentally unacceptable (Schütt, 1985). These chemicals can also be prohibitively costly in forests (Pearce *et al.*, 1995). There are also problems with their penetration into soil, and their injection deeper than 60 cm can, at best, decrease the number of foci rather than totally eradicating the disease (Guillaumin, 1988).

Some phenolic emulsions, such as 'Armillatox' and 'Bray's Emulsion', have been marketed for specific use against *Armillaria*, but their penetration into wood is minimal, control is not sustained, and there are phytotoxicity problems with these products. The fungus may even be stimulated into growth by the use of these chemicals (West, 1994).

A new generation of fungicides was investigated by Turner and Fox (1988) *in vitro* and in field trials. These included the ergosterol biosynthesis inhibitors (EBIs) hexaconazole and flutriafol (both triazoles) and fenpropidin (a piperidine). Though the EBIs were found effective protectants *in vitro*, no evidence of such an activity was found in the field trial using pear trees infected with *A. mellea*. High concentrations of the EBIs can cause serious phytotoxicity problems, especially on herbaceous plants such as strawberries, but also on young trees (Raziq, 1998).

Biological control of *Armillaria* root rot

It can be concluded from the preceding discussion on control options for *Armillaria* infections that there is a need for an alternative approach to the problem. Biological control, in itself or in integration with another control strategy, could offer a good choice.

Biological control agents (antagonists) that could be used to control *Armillaria* should be wood-inhabiting or rhizosphere-competent. They might function by inhibiting or preventing rhizomorph and mycelial development, by limiting the pathogen to substrate already occupied, by actively pre-empting the substrate, or by eliminating *Armillaria* (perhaps through replacement) from substrate already occupied (Hagle and Shaw, 1991). To perform these functions, an ideal antagonist should produce inoculum in excess; resist, escape, or tolerate other antagonists; germinate and grow rapidly; and invade and occupy organic substrates (Baker and Cook, 1974). The features of an antagonist to be useful in combination rather than in isolation are moderate production of antibiotics, the ability to invade the pathogen as judged from *in vitro* tests, and moderate growth rates. The possession of extremes of any of these features may act against the antagonist when introduced into the soil, resulting in its failure as an effective antagonist (Faull and Graeme-Cook, 1992).

Two important features may make the control of *Armillaria* by introduced antagonists difficult (Rishbeth, 1976). First, *Armillaria* in established infections has a positional advantage since it already occupies a considerable portion of the substrate. Introduced antagonists would have to reach these target sites in sufficiently active state to overcome the life processes of *Armillaria*. Production of antibiotics (Munnecke *et al.*, 1981) and formation of zonal lines and pseudosclerotial plates by *Armillaria* (Campbell, 1934) may protect it from invasion by antagonists. Second, *Armillaria* spreads quickly in the cambial region of freshly killed trees. Introduced antagonists may, therefore, not be able to prevent *Armillaria* from becoming established in stumps, but they may limit further stump colonization by restricting the available food base.

Perhaps the most thoroughly studied antagonists of *Armillaria* are *Trichoderma* species (Hagle and Shaw, 1991). *Trichoderma* is one of the commonest genera of fungi in the majority of soils (Aytoun, 1953). Interest in the genus was stimulated by Weindling (1932), who showed that *T. lignorum* could parasitize other soil fungi. Since then, it has been extensively studied for antagonistic activity against many plant pathogens in different parts of the world. Several reports have shown successful control of a number of diseases, both in glasshouse and field conditions, with the application of several species and strains of the antagonist (Papavizas, 1985; Chet, 1987). Isolates are obtained from suppressive soils and from a great variety of screening and selection procedures. Depending on the particular isolate, control may be achieved by competition, production of antibiotics, or by mycoparasitism (Campbell, 1989). The competitive ability of *Trichoderma* arises from its tolerance to changes in environmental conditions (Bliss, 1951; Munnecke *et al.*, 1981), ability to degrade various organic substrates in soil, metabolic versatility, and resistance to microbial inhibitors (Papavizas, 1985). Ishikawa *et al.* (1976) reported the isolation of trichodermin and another unidentified compound from *Trichoderma*. Species of *Trichoderma* are not only sources of various toxic metabolites, but also of various enzymes such as exo- and endoglucanases, cellobiases, chitinases (Papavizas, 1985), cellulases, and proteases. These enzymes permit *Trichoderma* to parasitize different structures of pathogenic fungi (van Driesche and Bellows, 1996). Commercial interest in the development of formulations based on *Trichoderma* has led to a range of products that are nowadays commercially available (van Driesche and Bellows, 1996) but, unfortunately, there is none in use for controlling *Armillaria*.

Bliss (1941) reported that his attempts to isolate *Armillaria* from mixtures of *Armillaria* and *Trichoderma* failed because the rapidly growing mycelium of *Trichoderma* overran and quickly suppressed *Armillaria*. He also noticed 'a greenish yellow pigment' which diffused through the agar and exerted a toxic effect on *Armillaria*. The mycelium of *Armillaria* became apparently non-viable in the presence of *Trichoderma*, but within citrus root segments incubated at 20–30°C in a dark room, *Armillaria* remained viable within its pseudosclerotium for seven years, despite the presence of *Trichoderma* on the surface of the segments for the full period (Bliss,

1951). However, towards the end of the period, *Armillaria* appeared to be losing its ability to prevent the invasion of *Trichoderma* through the pseudosclerotium. This weakening of viability accompanied the exhaustion of the food reserves and perhaps also the accumulation of toxic metabolic products. *Trichoderma,* however, showed no signs of weakening, perhaps due partly to spore formation (Bliss, 1951).

Bliss (1951) also conducted experiments on viability of *Armillaria* in artificially infected citrus root segments buried in moist, non-sterile soil containing *Trichoderma*. The segments contained viable mycelium after 76 months. Other artificially inoculated segments, buried in moist, non-sterile peat moss in galvanized iron containers, had viable mycelium of *Armillaria* after 82 months. In a further test, *Armillaria* was found viable after 107 months, with the pseudosclerotium still intact. Bliss concluded that the established inoculum of *Armillaria* can remain viable for six years or more in moist, non-sterile soil containing *Trichoderma*. The period of viability is probably affected by the amount of food reserve within the pseudosclerotium. Though these studies were the first thorough investigation into the interaction between *Armillaria* and *Trichoderma*, and the conclusions drawn may still be valid, it is not clear whether other isolates of *Armillaria* would resist this or other isolates of *Trichoderma* in a similar way. Raziq (1998) found significant differences in growth rates of 5 isolates of *Armillaria* challenged with a range of potential fungal antagonists, including several strains of *T. harzianum*, *T. viride*, and *T. hamatum.* Isolate 1 of *A. mellea* (Am1), which produced rhizomorphs abundantly, was found more vigorous. The rhizomorphs of Am1 continued to grow in the presence of the antagonistic fungi, while its mycelial growth was checked within a week. Isolates of *Armillaria* that produced rhizomorphs only rarely, such as isolate 2 (Am2), were less vigorous and their growth was inhibited in a shorter period of time, and by more antagonists compared to those against Am1.

Moreover, the populations of *Trichoderma* in the soils used by Bliss (1951) were probably very low as no artificial inoculations of the antagonist were carried out and, while *Armillaria* had an adequate food base (citrus root segments), the antagonist was not provided with organic substrates for growth and proliferation and production of antibiotics. Gindrat (1979) suggested mixing antagonists with suitable organic material for fast establishment in the soil environment. *Trichoderma* spp. are known to be decomposers of cellulose, and incorporation of this substrate into soil has been reported to have selectively increased the population density of *T. harzianum* (Liu and Baker, 1980). Production of antibiotics is also mediated by the composition of the growth medium (Wood and Tveit, 1955; Di Pietro *et al.*, 1992). Raziq (1998) found that the presence of organic substrates enhances antagonistic activities of *Trichoderma* spp. He demonstrated that the richness of the growth and carrier substrate has a direct relationship with the biological control efficacy of the antagonist. All of the potted strawberry plants in the glasshouse inoculated with *Armillaria* survived until the end of the experiment, lasting 560 days when treated with *T. harzianum* isolate Th2 on 50 g wheat bran or wheat germ per pot, while none of them survived that long when treated with Th2 on 50 g mushroom compost containing autoclaved mushroom mycelia. The mushroom compost inocula, however, induced greater growth responses in the strawberry plants during the first few months following inoculations. The non-sustainability of the mushroom compost inocula was thought to be due to shortage of essential nutrients resulting from considerable decomposition during phase I and II of the compost preparation. Residual carbohydrates are kept at a critical level at the end of phase I, sufficient to allow conversion of ammonia and amines to protein, but not enough to favour the development of moulds like *Trichoderma*. At the end of the decomposition process, the medium has 2.0–2.5% nitrogen, with less than 0.05% ammonia (Baker and Cook, 1974).

Aytoun (1953) studied *in vitro* interactions of *Trichoderma* and *Armillaria* and concluded that *Trichoderma* must be considered a possible controlling factor in the spread of pathogenic fungi. The degree of parasitism of *Armillaria* by *Trichoderma* varied with the pH of the medium and with the position at which the respective hyphae came into contact with one another. On a medium of pH 3.4, the *Trichoderma* colony invaded that of *Armillaria*. Coiling around and penetration of the host hyphae occurred. After 24 hours of contact, *Armillaria* hyphae began to show signs of disintegration, vacuolation of the protoplasm and, later, collapse of the hyphal wall. Parasitism by *Trichoderma* was less severe on a medium of pH 5.1. Though the antagonist stopped the growth of *Armillaria* almost immediately in the region of invasion, penetration was not observed. No parasitism was seen on a medium of pH 7.0. Aytoun (1953) also demonstrated that 4 days after the rhizomorphs had stopped growing because of the action of the antagonist on malt agar at pH 3.4, more than half of them were still found viable when washed free of the conidia and mycelia with sterile water and 50% alcohol, and placed on fresh medium. This was concluded to suggest little or no effect of *Trichoderma* on the rhizomorphs as they consist of material homologous to the pseudosclerotium. Perhaps further incubation would have resulted in complete disintegration of the rhizomorphs, as they were still subject to the antagonistic activities of *Trichoderma*.

Sokolov (1964) reported that several fungi, including *Trichoderma*, antagonized *Armillaria*. He recommended the use of *Trichoderma* for controlling *Armillaria* infections. Dubos *et al.* (1978) found that the medium in which *T. viride* was grown influenced the level of inhibition of rhizomorphs of *Armillaria* by the antagonist. *T. polysporum, T. harzianum* and *T. viride* have also been shown to be mycoparasites of the rhizomorphs of *A. gallica* (Dumas and Boyonoski, 1992). Using electron microscopy, they observed that hyphae of these *Trichoderma* species were able to invade the rhizomorphs of *A. gallica* after one week. Penetration of the rhizomorphs by the *Trichoderma* species was either direct or through the formation of a depressed area in which a small hole was formed. Once *Trichoderma* hyphae had penetrated the melanized tissues of the rhizomorphs, they parasitized the internal hyphae by forming enlarged tips on the surface, or penetrating directly, with the site of entry being collapsed. Coiling of the antagonist around the hyphae of *Armillaria* also occurred. Ultimately, after one week's exposure to the *Trichoderma* species, all rhizomorphs were devoid of any living hyphae. Raziq (1998) found a similar effect of *Trichoderma* species on *A. mellea*. The antagonists checked the mycelial growth of *Armillaria* within a week (*Figure 10.1*), but rhizomorphs continued to grow for another week in isolates producing these structures abundantly. Eventually, the rhizomorphs were disintegrated (*Figure 10. 2*). They appeared transparent and devoid of the core mycelium, and failed to melanize in due course. *T. harzianum* and *T. hamatum* have also been shown by Elad *et al.* (1983) to digest the cell walls of *Sclerotium rolfsii* and *Rhizoctonia solani* enzymatically. This caused the leakage of the cytoplasm from the host cells,

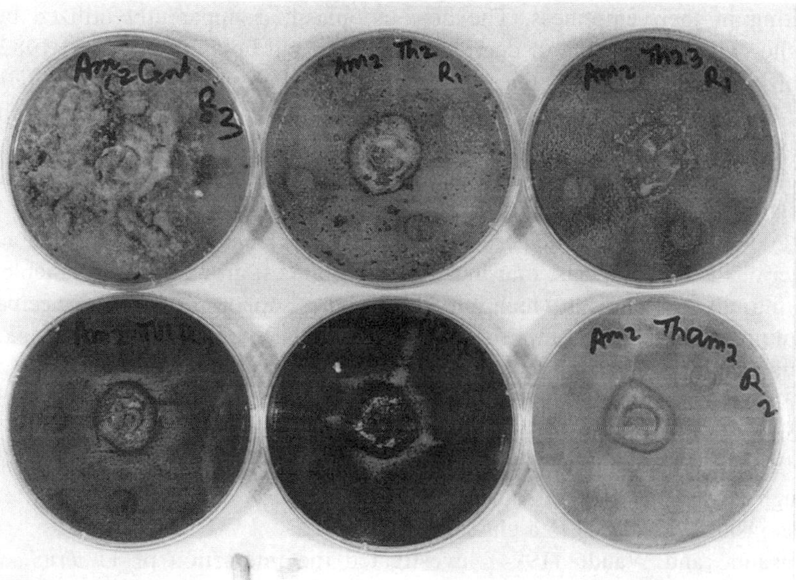

Figure 10.1. Inhibition of *Armillaria mellea* isolate 2 (Am2) by *Trichoderma* species. Clockwise from top left: Unchallenged growth of *Armillaria mellea* isolate 2 (Am2), growth of Am2 challenged with *Trichoderma harzianum* isolates 2 and 23 (Th2 and Th23), growth of Am2 challenged with *Trichoderma hamatum* isolate 2 (Tham2), growth of Am2 challenged with *Trichoderma viride* isolate 1 (Tv1). Other isolates of *Trichoderma harzianum*, *Trichoderma hamatum*, *Trichoderma viride* and an isolate of *Dactylium dendroides* had a similar effect on Am2.

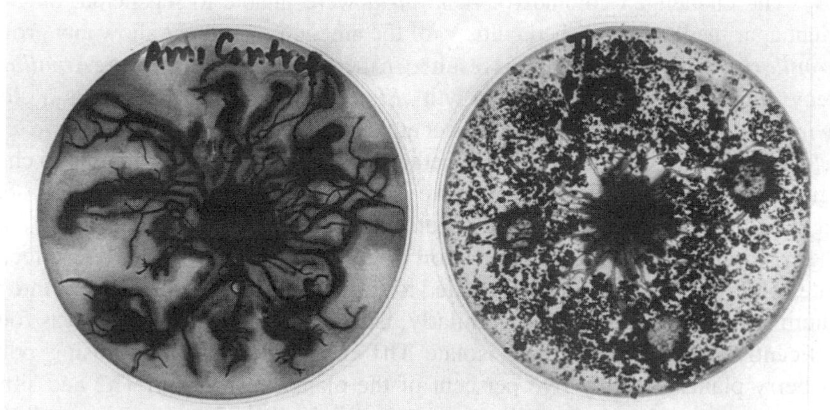

Figure 10.2. Disintegration of rhizomorphs of *Armillaria mellea* isolate 2 (Am1) by *Trichoderma harzianum* isolate 22 compared with its unchallenged growth, left. Other isolates of *Trichoderma harzianum*, *Trichoderma hamatum*, *Trichoderma viride* and an isolate of *Dactylium dendroides* had a similar effect on Am2.

resulting in their emptiness. The host cytoplasm is apparently utilized by the parasite, which is capable of degrading proteins and lipids (Elad *et al.*, 1982).

Rhizomorphs, like zonal lines and pseudosclerotial plates, enable *Armillaria* to resist antagonistic effects of *Trichoderma* and other antagonists. Rhizomorphs are highly differentiated structures conferring on *Armillaria* the ability to survive, grow, and infect roots in a hostile environment. An isolate's ability to incite disease is highly correlated with its production of rhizomorphs (Omdal *et al.*, 1995). Garraway *et al.* (1991) described the advantages of rhizomorphs to *Armillaria* as protection against deleterious external agents, translocation of resources, growth from a suitable food base into an environment which initially does not support growth, enhancement of inoculum potential, and amplification of individual hyphal sensitivity to external stimuli enabling directed growth responses. They also suggested that the outer cortex, composed of a ring of densely packed hyphae, is the main structure which protects the rhizomorhps in soil from being colonized by fungi and bacteria. Raziq (1998) reported that rhizomorphs of *A. mellea* were able to grow from inoculum source through carrier substrates of some *Trichoderma* strains to infect initially sterile hazel billets inside pots placed in a glasshouse.

Onsando and Waudo (1994) investigated the interaction of 11 *Trichoderma* isolates against *Armillaria* and found that vegetative structures and metabolites of *T. koningii* isolate 9, *T. longibrachiatum* isolate 3, and *T. harzianum* isolate 4 significantly reduced the radial mycelial and rhizomorph growth of *Armillaria* isolates M and N compared to the control. Raziq (1998) also noticed significant differences in the efficacy of different isolates of *T. harzianum, T. viride* and *T. hamatum* against isolates of *A. mellea*. *T. harzianum* isolates Th2, Th22 and Th23, *T. viride* isolates Tv3 (IMI 183289b) and Tv4 (IMI 170657), and *T. hamatum* isolate Tham1 (IMI 321194) were generally more effective. These isolates grew profusely over and around established colonies of *A. mellea*, greatly restricting their growth and disintegrating the rhizomorphs. This activity was maintained on different concentrations of potato dextrose agar (PDA), malt extract agar (MEA), and V-8 juice agar (VJA). The challenged colonies of *Armillaria* were unable to regenerate on fresh nutrient agar media. Established cultures of the antagonists did not allow any growth of *Armillaria*, while co-inoculations resulted in the antagonists outgrowing *Armillaria*, thereby completely inhibiting its growth (*Figure 10.3*). As *Armillaria* is a slow-growing fungus, and it was prone to direct attack of the antagonists in the above tests, hazel-billet inocula were used to see whether the antagonists were still able to check the growth of *Armillaria*. The inocula were placed on top of established cultures of the antagonists grown on 3% MEA. The effective antagonists named above did not allow any growth up to 5 months of incubation at 25°C in the dark. Other less effective isolates of *T. viride* and *T. hamatum* failed to stop the growth of rhizomorphs into the medium over the period of time. Similarly, isolate Th2 of *T. harzianum* was found significantly more effective than isolate Th1 in a glasshouse study using potted strawberry plants. Seventy five per cent of the plants treated with Th2 and simultaneously inoculated with *A. mellea* survived until the end of the experiment, lasting 413 days compared to no plant surviving that long when the treatment was Th1. However, in a field trial using apple trees, Raziq (1998) found no significant differences among the different isolates of *T. harzianum*, *T. viride*, and *T. hamatum*. None of the isolates was able to reduce the amount of infection on the tree roots

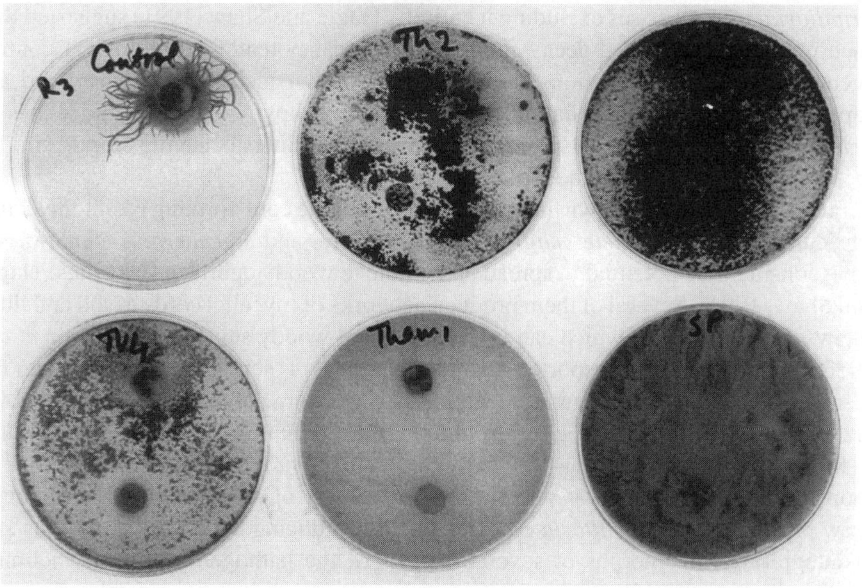

Figure 10.3. Prevention of growth of *Armillaria mellea* isolate 1 (Am1) in co-inoculations with the fungal antagonists. Clockwise from top left: Unchallenged growth of *Armillaria mellea* isolate 1 (Am1), Am1 co-inoculated with *Trichoderma harzianum* isolates 2 and 22 (Th2 and Th22), *Dactylium dendroides* (SP), *Trichoderma hamatum* isolate 1 (Tham1) and *Trichoderma viride* isolate 4 (Tv4).

compared to the control treatment of sterile mushroom compost with dead mycelia of *Agaricus bisporus*, the growth and carrier substrate of the antagonists. Nevertheless, the trees treated with isolate Th12 of *T. harzianum*, isolates Tv3 and Tv4 of *T. viride*, and isolate Tham1 of *T. hamatum* had significantly ($P<0.05$) less areas of their roots infected by *A. mellea*, compared to the control treatment of the pathogen without the mushroom compost.

Several other fungi have also been found antagonistic towards *Armillaria*. Leach (1937) observed that *Rhizoctonia lamellifera* prevented *Armillaria* from colonizing tea roots. Spruce stumps colonized by *Lenzites saepiaria* and *Phlebiopsis (Peniophora) gigantea* have been observed by Sokolov (1964) to have escaped attack by *Armillaria. Fomes pinicola* could be useful in controlling *Armillaria* (Orlos, 1957) because of its greater growth rate and ability to exclude the pathogen from occupied media. Cusson and LaChance (1974) reported that *Scytalidium lignicola* or scytalidin, the toxin it produces, halts the growth of *Armillaria* in culture. Two of the 10 basidiomycetes tested by Federov and Bobko (1989), *Phlebiopsis gigantea* and *Pleurotus ostreatus*, effectively prevented *Armillaria* growth in freshly cut stumps into which they had been inoculated.

Pearce and Malajczuk (1990) found that when *Coriolus versicolor, Stereum hirsutum,* and *Xylaria hypoxylon* were inoculated into karri thinning stumps simultaneously with *A. luteobubalina* each significantly reduced colonization by *Armillaria*. The eucalypt stumps were colonized both above and below ground by the competing fungi, but they were more effective antagonists above ground. A naturally occurring, cord-forming species of *Hypholoma* proved to be even more competitive with

Armillaria, in some cases excluding it entirely. Hagle and Shaw (1991) suggested that such cord-forming, wood-decay fungi have a great potential as biological control agents of *Armillaria* infections as they have a similar niche to *Armillaria* and are capable of subcortical mycelial growth in stumps, occupying the same initial sites as *Armillaria*. Rayner (1977) also reported some cord-formers behaving closely similar to *Armillaria*, except in pathogenicity.

Further studies have shown that several species of the cord-forming basidiomycetes, particularly *Phanerochaete velutina, H. fasciculare,* and *Steccherinum fimbriatum*, have considerable potential to spread and colonize woody debris in field sites (Hagle and Shaw, 1991). Several of them produce networks of mycelial cords in soil and litter (Dowson *et al.*, 1988a) which can infest additional woody substrates (Dowson *et al.*, 1988b). Rayner (1977) reported that populations of some cord-formers can be manipulated by chemically treating stumps. Ammonium sulphamate increased colonization by cord-formers of below ground portions of treated beech and birch stumps.

Raziq (1998) found that an isolate of *Dactylium dendroides* from Shiitake mushroom (*Lentinus edodes*) was an effective antagonist of *A. mellea.* This antagonist grew profusely over *Armillaria* colonies *in vitro*, reducing the mycelial growths and disintegrating rhizomorphs of several isolates of the pathogen. The overwhelming mode of action of this antagonist was competition, as it grew densely over soil compost under glasshouse conditions within a week of inoculation (*Figure 10.4*). Combinations of the *D. dendroides* isolate with isolates of *Trichoderma harzianum* and *T. viride* against *Armillaria* resulted in reduced mortality of potted strawberry

Figure 10.4. Growth of *Dactylium dendroides* isolate around strawberry plant grown in John Innes No. 2 Compost.

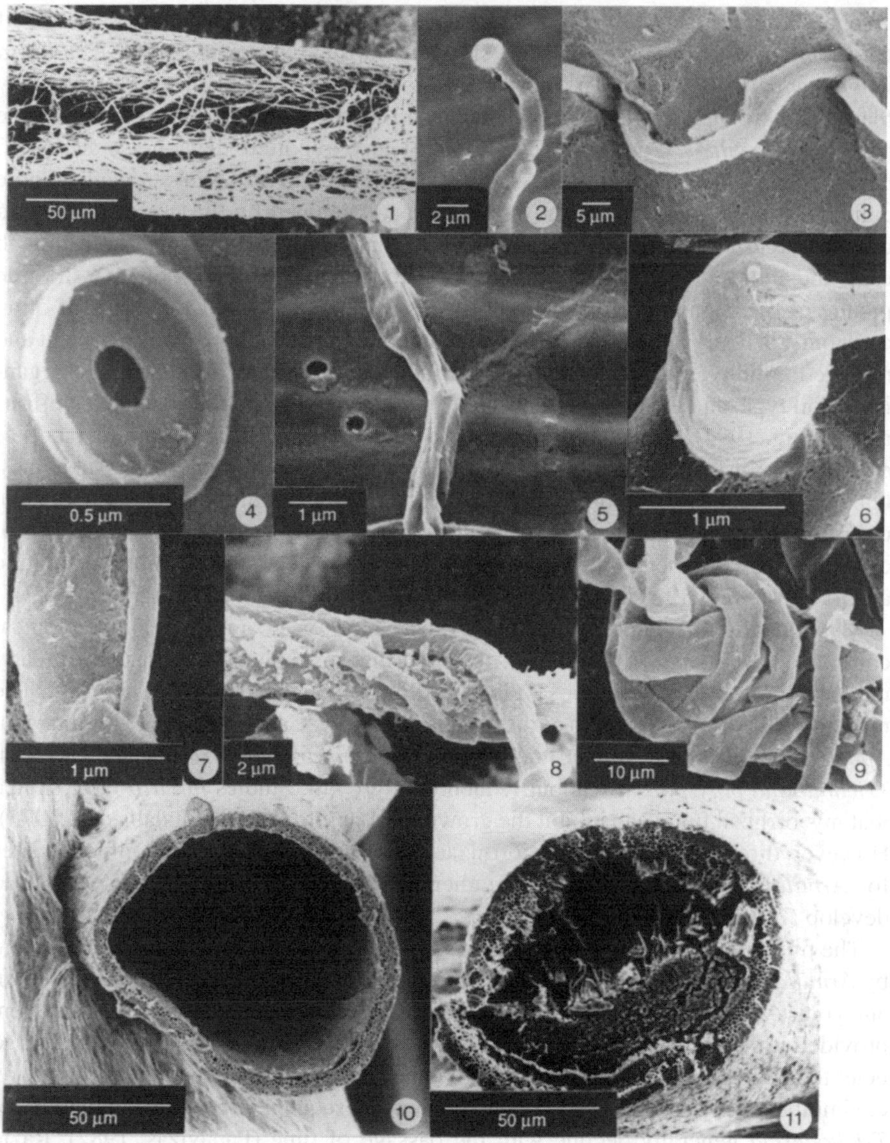

Figure 10.5. Stereo electron micrographs illustrating stages in mycoparasitism of *Armillaria mellea* hyphae by *Trichoderma* species.

plants in the glasshouse. The suppression of the disease was probably achieved by *D. dendroides* reinforcing the competitive ability of the *Trichoderma* isolates without interfering with their pathways of production of metabolites and mycoparasitic activities. The use of such combinations of antagonists is recommended for broad spectrum of activity and ecological adaptability (Gindrat, 1979). The mixtures should, however, be complementary and not competitive between themselves (Baker and Cook, 1974).

Raziq (1998) observed that the *in vitro* and *in vivo* or *in planta* activity of an antagonist may not necessarily be closely correlated. He found that an isolate of *Chaetomium olivaceum*, which did not show any substantial activity against *Armillaria* in the *in vitro* tests, reduced the mortality of potted strawberry plants more effectively than some isolates of *Trichoderma* that were able to suppress the growth of the pathogen *in vitro*. Wood and Tveit (1955) also reported that an isolate of *C. cochlioides*, which was not 'strikingly antagonistic' to *Fusarium nivale* in pure culture, gave field control of the same order as obtained by the use of a standard organomercurial seed dust. These results show that the *in vitro* screening tests, though easy and more economical to carry out (Campbell, 1989), may not give the true picture of the behaviour of an antagonist in soil. The physiological behaviour of an antagonist is likely to be affected by soil temperature, pH, aeration, water content and relative humidity, nutrient status, inoculum density, presence of other micro-organisms, soil type, and the physiological state of the pathogen and the host plant (Jeffries and Young, 1994). It is possible that some antagonists produce certain metabolites effective against pathogenic fungi in soils containing the required organic substrates, but not on solid agar media. On the other hand, metabolites produced under the optimized growth conditions of a laboratory may not be produced under natural soil conditions, or the metabolites may be produced but degraded by biotic or abiotic agents before they can work against the pathogen (Boudreau and Andrews, 1987).

Mycophagous nematodes and mycorrhizae have been implicated in biological control of *Armillaria*. Riffle (1973) observed that mycophagous nematodes greatly reduced mortality of ponderosa pine seedlings inoculated with *Armillaria*. The nematodes affected the growth of the fungus adversely. Cayrol *et al.* (1978) found that *Aphelenchus avenae* destroyed the hyphae of *Armillaria in vitro* but grew well on *Trichoderma polysporum* without reducing its growth. Studies *in vitro* have shown that mycorrhizal fungi can inhibit the growth of *Armillaria* (Eghbaltalab *et al.*, 1975). However, direct protection by mycorrhizae seems unlikely as the main infection sites for *Armillaria* are on coarse roots rather than the fine roots where mycorrhizae develop (Hagle and Shaw, 1991).

The time of application of an antagonist, in relation to the infection of a host plant by *Armillaria*, may be critical in the success of biological control of the disease. An antagonist well established in the rhizosphere will deny ports of entry to the pathogen, provided it maintains its antagonistic activities over a period of time long enough to deter the pathogen. This would suggest that the antagonist be added to the soil before, or simultaneously with, the pathogen to prevent subsequent infections. Populations of *Trichoderma* generally decline with the passage of time (Papavizas, 1982). Raziq (1998) reported that *T. harzianum* was more effective in reducing the mortality of potted strawberry plants when introduced simultaneously with *A. mellea*, but *Chaetomium olivaceum* had to be applied 85 days before the pathogen to achieve effective control of the disease in the glasshouse. Control was inadequate when the antagonists were added 85 days after inoculating the plants with *A. mellea*. The pathogen is likely to be more susceptible to the antagonists at the early stages of its growth because of the mycelium being young and, therefore, more sensitive. Turner (1991) and West (1994) found greater sensitivity of the young fresh mycelium of *Armillaria* to fungicides compared to the old mycelium. In addition to greater biomass, compact hyphae, and melanization, production of enough quantities of antibiotics by the established old mycelium of A*rmillaria* enables it to defend itself against the antagonists. In practical situations, antagonists would be expected to cure existing infections as well. Repeated applications of the antagonists, like fungicides, in better formulations may help to achieve this goal. Mycelial preparations resist fungistasis. Depending on the geographical location where control is attempted, antagonists should be applied at the time of the year when natural growth conditions are optimum for the antagonists but not for *Armillaria*. Optimum growth of *A. mellea* from woody inocula into soil takes place at a relatively low temperature of 22°C (Rishbeth, 1968). Growth of some isolates of the fungus may stop when temperatures reach 30°C (Turner, 1991). Conversely, growth of *Trichoderma* spp., especially *T. harzianum*, is favoured by high temperatures (Danielson and Davey, 1973; Munnecke *et al.*, 1981). Raziq (1998) observed that infection levels by *A. mellea* on the roots of apple trees in the field in southern England were lower when antagonists, including several species of *Trichoderma*, were applied in the second half of May 1995 than in the second half of June the same year. Exceptionally early start of the very warm weather that year provided longer duration of high temperatures for the establishment and proliferation of the antagonists applied earlier. *A. mellea* was probably also stressed to some extent by the high temperatures.

Integrated control of *Armillaria* root rot

In an effort to limit the use of chemicals in view of their harmful effects on the environment and human health, or where biological or chemical control alone has been ineffective, researchers have often tried to combine the two. In fact, biological control has been going on side by side with chemical control even though only the chemical control agent has often been credited with the measure of control achieved. Fungicides applied to the soil to control a disease affect the microflora therein. If the antagonistic fungi that constitute part of the microbial population are stimulated, remain unaffected, or are affected comparatively less than the pathogen at which the fungicide is targeted, they can dominate the treated soil by rapid growth in the absence of competing organisms. This can result in the exclusion of the pathogen from the soil. Therefore, Baker and Cook (1974) suggested that rather than killing the pathogen, it may only be necessary to weaken it and make it more vulnerable to the antagonism of the associated microflora.

Several examples of integrated control of *Armillaria* have been reported in the literature. Bliss (1951) demonstrated the ability of *T. viride* to replace *Armillaria* in artificially infected root segments fumigated with carbon disulphide. He reported that when root pieces infected with *A. mellea* were fumigated with the chemical and incubated either without soil, or in soil that was previously sterilized, the fungus survived, but when the infected root pieces were buried after fumigation in unsterilized soil or in soil amended with *Trichoderma, A. mellea* was killed and replaced by the antagonist. Pure soil cultures of *Trichoderma* were also able to kill *Armillaria* in unfumigated root inocula. Bliss concluded that *Armillaria* was not killed by direct fungicidal action of carbon disulphide but by the increase in the populations of *Trichoderma* spp. He explained that, normally, the pseudosclerotium formed by *Armillaria* inside woody substrates resists attack by *Trichoderma* in the presence of other micro-organisms in the soil, and the longevity of *Armillaria* in the pseudosclerotial phase of its life cycle is determined largely by the availability of its food reserve and by other environmental conditions. When the normal biological equilibrium was disturbed by the fumigation, *Trichoderma* developed rapidly and overcame the protective mechanism of *Armillaria*. After gaining entrance to the pseudosclerotium, *Trichoderma* requires only time to reach all parts of the fungus body and to kill the mycelium by antibiotic action.

Filip and Roth (1977) frequently isolated *T. viride* from pine stumps in which *Armillaria* was no longer viable after fumigation. Munnecke *et al.* (1973) suggested that after fumigation with carbon disulphide or methyl bromide, a lag period for *Armillaria* growth occurred, indicating a weakening of the fungus. As *T. viride* is more tolerant of the chemicals, it was able to exploit the lag period and exert an antagonistic action on *Armillaria*. Ohr and Munnecke (1974) showed that sub-lethal methyl bromide fumigation prevented the production of antibiotics by *Armillaria*. A similar effect may be caused by heating or drying (Munnecke *et al.*, 1976). These stress factors may be very critical and may concurrently stimulate antagonistic organisms, resulting in further damage to the already weakened *Armillaria* (Hagle and Shaw, 1991).

Garrett (1957) confirmed some of the findings of Bliss (1951), but also demonstrated that fumigation of the soil did not stimulate *Trichoderma* population increases to levels sufficient to kill unfumigated *A. mellea*. He reasoned that *A. mellea* must also be affected in some way. Ohr *et al.* (1973) found that, while almost 100% of *Armillaria* inoculum treated with sub-lethal but high concentrations of methyl bromide survived in sterilized field soil, only a small fraction of that stored in the unsterilized soil was viable after 30 days. A decrease in the viability of *A. mellea* was positively correlated with an increase in *Trichoderma* spp. in the fumigated root pieces incubated in the unsterilized field soil. *Trichoderma* spp. were the only organisms directly correlated with the increase in viability of *A. mellea*, although species of *Fusarium, Rhizopus*, and *Mucor*, and various unidentified bacteria, were also found frequently. Munnecke *et al.* (1981) suggested that fumigation of soil weakens the defence mechanisms of *Armillaria*. Metabolism of the fungus is presumably affected and repair of ruptures in, or formation of, pseudosclerotial walls is hindered. This breach in the defence mechanisms, cessation of antibiotic production, and simultaneous increase in the growth of *Trichoderma* result in the death of *Armillaria*. The stressed mycelium of *Armillaria* was also found to leak substances that could be attractive or stimulatory to *Trichoderma* spp. and other antagonists of *A. mellea*.

In the interactions between *Armillaria* and *Trichoderma*, the critical factor is the tolerance of *Trichoderma* to stress factors that inhibit the growth and metabolism of *Armillaria*. Munnecke *et al.* (1981) reported that *Trichoderma* spp. were 1.9–3.2 times more resistant to methyl bromide than *A. mellea*. Thus, *Trichoderma* spp. were able to survive much higher concentrations of methyl bromide than *A. mellea*, and populations are presumably unaffected or even increased by field fumigations with methyl bromide that are lethal to *A. mellea*. Munnecke *et al.* (1981) also found that *Trichoderma* spp. were much more tolerant of environmental conditions than *A. mellea*. The responses of the two fungi to temperatures of 30 to 39°C were particularly noteworthy. While growth of *Trichoderma* increased as temperatures increased, *A. mellea* was severely affected and its growth ceased either temporarily (10–30 days) or permanently, depending on the temperature and exposure time. Indirect kill, by storage in natural soil for 15 days following the heat treatment, required less exposure time and lower temperatures than the direct kill. They concluded that the fact that treatment with a physical agent like heat elicits responses in the pathogen and its antagonist, both *in vitro* and *in vivo*, similar to those elicited by the treatment with fumigants, indicates that the stressing effect is not due to residual effect of the fumigants.

Nelson *et al.* (1995) found that spore inoculum of a mixture of 5 isolates belonging to *T. harzianum, T. hamatum*, and *T. virens* placed in drill holes in the stumps of karri (*Eucalyptus diversicolor*) resulted in a significant reduction in root colonization by *A. luteobubalina*. Spore inoculum was found more effective than calcium alginate bead inoculum, and treatments of the stumps with 40% ammonium sulphamate (AMS), below or above ground level, resulted in increased colonization of the roots by *A. luteobubalina*. In the wood decayed by *Armillaria* or other fungi, the occurrence or relative frequency of *Trichoderma* spp. was higher in the AMS-treated than non-treated stumps. Laboratory tests by Nelson *et al.* (1995) indicated higher tolerance to AMS by the isolates of *Trichoderma* spp. than by the isolate of *A. luteobubalina*.

Studies by Pearce *et al.* (1995) showed that the cord-forming saprotrophs *Hypholoma australe* and *Phanerochaete filamentosa* reduced the colonization of karri stumps by *A. luteobubalina* significantly, but only when the stumps were treated with 40% AMS. *P. filamentosa* was more effective than *H. australe* in reducing colonization by *Armillaria*, and inoculations of both antagonists were more effective below ground level than above ground level. *P. filamentosa* all but eliminated *Armillaria* when inoculated below ground level. AMS alone significantly enhanced colonization by *Armillaria* below ground level, whereas at ground level or above ground level, colonization by the pathogen was reduced. The naturally occurring fungi *Stereum hirsutum* and *Trametes versicolor* fruited only on the AMS-treated stumps, whereas *Chondrostereum purpureum* fruited only on those not treated with the chemical. AMS enhanced the fruiting of *H. australe* and suppressed that of *A. luteobubalina.*

The cord-forming saprotrophs may have a role in limiting the saprophytic spread of *Armillaria* to new host plants by occupying the available substrates and taking hold of the niches that would have otherwise been colonized by the pathogen. Preventing or eradicating infections in living host plants with these fungi is, however, not advisable as they themselves cause decay. Control with *Trichoderma* is, therefore, desirable as it can not only work as an antagonist towards the pathogen, but is also safe to the host plant, and can even stimulate growth responses in it (Chet, 1987). Other non-decay antagonists may also be considered for integration with suitable fungicides.

In the case of integrated control with an introduced antagonist, consideration as to whether the antagonist or the fungicides should be applied first is very important. If the antagonist is likely to be adversely affected by the fungicide, then the fungicide would have to be applied first. The antagonist could then be applied after the toxic effects of the fungicide have diminished. This may also help in the establishment of the antagonist as part of the soil microflora will have been killed by the fungicide, thus reducing competition to the introduced antagonist, but if the antagonist is not likely to be adversely affected by the fungicide , then it could be applied before, or simultaneously with, the fungicide. This could have a combined effect on the pathogen resulting, possibly, in its death.

Extensive studies by Raziq (1998) on integration of selected antagonists with two systemic fungicides, fosetyl-Al and fenpropidin, against *A. mellea* in the laboratory, glasshouse and in the field showed that the antagonists were more sensitive to fenpropidin than to fosetyl-Al. Consequently, the antagonists were generally more effective when applied in the glasshouse and field 40 days after fenpropidin or before fosetyl-Al. Fenpropidin binds to the soil particles and is rapidly and extensively degraded (Tomlin, 1994). Therefore, the antagonists were not adversely affected when applied 40 days after using a moderate dose rate of 2000 mg/l of the fungicide as drenches. High dose rates of the fungicide were, however, phytotoxic to strawberry plants in the glasshouse and apple trees in the field. Fosetyl-Al, on the other hand, was safe even when used at the highest dose rate of 8000 mg/l. The strawberry plants treated with fosetyl-Al had significantly ($P<0.05$) more leaves and grew more vigorously than those treated with fenpropidin. Similarly, the apple trees drenched with fosetyl-Al produced significantly more branches than those treated with fenpropidin. *In vitro*, growth of *A. mellea* on MEA amended with 200 mg/l of fenpropidin was completely inhibited, while that of the antagonists was either inhibited or severely restricted. Fosetyl-Al, however, allowed growth of both *Armillaria* and the antagonists, even when applied at the higher dose rate of 2000 mg/l. The general theme that emerges from these studies is that, while the individual role of the antagonists and the fungicides is important in stressing the pathogen at the early crucial stages of its development, the time of application of one control strategy in relation to the other can be a determining factor in the success, or otherwise, of controlling the pathogen.

Acknowledgements

The author would like to thank the Federal Ministry of Education, Government of Islamic Republic of Pakistan, for sponsoring his Ph.D. studies on *Armillaria* at The University of Reading (U.K.) and the N-W.F.P. Agricultural University, Peshawar (Pakistan), for study leave during the course.

References

AYTOUN, R.S.C. (1953). The genus *Trichoderma*: its relationship with *Armillaria mellea* (Vahl ex Fries) Quél. and *Polyporus schweinitzii* Fr., together with preliminary observations on its ecology in woodland soils. *Transactions and Proceedings of the Botanical Society of Edinburgh* **36**, 99–114.

BAKER, K.F. AND SNYDER , W.C. (eds.) (1965). *Ecology of soil-borne plant pathogens: Prelude to biological control.* Berkeley: University of California Press.

BAKER, K.F. AND COOK, R.J. (1974). *Biological control of plant pathogens.* San Francisco: W.H. Freeman and Company.

BLISS, D.E. (1941). Artificial inoculation of plant with *Armillaria mellea* (Abstract). *Phytopathology* **31**, 859.

BLISS, D.E. (1951). The destruction of *Armillaria mellea* in citrus soils. *Phytopathology* **41**, 665–683.

BOUDREAU, M.A. AND ANDREWS, J.H. (1987). Factors influencing antagonism of *Chaetomium globosum* to *Venturia inaequalis*: a case study in failed biocontrol. *Phytopathology* **77**, 1470–1475.

CAMPBELL, A.H. (1934). Zone lines in plant tissues. II. The black lines formed by *Armillaria mellea* (Vahl) Quél. *Annals of Applied Biology* **21**, 1–22.

CAMPBELL, R. (1989). *Biological control of microbial plant pathogens.* Cambridge: Cambridge University Press.

CAYROL, J.C., DUBOS, B. AND GUILLAUMIN, J.J. (1978). Etude préliminaire in vitro de l'agressivité de quelque nématodes mycophages vis-à-vis de *Trichoderma viride* Pers., *T. polysporum* (Link. ex. Pers.) Rifaï et *Armillaria mellea* (Vahl) Karst. *Annales de Phytopathologie* **10**, 177–185.

CHET, I. (1987). *Trichoderma*–application, mode of action, and potential as a biocontrol agent of soilborne plant pathogenic fungi. In: *Innovative approaches to plant disease control*. Ed. I. Chet, pp 137–160. New York: John Wiley.

COOK, R.J. (1991). Biological control of plant diseases: broad concepts and applications. In: *The biological control of plant diseases*. Proceedings of the International Seminar on Biological Control of Plant Diseases and Virus Vectors. FFTC, Taipei. Ed. J. Bay-Peterson, pp 1–29.

CUSSON, Y. AND LACHANCE, D. (1974). Antagonism between *Scytalidium lignicola* and two root rot fungi. *Phytoprotection* **55**, 17–28.

DANIELSON, R.M. AND DAVEY, R.B. (1973). The abundance of *Trichoderma* propagules and the distribution of species in forest soils. *Soil Biology and Biochemistry* **5**, 485–494.

DI PIETRO, A., GUT-RELLA, M., PACHLATKO, J.P. AND SCHWINN, F.J. (1992). Role of antibiotics produced by *Chaetomium globosum* in biocontrol of *Pythium ultimum*, a causal agent of damping-off. *Phytopathology* **82**, 131–135.

DOWSON, C.G., RAYNER, A.D.M. AND BODDY, L. (1988a). Inoculation of mycelial cord-forming basidiomycetes into woodland soil and litter, I. Initial establishment. *New Phytologist* **109**, 335–341.

DOWSON, C.G., RAYNER, A.D.M. AND BODDY, L. (1988b). Inoculation of mycelial cord-forming basidiomycetes into woodland soil and litter, II. Resource capture and persistence. *New Phytologist* **109**, 343–349.

DUBOS, B., GUILLAUMIN, J.J. AND SCHUBERT, M. (1978). Action du *Trichoderma viride* Pers., approté avec divers substrats organiques, sur l'initiation et la croissance des rhizomorphes *d'Armillaria mellea* (Vahl) Karst. dans deux types de sols. [The action of *Trichoderma viride* Pers., applied with various organic substrates, on the initiation and growth of rhizomorphs of *Armillaria mellea* (Vahl) Karst. in two types of soils.] *Annales de Phytopathologie* **10**, 187–196.

DUMAS, M.T. AND BOYONOSKI, N.W. (1992). Scanning electron microscopy of mycoparasitism of *Armillaria* rhizomorphs by species of *Trichoderma*. *European Journal of Forest Pathology* **22**, 379–383.

EGHBALTALAB, M., GAY, G. AND BRUCHET, G. (1975). Antagonisme entre 15 espèces de basidiomycètes et 3 champignons pathogènes de racines d'abres. *Bulletin de la Société Linnéenne de Lyon* **44**, 203–229.

ELAD, Y., CHET, I. AND HENIS, Y. (1982). Degradation of plant pathogenic fungi by *Trichoderma harzianum*. *Canadian Journal of Microbiology* **28**, 718–725.

ELAD, Y., BARAK, R., CHET, I. AND HENIS, Y. (1983). Ultrastructural studies of the interaction between *Trichoderma* spp. and plant pathogenic fungi. *Phytopathologische Zeitschrift* **107**, 168–175.

FAULL, J.L. AND GRAEME-COOK, K. (1992). Characterization of mutants of *Trichoderma harzianum* with altered antibiotic production characteristics. In: *Biological control of plant diseases: progress and challenges for the future*. Eds. E.S. Tjamos, G.C. Papavizas and R.J. Cook, pp 345–351. New York: Plenum Press.

FEDEROV, N.I. AND BOBKO, I.N. (1989). *Armillaria* root rot in Byelorussian forests. In: *Proceedings of the 7th International Conference on Root and Butt Rots*, August 9–16, 1988, Vernon and Victoria, B.C. Ed. D.J. Morrison, pp 469–476. Victoria, B.C.: International Union of Forestry Research Organizations.

FILIP, G.M. AND ROTH, L.F. (1977). Stump infections with soil fumigants to eradicate *Armillaria mellea* from young-growth ponderosa pine killed by root rot. *Canadian Journal of Forest Research* **7**, 226–231.

GARRAWAY, M.O., HÜTTERMANN, A. AND WARGO, P.M. (1991). Ontology and physiology. In: *Armillaria Root Disease*. Eds. C.G. Shaw, III and G.A. Kile, pp 21–47. *Forest Service Agriculture Handbook No. 691*. Washington, D.C.: USDA.

GARRETT, S.D. (1957). Effect of soil microflora selected by carbon disulphide fumigation on survival of *Armillaria mellea* in woody host tissues. *Canadian Journal of Microbiology* **3**, 135–149.

GINDRAT, D. (1979). Biocontrol of plant diseases by inoculation of fresh wounds, seeds, and soil with antagonists. In: *Soil-borne plant pathogens*. Eds. B. Schippers and W. Gams, pp 536–551. New York: Academic Press.

GREGORY, S.C., RISHBETH, J. AND SHAW, C.G., III (1991). Pathogenicity and virulence. In: *Armillaria Root Disease*. Eds. C.G. Shaw, III and G.A. Kile, pp 76–87. *Forest Service Agriculture Handbook No. 691*. Washington, D.C.: USDA.

GUILLAUMIN, J.J. (1988). The *Armillaria mellea* complex. *Armillaria mellea* (Vahl) Kummer *sensu stricto*. In: *European handbook of plant diseases*. Eds. I.M. Smith, J. Dunez, D.H. Phillips, R.A. Lelliot and S.A. Archer, pp 520–523. Oxford: Blackwell Scientific Publications.

HAGLE, S.K. AND SHAW, C.G., III (1991). Avoiding and reducing losses from Armillaria root disease. In: *Armillaria Root Disease*. Eds. C.G. Shaw, III and G.A. Kile, pp 157–173. *Forest Service Agriculture Handbook No. 691*. Washington, D.C.: USDA.

ISHIKAWA, H., OKI, T. AND KIRIYAMA, H. (1976). (The toxic function of the antifungal compounds prepared by some *Hypocrea* species to wood-rotting fungi.) (In Japanese.) *Reports of the Tottori Mycological Institute* **14**, 105–110. (*Review of Applied Mycology* **56**, 2249).

JEFFRIES, P. AND YOUNG, T.W.K. (1994). *Interfungal parasitic relationships*. Wallingford, U.K.: CAB International.

LAMANA, C. AND MALETTE, M.F. (1965). *Basic bacteriology: its biological and chemical background*. 3rd Edition. Baltimore: Williams and Wilkins.

LEACH, R. (1937). Observations on the parasitism and control of *Armillaria mellea*. *Proceedings of the Royal Society of London, Series B* **121**, 561–573.

LIU, S. AND BAKER, R. (1980). Mechanism of biological control in soil suppressive to *Rhizoctonia solani*. *Phytopathology* **70**, 404–412.

MASER, C., TRAPPE, J.M. AND LI, C.Y. (1984). Large woody debris and long-term forest productivity. In: *Pacific Northwest bioenergy systems: policies and application*. Proceedings of a Conference Sponsored by the U.S. Department of Energy and the Pacific Northwest and Alaska Bioenergy Program, May 10–11, 1984, Portland, Oregon.

MORRISON, D.J. (1976). Vertical distribution of *Armillaria mellea* rhizomorphs in soil. *Transactions of the British Mycological Society* **66**, 393–399.

MUNNECKE, D.E., KOLBEZEN, M.J. AND WILBER, W.D. (1973). Effects of methyl bromide or carbon disulfide on *Armillaria* and *Trichoderma* growing on agar medium and relation to survival of *Armillaria* in soil following fumigation. *Phytopathology* **63**, 1352–1357.

MUNNECKE, D.E., WILBER, W.D. AND DARLEY, E.F. (1976). Effects of heating or drying on *Armillaria mellea* and *Trichoderma viride* and the relation to survival of *A. mellea* in soil. *Phytopathology* **66**, 1363–1368.

MUNNECKE, D.E., KOLBEZEN, M.J., WILBUR, W.D. AND OHR, H.D. (1981). Interactions involved in controlling *Armillaria mellea*. *Plant Disease* **65**, 384–389.

NELSON, E.E., PEARCE, M.H. AND MALAJCZUK, N. (1995). Effects of *Trichoderma* spp. and ammonium sulphamate on establishment of *Armillaria luteobubalina* on stumps of *Eucalyptus diversicolor*. *Mycological Research* **99**, 957–962.

OHR, H.D., MUNNECKE, D.E. AND BRICKER, J.L. (1973). Interaction of *Armillaria mellea* and *Trichoderma* spp. as modified by methyl bromide. *Phytopathology* **63**, 965–973.

OHR, H.D. AND MUNNECKE, D.E. (1974). Effects of methyl bromide on antibiotic production by *Armillaria mellea*. *Transactions of the British Mycological Society* **62**, 65–72.

OMDAL, D.W., SHAW, C.G., III, JACOBI, W.R. AND WAGER, T.C. (1995). Variation in pathogenicity and virulence of isolates of *Armillaria ostoyae* on eight tree species. *Plant Disease* **79**, 939–944.

ONSANDO, J.M. AND WAUDO, S.W. (1994). Interaction between *Trichoderma* species and *Armillaria* root rot fungus of tea in Kenya. *International Journal of Pest Management* **40**, 69–74.

ORLOS, H. (1957). Badania nad zwalczaniem opienki miodowej (*Armillaria mellea* Vahl) metoda biologiczna. [Investigations on the biological control of *Armillaria mellea*.] (In Polish.) *Roczniki Nauk lesn.* **15**, 195–236. (*Forestry Abstracts* **19**, 4401).

PAPAVIZAS, G.C. (1982). Survival of *Trichoderma harzianum* in soil and in pea and bean rhizospheres. *Phytopathology* **72**, 121–125.

PAPAVIZAS, G.C. (1985). *Trichoderma* and *Gliocladium*: biology, ecology, and potential for biocontrol. *Annual Review of Phytopathology* **23**, 23–54.

PEARCE, M.H. AND MALAJCZUK, N. (1990). Inoculation of *Eucalyptus diversicolor* thinning stumps with wood decay fungi for control of *Armillaria luteobubalina. Mycological Research* **94**, 32–37.

PEARCE, M.H., NELSON, E.E. AND MALAJCZUK, N. (1995). Effects of the cord-forming saprotrophs *Hypholoma australe* and *Phanerochaete filamentosa* and of ammonium sulphamate on establishment of *Armillaria luteobubalina* on stumps of *Eucalyptus diversicolor. Mycological Research* **99**, 951–956.

POWELL, K.R., FAULL, J.L. AND RENWICK, A. (1990). The commercial and regulatory challenge. In: *Biological control of soil-borne plant pathogens.* Ed. D. Hornby, pp 445–463. Wallingford, U.K.: CAB International.

RAYNER, A.D.M. (1977). Fungal colonization of hardwood stumps from natural sources, I. Basidiomycetes. *Transactions of the British Mycological Society* **69**, 303–312.

RAZIQ, F. (1998). *Biological and integrated control of the root rot caused by Armillaria mellea.* Ph.D. Thesis. U.K.: University of Reading.

REDFERN, D.B. (1970). The ecology of *Armillaria mellea*: rhizomorph growth through soil. In: *Root diseases and soil-borne pathogens.* Proceedings of the Symposium, July, 1968. Imperial College, London. Eds. T.A. Toussoun, R.V. Bega and P.E. Nelson. Berkeley: University of California Press.

RIFFLE, J.W. (1973). Effect of two mycophagous nematodes on *Armillaria mellea* root rot of *Pinus ponderosa* seedlings. *Plant Disease Reporter* **57**, 355–357.

RISHBETH, J. (1968). The growth rate of *Armillaria mellea. Transactions of the British Mycological Society* **51**, 575–586.

RISHBETH, J. (1976). Chemical treatment and inoculation of hardwood stumps for control of *Armillaria mellea. Annals of Applied Biology* **82**, 57–70.

SCHÜTT, P. (1985). Control of root and butt rots: limits and prospects. *European Journal of Forest Pathology* **15**, 357–363.

SHAW, C.G., III AND ROTH, L.F. (1978). Control of Armillaria root rot in managed coniferous forests. *European Journal of Forest Pathology* **8**, 163–174.

SOKOLOV, D.V. (1964). *Kornevaya gnil' ot openki i bor'ba s nei.* [Root rot caused by *Armillaria mellea* and its control.] (In Russian.) Moscow, Izadatel'stvo Lesnaya Promýshlennost'. (Canada Department of Forestry. 235pp.)

TOMLIN, C. (ed.) (1994). *The pesticide manual.* 10th Edition. Farnham, U.K.: The British Crop Protection Council and Royal Society of Chemistry.

TURNER, J.A. (1991). *Biology and control of Armillaria.* Ph.D. Thesis. U.K.: University of Reading.

TURNER, J.A. AND FOX, R.T.V. (1988). Prospects for the chemical control of *Armillaria* species. In: *Proceedings of the Brighton Crop Protection Conference: Pests and Diseases 1*, 1988, pp 235–240.

VAN DRIESCHE, R.G. AND BELLOWS, T.S., JR. (1996). *Biological control.* New York: Chapman and Hall.

WEINDLING, R. (1932). *Trichoderma lignorum* as a parasite of other soil fungi. *Phytopathology* **22**, 837.

WEST, J.S. (1994). *Chemical control of Armillaria.* Ph.D. Thesis. U.K.: University of Reading.

WOOD, R.K.S. AND TVEIT, M. (1955). Control of plant diseases by use of antagonistic organisms. *The Botanical Review* **21**, 441–492.

SECTION 5
Future Possibilities

11
Answering All the Questions About *Armillaria*

ROLAND T.V. FOX

The University of Reading, School of Plant Sciences, Department of Horticulture and Landscape (Crop Protection), 2 Earley Gate, Reading, Berkshire RG6 6AU, U.K.

Synopsis

In the future, control of *Armillaria* root rot under commercial forest conditions would probably only become as worthwhile as in amenity and commercial horticulture if cheaper and more effective ways can be found to treat or destroy tree stumps, infected roots and soil in order to speed replanting. Carbon disulphide and methyl bromide have been used as soil fumigants for many years, but both require specialist handling. The use of such fumigants may be banned as they have been implicated in global warming, so eradication would have to involve more specific and less hazardous soil fumigants or drenches. Several agricultural fungicides have shown both eradicant and protectant properties in laboratory tests without serious phytotoxic effects on the range of tree species tested so far, but results in the field have been disappointing. However, new toxophores are continually being developed that would be worth testing against *Armillaria*. A suitable fungicide drench, perhaps integrated with cultures of antagonistic, possibly genetically modified, micro-organisms, such as *Trichoderma harzianum* or *Dactylium dendroides*, probably incorporated on a sustaining substrate, could protect recently planted trees while they become established. New ways may be found to manipulate *Armillaria* outbreaks by altering the environmental factors which favour the pathogen. It may also eventually be possible to replant previously infected areas with trees or herbaceous plants which have been selected or genetically modified to tolerate *Armillaria* infection, as well as injury from exposure to pollutants and other factors that encourage infection. Since correct diagnosis is an essential prerequisite for deciding on the most effective control method to choose, or indeed whether any form of control should be practised, a rapid, simple, cheap, portable, qualitative test kit is needed for use in the field. Infection in the roots of a tree by pathogenic *Armillaria* species could then be confirmed unambiguously and quickly, allowing more time for treatment or destruction long before it develops into a dangerous rot or spreads to other trees.

Armillaria *Root Rot: Biology and Control of Honey Fungus*

Opportunities for further basic research

There are several unresolved future research topics on *Armillaria* among the current subjects being explored. Features of the biology of the many different *Armillaria* species should continue to provide abundant opportunities for study, such as understanding the mechanisms of homothallism. Now that sexual and parasexual crosses are available in the laboratory, it may even be possible to identify the determinants of pathogenicity. Because of the considerable experience of the breeding relationships, morphology, ecology, and distribution of well delineated species, *Armillaria* offers an excellent opportunity to use molecular characters to reconstruct phylogenetic relationships and to assess the relative roles of geographic isolation and intersterility in fungal speciation.

Because species, and even individual genotypes, can now be accurately identified, we can expect better resolution of epidemiological patterns, from long-range dispersal through local spread and infection in forests. Despite much progress, there remain abundant opportunities for research into the epidemiology of the numerous diverse *Armillaria* species, as these may be found almost anywhere in the world where there is a suitable substrate or host, even remote oceanic islands (Raabe and Trujillo, 1963). As honey fungus damages forests, destroys plantations and orchards, and spoils gardens and other amenity sites, there should be an impetus for further research into host susceptibility. Although trees and shrubs are attacked by one or more virulent species of *Armillaria* almost anywhere, and few, if any, woody plants are immune (Raabe, 1962), herbaceous plants show a much wider spectrum of susceptibility. It is not yet clear if, or how, this diverse range of specific responses can be exploited in order to avert future losses, but understanding the physiological bases for pathogenesis and the interactions of *Armillaria* species with their hosts is the key to understanding the variation in virulence among and within them. Some of the morphological and anatomical reactions have been characterized but Wargo and Harrington (1991) considered this area of research is ripe for many more studies to elucidate the physiology of host-pathogen interactions. Several modelling/quantitative aspects of the epidemiology of *Armillaria* in the field deserve attention, such as a comparison of the spread of the disease to adjacent trees by direct root-to-root contact with that by rhizomorphs.

Although it is already clear that the genus *Armillaria* has a worldwide distribution, and many species have now been described, probably others exist. Species which cause root and butt rot of trees and infect shrubs and some herbs are more readily recognized than the numerous species of *Armillaria* which are non-parasitic. If any species have lost the ability to produce basidiocarps and rhizomorphs, the resultant sterile mycelium might not be easily recognized in rotten wood or soil, particularly in the tropics. Doubtless, if such fungi exist, molecular methods would show their affinities. The problem is who, if anyone, is looking for these mycelia?

Immunological assays are also entering a period of great change as relatively cheap, easy to use kits are being developed which allow low levels of disease to be monitored on the spot under field conditions. This may enable the initial stages of infection by basidiospores to be observed (Shaw and Kile, 1991). Since these kits are so sensitive, not only could the life cycle (Wargo and Shaw, 1985) and patterns of spread (Rishbeth, 1986) be validated, but it would be possible to treat lower inoculum levels of pathogens than previously. Consequently, if *Armillaria* can be detected earlier, this should permit more effective control, and the use of fungicides may be avoided where no pathogens are detected. Also, formerly difficult to diagnose pathogens, such as *Armillaria* and other soil-borne organisms which are traditionally rarely easily quantified even when recognized, could become a more commercially viable market for fungicides and other control measures.

In future, experimental designs intended to elucidate the resistant and susceptible reactions may be improved by the increased use of clonal host material and known species, together with known genotypes of *Armillaria* species and stressed and non-stressed systems.

Improving diagnosis

Now that *Armillaria* root rot is known to be caused by over thirty distinct species with quite distinct patterns of virulence and somewhat varied morphology rather than a single species, *A. mellea*, their identity, as well as the extent of current damage, determines whether it is worthwile to employ control measures. Symptoms are non-specific and too late, as are the presence of signs such as mycelial fans, rhizomorphs, toadstools and decay *in situ*, or when seen *in vitro* after culturing it onto agar, even though cultures of *Armillaria* have several distinctive characteristics. In future, it is likely that the methods outlined in Chapter 5 will be improved further to allow the identification and separation of the similar species of *Armillaria* that co-habit the same environment.

The earlier that diagnosis can be confirmed, the greater the opportunity for successful control of the disease. Now that it is possible to recognize the presence of infection before the appearance of above-ground symptoms, using enzyme-linked immunosorbent assays (ELISA) and nucleic acid techniques should make control of the disease more feasible than when the fungus had to be isolated onto agar (Fox and Hahne, 1989; Manley, 1996).

Improved husbandry practices

Control by cultural practices has been recommended for a long time but, since viable mycelium can be found on roots at a depth of several metres (Bliss, 1951), removal must be total to be completely effective. The process of extraction of wood buried in the soil is always laborious and usually impracticable, since it is not easy to know whether any roots still remain deep in the soil. However, in the future it may be possible to detect and remove buried wooden inoculum more effectively than at present (Roth *et al.*, 1980). Also, new methods of destroying stumps and root remnants may be developed (Morrison *et al.*, 1988), perhaps by turning the earth over more thoroughly and deeply (Heaton and Dullahide, 1989). Trenching (Guillaumin, 1988) might be made more effective by incorporating a fungicide impregnated plastic barrier in the trench before it is backfilled with soil.

Local tree species, grown in typical mixtures and densities in natural forests, may tolerate the local species of *Armillaria* root disease, even though individually each may be a host for it (Hagle and Shaw, 1991). This technique of matching indigenous species with suitable sites as a way to minimize disease hazard could be tried more widely than in the coniferous production forests of western North America. There, only indigenous tree species and the species, seed sources and cultural methods are carefully selected to fit local habitat and site conditions (Daubenmire, 1952; Corns and Annas, 1986; Hadfield *et al.*, 1986; Morrison, 1981; Williams *et al.*, 1989). If locally adapted seed sources are not available, the genetic differences within a population of naturalized trees should be minimal (Lung-Escarmant and Taris, 1989).

Assessing planting sites

Existing knowledge of the ecology of *Armillaria* is rather restricted (Hagle and Shaw, 1991), but in the future it may be possible to discover more accurately beforehand whether a potential planting site is unsuitable due to heavy levels of inoculum of a virulent species of *Armillaria* remaining from numerous infections in previous stands, harmful operations by humans or other factors or conditions that greatly favour disease development. Soil type and situation is important (Sokolov, 1964; Whitney, 1984). These differences affect the type of plants growing prior to establishing orchards or plantations, and so may predict the risk of disease.

Although it is feasible to partly modify the pH, organic content, and nutrient status of the soil in orchards or forests, there is insufficient direct evidence that the application of fertilizers or other soil amendments can influence *Armillaria* root disease, particularly in forest trees whose nutritional requirements are not yet well understood.

Possible future trends in chemical control

At present, the active fumigants face a possible ban on their use, yet no replacement fumigant chemicals have yet been identified, nor have adequately effective non-phytotoxic fungicide drenches or other formulations been discovered (Hagle and Shaw, 1991). Only Armillatox, a phenolic emulsion containing 48% cresylic acid as the active ingredient is approved in the U.K. and marketed for use against *Armillaria* (Pawsey and Rahman, 1974; Rahman, 1974). However, Redfern (1971) obtained little benefit from its use, since its penetration into wood is limited. A comparable mixture of cresylic acids (Bray's Emulsion) behaved similarly. West (1994) found cresylic acid was responsible for unacceptable phytotoxicity and also eventually stimulated the mycelium to grow. Efforts to find a superior non-phytotoxic chemical treatment have not yet been successful.

Any fungicide applied as a drench requires substantial volumes of a solvent, often water. Highly absorbent soils pose problems. All soil-borne pathogens are difficult to control because of the nature of soil and the presence of absorbent matter like organic substances. However, *Armillaria* has both extensive mycelium inside host tissue and rhizomorphs covered in a protective rind. After becoming established in the roots, the vegetative mycelium of *Armillaria* develops a protective layer of thick-walled fungus cells, the pseudosclerotial envelope which covers the bulk of the infected wood. The rhizomorphs growing from the apices of the pseudosclerotium are also covered by this protective layer. In order to eradicate the fungus, a chemical or biological control agent will have to enter these resistant structures and maintain its eradicant activity. Any novel method of achieving this would be valuable. Possibly, mycoviruses could be effective.

After a time, the linked *Armillaria* infections become widely distributed in the soil. Species have now been described which consist of an enormous long-lived biomass at least as large as any organism on Earth (Smith *et al.*, 1992). Since the target is so enormous, it is hardly surprising that no non-systemic chemicals (Pawsey and Rahman, 1976; Filip and Roth, 1987; Fedorov and Bobko, 1989), or systemic chemicals (Turner, 1991), proved adequate in the field. It is likely that chemicals applied as drenches fail to reach sufficient of their target, the mycelium of the pathogen. Apart from its huge extent, this may also be due to the chemicals becoming bound to the soil particles or wood surfaces, thereby making them unavailable even for localized fungicidal action. Alternatively, a water soluble chemical is often leached away. Even if the currently available fungicides were sufficiently active, it would logically be both costly and extremely difficult to achieve long term control of many well-established outbreaks.

Several possibilities have been tried to improve matters, including the use of high presssure soil and tree injection apparatus, as well as systems for the gradual uptake and accumulation of non-phytotoxic systemic fungicides, initially through the tips of young roots, and their subsequent translocation even into mature woody root tissue.

None of the major agrochemical companies is giving priority to research on new chemicals whose main target is to control *Armillaria*, but several target soil-borne pathogens of arable crops, such as *Gaeumannomyces*. New antifungal toxophores are continually being created synthetically, or modified from natural products (Clough *et al.*, 1996). Controlled experiments should also be conducted once a potential chemical is identified. Its efficacy should be examined in the field against different species and genets of *Armillaria mellea*. In order to understand any limitations to the performance of the chemicals so that improvements may be made, experiments should concentrate on controlling infection by rhizomorphs and root-to-root contact. In order to make trials more uniform and reliable, artificial inoculum should be used where possible. Both the protectant and eradicant properties of each chemical (against rhizomorphs growing in soil and mycelium within wooden inocula) should be investigated, as well as its phytotoxicity (which might vary between seasons), behaviour in different soil types, temperatures, levels of wetness, and the probability of resistance by fungi or premature breakdown. Even though many new fungicides have been investigated *in vitro* and in field trials by Turner and Fox (1988) and West (1994), a thin layer of bark prevented any of the chemicals from eradicating the underlying mycelium.

Since penetration of either living or dead rots by soil acting fungicides is still neither straightforward nor well understood, one novel approach to the effective control of these destructive root rot that still appears feasible would be the development of a safe fungicide that is freely mobile from the foliage through the phloem to the root cambium and other susceptible tissues. A downwardly mobile systemic chemical would also be desirable, as it should cause less disruption to ectomycorrhizae. Both phosphonic acid, the active ingredient in fosetyl-Al (West, 1994), and N-(phosphomethyl)glycine, the active ingredient in glyphosate (Zolciak, 1998), appear to control *Armillaria,* so it is possible more active, novel, downwardly mobile toxophores may be developed in the future. Once any potential chemical is identified, its efficacy in the field should be examined, with consideration to a number of factors such as: phytotoxicity (which may be less after an application in any dormant season), soil type, effect of temperature and wetness, and the potential for fungicide resistance or premature breakdown.

In order to test potential chemicals indicated by previous work for activity in field conditions against other soil-borne diseases, experiments should concentrate on controlling *Armillaria* infection by rhizomorphs and root-to-root contact that occurs in an established area of infection. In order to make trials more uniform and reliable, artificial inoculum should be used where possible. Both the protectant and eradicant properties of each chemical should be investigated. Such controlled experiments should also be conducted to understand any limitations to the performance of the chemicals so that improvements may be made.

Alternatively, a fungicide that could penetrate, but eventually disrupt, the hyphae might be useful to curtail the spread of the disease to adjacent trees through the rhizomorphs. Spread by direct root-to-root contact may still require control by cultural procedures but, unlike those currently recommended, these will need to be more effective, less laborious, and more practicable in real situations, possibly involving electricity (Fox and Sanson, 1996). By labelling with radioactive or immunological markers, it would be possible to map the uptake patterns of the most active fungicides into the root systems of different woody plants treated in different ways. These, coupled with parallel bioassays, should eventually point the way to developing a practical system of control, thus ensuring for the first time an adequate level of protection to those healthy trees and shrubs in danger from this most destructive disease, in a way which has previously been impossible.

In future, such antibiotics as the strobilurins produced naturally by various forest, soil-dwelling fungi and bacteria which inspired azostrobin (Clough *et al.*, 1996), might be delivered by robust rhizoplane organisms genetically modified to secrete them around the roots, where they could not otherwise be placed.

Another possible development that could be of great value is to use the decomposition of green manures, rapidly growing crop plants like mustard or other brassicas grown to be ploughed under prematurely to improve soil structure and health, which release antifungal volatiles that might inhibit *Armillaria*.

Biological and integrated control

Baker and Cook (1974) defined biological control as the reduction of inoculum density or disease-producing activities of a pathogen or parasite in its active or dormant state, by one or more organisms, accomplished naturally or through manipulation of the environment, host, or antagonist, or by mass introduction of one or more antagonists. They set out the objectives of biological control of plant pathogens as the reduction of disease. Each of these could be improved by interactions that reduce inoculum of the pathogen through more effective methods to eradicate inoculum between crops, decrease production or release of viable propagules, or decrease spread by mycelial growth, better methods to impede infection of the host by the pathogen and reduce the severity of attack by the pathogen. This could be done by improving the three types of activity antagonism identified by Baker and Cook (1974). Novel antibiotic metabolic products could be created, and antibiotic production could be improved by the mutation or genetic modification of antagonistic organisms, leading to improved lysis to destroy, disintegrate or decompose *Armillaria*.

More competitive organisms could be found or genetically modified which deny an adequate supply of a substrate to the pathogen (Baker and Snyder, 1965). Also, superior antagonists may be found or created that make better use of enzymes such as chitinases and cellulases to break down the walls of the pathogen as a food source (Campbell, 1989).

Improved antagonists should produce inoculum in excess; resist, escape, or tolerate other antagonists; germinate and grow rapidly; and invade and occupy organic substrates (Baker and Cook, 1974). Faull and Graeme-Cook (1992) suggested that the features of a biological control agent to be useful in combination rather than in isolation are moderate production of antibiotics, the ability to invade the pathogen as judged from *in vitro* tests, and moderate growth rates. They reasoned that the possession of extremes of any of these features appears to act against the strain in the soil, and may lead to its failure as a biological control agent.

To control *Armillaria*, a rhizosphere or wood-inhabiting organism might function by inhibiting or preventing rhizomorph and mycelial development, by limiting the pathogen to substrate already occupied, by actively pre-empting the substrate, or by eliminating *Armillaria* (perhaps through replacement) from substrate already occupied (Hagle and Shaw, 1991). Rishbeth (1976) suggested that antagonistic organisms might not be able to prevent *Armillaria* from becoming established in stumps, but they may restrict further stump colonization and thus limit the available food base. The main objectives of further studies needed to fulfil the goals of biological control are to investigate the eradicant and protectant capability of potential fungal antagonists, including *Trichoderma* spp., against several isolates of *A. mellea in vitro*, to find a suitable carrier substrate for the delivery of the fungal antagonists to soil. The biocontrol efficacy of these antagonists would need to be tested in the glasshouse and field, both alone and in mixture with other antagonists or fungicide chemicals, and find the best time for introducing the antagonists in relation to the occurrence of the disease.

Although Sokolov (1964) found that fungi in six genera, including *Trichoderma*, *Penicillium*, and *Peniophora* (*Phlebiopsis*), antagonized *Armillaria*, he only recommended using *T. viride* as a control for *Armillaria* root disease. Commercial interest in the development of formulations based on *Trichoderma* has led to a range of products that are available commercially (van Driesche and Bellows, 1996), but there is none currently in use for controlling *Armillaria.* Since Dubos *et al.* (1978) found that the medium in which the inoculum of *T. viride* was grown altered the degree to which the antagonist inhibited production of rhizomorphs by *Armillaria*, the introduction of more novel substrates might be merited. There may also be some advatages in using mixtures of fungi as inoculum. Nelson *et al.* (1995) found that spore inoculum of a mixture of five *Trichoderma* isolates belonging to *T. harzianum*, *T. hamatum,* and *T. virens* placed in drill holes in the stumps of karri (*Eucalyptus diversicolor*) resulted in a significant reduction in root colonization by *A. luteobubalina*. Mixtures with other antagonistic fungi can improve the eradicant and protectant capability of a number of isolates of *Trichoderma* spp. (Raziq, 1998). These, and several other potential fungal antagonists isolated from mushroom, showed promising activity against *A. mellea*, alone and in combination, when dead mushroom compost was used as substrate to convey the fungal antagonists to soil (Raziq, 1998).

There is also scope for testing other fungi for antagonism towards *Armillaria.* Leach (1937) observed that *Rhizoctonia lamellifera* prevented *Armillaria* from colonizing tea roots. Cusson and LaChance (1974) reported that *Scytalidium lignicola* or scytalidin, the toxin it produces, halts the growth of *Armillaria* in culture. Two of the ten basidiomycetes capable of excluding *Armillaria* from occupied substrates tested by Federov and Bobko (1989), *Phlebiopsis* (*Peniophora*) *gigantea* and *Pleurotus ostreatus*, also effectively prevented the growth of *Armillaria* in freshly cut stumps into which they had been inoculated. Pearce and Malajczuk (1990) found that *Coriolus versicolor*, *Stereum hirsutum*, and *Xylaria hypoxylon* inoculated into karri thinning stumps simultaneously with *A. luteobubalina* each significantly reduced colonization by *Armillaria*. Hagle and Shaw (1991) suggested that such cord-forming, wood-decay fungi have a great potential as biological control agents of *Armillaria* infections as they have a similar niche to *Armillaria* and are capable of subcortical mycelial growth in stumps, occupying the same initial sites as *Armillaria*. Rayner (1977) also reported some cord-formers behaving closely similar to *Armillaria*, except in pathogenicity.

Further studies have indicated that several species of the cord-forming basidiomycetes, particularly *Phanerochaete velutina, H. fasciculare,* and *Steccherinum fimbriatum*, have considerable potential to spread and colonize woody debris in field sites (Hagle and Shaw, 1991). Several of them produce networks of mycelial cords in soil and litter (Dowson *et al.*, 1988a) which can infest additional woody substrates (Dowson *et al.*, 1988b).

Rayner (1977) suggested that populations of some cord-formers can be manipulated by chemically treating stumps. Ammonium sulphamate increased colonization by cord-formers of below-ground portions of treated beech and birch stumps. Other herbicides and silvicides might be even more effective, if tested.

Mycophagous nematodes and mycorrhizae have been implicated in biological control of *Armillaria*. Riffle (1973) observed that mycophagous nematodes greatly reduced mortality of ponderosa pine seedlings inoculated with *Armillaria* as they affected the growth of the fungus adversely. Cayrol *et al.* (1978) found that *Aphelenchus avenae* destroyed the hyphae of *Armillaria in vitro*, but grew well on *Trichoderma polysporum* without reducing its growth. Studies *in vitro* have shown that mycorrhizal fungi can inhibit the growth of *Armillaria* (Eghbaltalab *et al.*, 1975). However, direct protection by mycorrhizae seems unlikely as the main infection sites for *Armillaria* are on coarse roots, rather than the fine roots where mycorrhizae develop (Hagle and Shaw, 1991). However, it might be possible that some mycorrhizae at the root tips can produce antimicrobial substances that could be translocated to the base of the roots, butt and collar region.

It is rare that integrated control leads to ideas for biological control rather than *vice versa*, but several examples have already been reported since Bliss (1951) demonstrated the ability of *T. viride* to replace *Armillaria* in artificially infected root segments fumigated with carbon disulphide, and Ohr and Munnecke (1974) found that the fumigant methyl bromide disrupted the antibiotic production of the fungus (as with carbon disulphide), allowing *Trichoderma viride* to attack it. Integrated control could be a wise choice for future research (Onsando and Waldo, 1994).

Control of adverse environmental conditions

It is likely that improved husbandry techniques and technology will continue to reduce stress due to factors such as drought, waterlogging, shading, defoliation, and root damage, which make the host plant prone to infection. Hagle and Shaw (1991) recommend this could be achieved by continuing to investigate the relative resistance of different species and genotypes under a variety of conditions.

Resistant/tolerant hosts

Although there are no incontrovertible examples of truly woody hosts that are completely resistant to all *Armillaria* species, some woody hosts are considered more tolerant than others, even to species of *Armillaria* like *A. mellea* which are known to have a very wide range of tree and shrub hosts. Several species with a more restricted host range, such as *A. ostoyae*, are known to be more virulent to conifers than maples. Also, resistant rootstocks have been developed for some fruit and fibre species (Raabe, 1966). Whereas no woody plant has yet been demonstrated to be immune to infection by *Armillaria* among the herbaceous plants, there are numerous non-hosts, as well as many hosts, yet the basis of the resistance or tolerance of particular hosts is still uncertain. There is also an apparent higher level of tolerance in some species when they are seedlings, which reduces with maturity. Since the mechanism of neither genetic nor phenotypic resistance or tolerance is fully understood, it is not at all clear whether the selection or breeding of plants with enhanced resistance or tolerance will ever be worthwhile.

Once we have evaluated more herbaceous plants for susceptibility, it may be possible to construct a garden less likely to become diseased by choosing only the more resistant plants. Some susceptible plants like strawberry that are killed within a few weeks, are valuable in a different way because they can be used as 'live bait' to reveal where *Armillaria* is present or absent in the soil. Since such differences in resistance can readily be observed even within a genus, there may be more opportunities for making interspecific crosses for *Armillaria* resistance. In the field, host resistance to *Armillaria* root rot involves the genetics of both the host and the pathogen, as well as environmental influences. It can also involve managing mixtures of genotypes with varying levels of resistance. Some host species with superior resistance or tolerance to infection in one location may be more susceptible in other locations. Even within a locality, or even in a hedge or orchard, discrepancies in resistance can be seen, and intraspecific, or even individual, variation in adaptation to sites may be as great as interspecific variation within the natural ranges of any two or more species.

Genetic modification of more resistant cultivars

It may be possible to breed or transfer genes for resistance in which the host's biochemistry is altered to produce more preformed constituents in the bark, or mobilized constituents in response to penetration by the fungus. Alternatively, cultivars might be produced in which the detoxification of the host's antifungal chemicals, such as gallic acid and gallotannins, by phenol oxidases, peroxidases, tyrosinases, or laccases produced by *Armillaria*, could be inhibited, or the host's cell walls reinforced with melanin and other phenolic substances.

To make further progress, additional information is essential on physiological, genetic, or environmental resistance, and how they interact with different species and genotypes of *Armillaria*. Cultivars may be developed in which biochemical changes are induced to counteract the influences of stress leading to host susceptibility. As well as increasing the capacity of the host tissues to tolerate or control the metabolites produced by the fungus, the influences which could be reduced include the repression of fungal inhibitors, release of nutrients and metabolites required by the fungus for pathogenesis, and the stimulation of the growth of *Armillaria*, allowing it to overwhelm the capacity of the host root system to resist harmful fungal metabolites. Cultivars may also be produced with a low incidence of mortality from *Armillaria* infections normally only attributed to increased host resistance with age, associated with physiological or biochemical changes in the host.

Although much of the physiology of the pathogenic species is of fundamental importance, it is still not sufficiently understood (Garraway *et al.*, 1991). Further molecular research is required to establish whether or not a common mechanism is involved in the response of *Armillaria* to the various growth factors. This is complex as it involves the genetics of both host and pathogen, as well as environmental influences.

At present, mixtures of genotypes with varying levels of resistance or, in the future, species with superior resistance or tolerance to infection, may be included. However, judging by experience with other diseases, resistant cultivars in one location may be more susceptible elsewhere, but this is likely to be less marked than with fungi that are predominately spore dispersed.

Hubbes (1987) has suggested that gene manipulation techniques can improve the adaptability of a host species and its physiological resistance to disease, produce populations or clones immune to *Armillaria*, or improve the economic qualities of endemic, resistant species. Hagle and Shaw (1991) propose that when resistance genes known to produce successful resistance reactions in other species or genotypes can be identified, they should be transferred first to the genome of fruit tree rootstocks with superior growth and compatibility characteristics. In practice, species or cultivars which are resistant to *Armillaria*, although valuable, might be more worthwhile in combination with other control procedures.

Planting mixtures of species with differing resistance to *Armillaria* may be more widely adopted, as it can reduce secondary spread of disease in a plantation if rows of the more resistant species alternate with a highly susceptible species (Morrison *et al.*, 1988).

References

BAKER, K.F. AND SNYDER, W.C. (eds.) (1965). *Ecology of soil-borne plant pathogens. Prelude to biological control.* Berkeley: University of California Press.

BAKER, K.F. AND COOK, R.J. (1974). *Biological control of plant pathogens.* San Francisco: W.H. Freeman and Company.

BLISS, D.E. (1951). The destruction of *Armillaria mellea* in citrus soils. *Phytopathology* **41**, 665–683.

CAMPBELL, R. (1989). *Biological control of microbial plant pathogens.* Cambridge: Cambridge University Press.

CAYROL, J.C., DUBOS, B. AND GUILLAUMIN, J.J. (1978). Etude préliminaire in vitro de l'agressivité de quelque nématodes mycophages vis-à-vis de *Trichoderma viride* Pers., *T. polysporum* (Link. ex. Pers.) Rifaï et *Armillaria mellea* (Vahl) Karst. *Annales de*

Phytopathologie **10** (2), 177–185.

CLOUGH, J.M., GODFREY, C.R.A., GODWIN, J.R., JOSEPH, R.S.I. AND SPINKS, C. (1996). Azostrobin: a novel broad-spectrum systemic fungicide. *Pesticide Outlook*, August.

CORNS, I.G.W. AND ANNAS, R.M. (1986). *Field guide to forest ecosystems of west-central Alberta*. 251pp. Edmonton, Alberta: Canadian Forestry Service, Northern Forestry Centre.

CUSSON, Y. AND LACHANCE, D. (1974). Antagonism between *Scytalidium lignicola* and two root rot fungi. *Phytoprotection* **55**, 17–28.

DAUBENMIRE, R. (1952). Forest vegetation in Idaho and adjacent Washington, and its bearing on concepts of vegetation classification. *Ecological Monographs* **22**, 301–330.

DOWSON, C.G., RAYNER, A.D.M. AND BODDY, L. (1988a). Inoculation of mycelial cord-forming basidiomycetes into woodland soil and litter, I. Initial establishment. *New Phytologist* **109**, 335–341.

DOWSON, C.G., RAYNER, A.D.M. AND BODDY, L. (1988b). Inoculation of mycelial cord-forming basidiomycetes into woodland soil and litter, II. Resource capture and persistence. *New Phytologist* **109**, 343–349.

DUBOS, B., GUILLAMIN, J.J. AND SCHUBERT, M. (1978). Action du *Trichoderma viride* Pers., apporté avec divers substrats organiques, sur l'initiation et la croissance des rhizomorphes d'*armillariella mellea* (Vahl) Karst. dans deux types de sols. [The action of *Trichoderma viride* Pers., applied with various organic substrates, on the initiation and growth of rhizomorphs of *Armillariella mellea* (Vahl) Karst. in two types of soils.] *Annales de Phytopathologie* **10**, 187–196. [English translation: P. Aukland.]

EGHBALTALAB, M., GAY, G. AND BRUCHET, G. (1975). Antagonisme entre 15 espèces de basidiomycetes et 3 champignons pathogènes de racines. *Bulletin de la Société Linnéenne de Lyon* **44**, 203–229.

FAULL, J.L. AND GRAEME-COOK, K. (1992). Characterization of mutants of *Trichoderma harzianum* with altered antibiotic production characteristics. In: *Biological control of plant diseases: progress and challenges for the future*. Eds. E.S. Jamos, G.C. Papavizas and R.J. Cook. New York: Plenum Press.

FEDOROV, N.I. AND BOBKO, I.N. (1989). Armillaria root rot in Byelorussian forests. In: *Proceedings of the 7th international conference on root and butt rots*, August 9–16, 1988, Vernon and Victoria, B.C. Ed. D.J. Morrison, pp 469–476. Victoria, B.C.: International Union of Forestry Research Organizations.

FILIP, G.M. AND ROTH, L.F. (1987). *Seven chemicals fail to protect ponderosa pine from Armillaria root disease in central Washington*. Res. Note PNW-RN-460. 8pp. Portland, OR: USDA, Forest Service, Pacific Northwest Forest and Range Experiment Station.

FOX, R.T.V. (1990). Diagnosis and control of *Armillaria* honey fungus root rot of trees. *Professional Horticulture* **4**, 121–127.

FOX, R.T.V. AND HAHNE, K. (1989). Prospects for the rapid diagnosis of *Armillaria* by monoclonal antibody ELISA. In: *Proceedings of the 7th international conference on root and butt rots*, August 9–16, 1988, Vernon and Victoria, B.C. Ed. D.J. Morrison, pp 458–468. Victoria, B.C.: International Union of Forestry Research Organizations.

FOX, R.T.V. AND SANSON, S. (1996). Lethal effects of an electrical field on *Armillaria mellea*. *Mycological Research* **100** (3), 318–320.

GARRAWAY, M.O., HÜTTERMANN, A. AND WARGO, P.M. (1991). Ontogeny and Physiology. In: *Armillaria Root Disease*. Eds. C.G. Shaw, III and G.A. Kile, pp 21–47. *Forest Service Handbook No. 691*. Washington, D.C.: USDA.

GUILLAUMIN, J.J. (1988). The *Armillaria mellea* complex. *Armillaria mellea* (Vahl) Kummer *sensu stricto*. In: *European Handbook of Plant Diseases*. Eds. I.M. Smith, J. Dunez, D.H. Phillips, R.A. Lelliot and S.A. Archer, pp 520–523. Oxford: Blackwell Scientific Publications.

HADFIELD, J.S., GOHEEN, D.J. AND FILIP, G.M. (1986). *Root diseases in Oregon and Washington conifers*. R6-FPM25086. 27pp. Portland, OR: USDA, Forest Service, Pacific Northwest Region, Forest Pest Management.

HAGLE, S.K. AND SHAW, C.G., III (1991). Avoiding and reducing losses from *Armillaria* root disease. In: *Armillaria Root Disease*. Eds. C.G. Shaw, III and G.A. Kile, pp 157–173. *Forest Service Handbook No. 691*. Washington, D.C.: USDA.

HEATON, J.B. AND DULLAHIDE, S.R. (1989). Overcoming Armillaria root rot in Granite Belt orchards. *Queensland Agricultural Journal* January–February 1989, 25–27.

HUBBES, M. (1987). Influence of biotechnology on forest disease research and disease control. *Canadian Journal of Plant Pathology* **9**, 343–348.

LEACH, R. (1937). Observations on the parasitism and control of *Armillaria mellea*. *Proceedings of the Royal Society of London, Series B* **121**, 561–573.

LUNG-ESCARMANT, B. AND TARIS, B. (1989). Methodological approach to assess host response (resinous and hardwood species) to *Armillaria obscura* infection in the southwest French pine forest. In: *Proceedings of the 7th international conference on root and butt rots*, August 9–16, 1988, Vernon and Victoria, B.C. Ed. D.J. Morrison, pp 226–236. Victoria, B.C.: International Union of Forestry Research Organizations.

MANLEY, H. (1996). *Inter- and intraspecific genetic variation in European Armillaria species.* Ph.D. Thesis. U.K.: University of Reading.

MORRISON, D.J. (1981). *Armillaria root disease. A guide to disease diagnosis, development and management in British Columbia*. Information Report BC-X-203, pp 1–16. Environment Canada, Canadian Forestry Service.

MORRISON, D.J., WALLIS, G.W. AND WEIR, L.C. (1988). *Control of Armillaria and Phellinus root diseases: 20-year results from the Skimikin stump removal experiment*. Information Report BC-X-302. 16pp. Canadian Forestry Service, Pacific Forestry Centre.

NELSON, E.E., PEARCE, M.H. AND MALAJCZUK, N. (1995). Effects of *Trichoderma* spp. and ammonium sulphamate on establishment of *Armillaria luteobubalina* on stumps of *Eucalyptus diversicolor*. *Mycological Research* **99** (8), 957–962.

OHR, H.D. AND MUNNECKE, D.E. (1974). Effects of methyl bromide on antibiotic production by *Armillaria mellea*. *Transactions of the British Mycological Society* **62**, 65–72.

ONSANDO, J.M. AND WALDO, S.W. (1994). *Trichoderma* species and *Armillaria* root disease of tea in Kenya. *International Journal of Pest Management* **40** (1), 69–74.

PAWSEY, R.G. AND RAHMAN, M.A. (1974). Armillatox field trials. *Gardeners Chronicle* **175**, 29–31.

PAWSEY, R.G. AND RAHMAN, M.A. (l976). Chemical control of infection by honey fungus, *Armillaria mellea*: a review. *Arboricultural Journal* **2**, 468–479.

PEARCE, M.H. AND MALAJCZUK, N. (1990). Inoculation of *Eucalyptus diversicolor* thinning stumps with wood decay fungi for control of *Armillaria luteobubalina*. *Mycological Research* **94**, 32–37.

RAABE, R.D. (1962). Host list of the root rot fungus, *Armillaria mellea*. *Hilgardia* **33**, 25–88.

RAABE, R.D. (1966). Testing plants for resistance to oak root fungus. *California Agriculture* **20**, 12.

RAABE, R.D. AND TRUJILLO, E.E. (1963). *Armillaria mellea* in Hawaii. *Plant Disease Reporter* **47**, 776.

RAHMAN, M.A. (1974). *Studies on the Effect of Armillatox, a Proprietary Phenolic Emulsion, on the Growth and Infection by Armillaria mellea*. M.Sc. Thesis. U.K.: Oxford University.

RAYNER, A.D.M. (1977). Fungal Colonisation of Hardwood Stumps from natural sources. *Transactions of the British Mycological Society* **69** (2), 303–312.

RAZIQ, F. (1998). *Biological and integrated control of the root rot caused by Armillaria mellea.* Ph.D. Thesis. U.K.: University of Reading.

REDFERN, D.B. (1971). Chemical control of honey fungus (*Armillaria mellea)*. In: *Proceedings of the Brighton Crop Protection Conference: Pests and Diseases 1*, 1971, pp 469–474.

RIFFLE, J.W. (1973). Effet of two mycophagous nematodes on *Armillaria mellea* root rot of *Pinus ponderosa* seedlings. *Plant Disease Reporter* **57**, 355–357.

RISHBETH, J. (1976). Chemical treatment and inoculation of hardwood stumps for control of *Armillaria mellea*. *Annals of Applied Biology* **82** (1), 57–70.

RISHBETH, J. (1986). Some characteristics of English *Armillaria* species in culture. *Transactions of the British Mycological Society* **85**, 213–218.

ROTH, L.F., ROLF, L. AND COOLEY, S. (1980). Identifying infected ponderosa pine stumps to reduce costs of controlling *Armillaria* root rot. *Journal of Forestry* **78**, 145–151.

SMITH, M.L., BRUHN, J.N. AND ANDERSON, J.B. (1992). The fungus *Armillaria bulbosa* is among the largest and oldest living organisms. *Nature (London)* **256**, 428–431.

SHAW, C.G., III AND KILE, G.A. (1991). *Armillaria Root Disease*. xi+239pp. *Forest Service Agriculture Handbook No. 691*. Washington, D.C.: USDA.

SOKOLOV, D.V. (1964). *Kornevaya gnil' ot Openki bor'ba s nel*. [Root rot caused by *Armillaria mellea* and its control.] (In Russian.) Moscow, Izadatel'stvo Lesnaya Promyshlennost'. [Canada Department of Forestry. 235pp.]

TURNER, J.A. (1991). *Biology and control of Armillaria*. Ph.D. Thesis. U.K.: University of Reading.

TURNER, J.A. AND FOX, R.T.V. (1988). Prospects for the chemical control of *Armillaria* species. In: *Proceedings of the Brighton Crop Protection Conference: Pests and Diseases 1*, 1988, 235–240.

VAN DRIESCHE, R.G. AND BELLOWS, T.S. (1996). *Biological control*. New York: Chapman and Hall.

WARGO, P.M. AND SHAW, C.G., III (1985). *Armillaria* root rot: the puzzle is being solved. *Plant Disease* **69** (10), 826–832.

WARGO, P.M. AND HARRINGTON, T.C. (1991). Host stress and susceptibility. In: *Armillaria Root Disease*. Eds. C.G. Shaw, III and G.A. Kilc, pp 88–101. *Forest Service Agriculture Handbook No. 691*. Washington, D.C.: USDA.

WEST, J.S. (1994). *Chemical control of Armillaria*. Ph.D. Thesis. U.K.: University of Reading.

WHITNEY, R.D. (1984). Site variation of *Armillaria* in three Ontario conifers. In: *Proceedings of the 6th international conference on root and butt rots of forest trees*, August 25–31, 1983, Melbourne, Victoria, Gympie, Queensland, Australia. Ed. G.A. Kile, pp 122–130. Melbourne: International Union of Forestry Research Organizations.

WILLIAMS, R.E., SHAW, C.G., III AND WARGO, P.M. (1989). *Armillaria Root Disease*. Forest Insect and Disease Leaflet 78 (rev.). 8pp. Washington, D.C.: USDA, Forest Service.

ZOLCIAK, A. (1998). Refraining the regeneration of *Armillaria ostoyae* Romagn. rhizomorphs with Roundup. In: *Proceedings of the 9th international conference on root and butt rots of forest trees*, September 1–7, 1997, Carcans-Maubuisson (France). Eds. C. Delatour, J.J. Guillaumin, B. Lung-Escarmant and B. Marchais, p 446. France: International Union of Forestry Research Organizations.

Index